Theory of the Image

Theory of the Image

Thomas Nail

OXFORD
UNIVERSITY PRESS

OXFORD
UNIVERSITY PRESS

Oxford University Press is a department of the University of Oxford. It furthers the University's objective of excellence in research, scholarship, and education by publishing worldwide. Oxford is a registered trade mark of Oxford University Press in the UK and certain other countries.

Published in the United States of America by Oxford University Press
198 Madison Avenue, New York, NY 10016, United States of America.

CIP data is on file at the Library of Congress

ISBN 978–0–19–005008–5 (pbk)
ISBN 978–0–19–092403–4 (hbk)

Hardback printed by Bridgeport National Bindery, Inc., United States of America

CONTENTS

LIST OF FIGURES

ACKNOWLEDGMENTS

I am indebted to a number of people for their support and encouragement of this project.

I greatly value my discussions about classical Greek art and aesthetics with Robert Urquhart and my long late-night philosophy strolls around Harvard Gulch park with Chris Gamble and Josh Hanan.

I am grateful to my research assistants, who helped me with various aspects of the project: Adam Loch and Kevin Buskager.

I thank my anonymous peer reviewers for their constructive feedback and everyone I worked with at Oxford University Press for guiding this book to publication. I would also like to thank my copy editor, Dan Thomas for his excellent close corrections and suggestions.

Thanks also to the University of Denver Rosenberry Fund for its financial support.

Finally, I would like to thank my family, and especially my wife, Katie, for her continued support and direction on parts of this project. These are the material conditions without which this work would not have been possible.

Introduction

The Age of the Image

We live in the age of the image. Today, there are more images in wider circulation than ever before in human history. This is due in part to the worldwide increase in mechanical reproduction technologies, global transportation methods, and distribution circuits during the latter part of the twentieth century. There are now more written, spoken, and visual images moving around the world faster and farther than anyone could ever have anticipated.

However, perhaps the single greatest source of this massive circulation of images has been the advent of the digital image, which began around the same time. Just before the turn of the twenty-first century, a host of digital-media technologies (computers, internet, video games, mobile devices, and many others) unleashed the largest flow of digitally reproduced words, images, and sounds the world has ever witnessed. No other kind of aesthetic medium or method of mechanical reproduction could possibly compete with what digital media have done to the image in the past twenty years. All that was solid has melted into the electromagnetic field. The digital image thus gave mobility to the image on a scale never before witnessed in human history.

Our world is now saturated with moving images of all kinds, both analog and digital. This sea change in image production and circulation is nothing less than the equivalent of a Copernican revolution in our time. The centrality of the movement and the mobility of the image have never been more dramatic. And just like the Copernican revolution, the aesthetic revolution of the image has consequences not only for the way we think about the contemporary image but for the way we think about all previous images as well. The contemporary mobility of the image lets us see something new about the nature of all hitherto existing images. Only now are we able to understand that movement and mobility have always been at the heart of the image. The core argument of this book, therefore, is that it is this contemporary insight that makes possible a complete reinterpretation of aesthetics and art history from a new perspective. A specter is haunting the twenty-first century, and it is the specter of the image.

THE RISE OF THE IMAGE

The advent of radio and television in the twentieth century marked the beginning of the image revolution. They gave birth to its earliest electromagnetic and increasingly mobile form. However, they also restricted it to relatively centralized, homogenized, and unidirectional forms of "programming." By contrast, the new interactive and bidirectional nature of digital media has expanded the mobility and mutability of the image in completely new ways. With popularization of the internet and mobile devices—cellphones, smartphones, tablets, and laptops—at the turn of the twenty-first century, the image has become something not only ubiquitous but also increasingly portable. As of 2014, there were more active mobile devices than there were people on the planet. The mobile phone is probably the single, fastest-growing human sensory technology ever made, increasing from zero to 7.2 billion in a mere three decades.[1] What is more, the digital image has incited a huge revolution in publishing, journalism, entertainment, education, commerce, and politics unmatched by that of radio and television. The digital image has both integrated and carried forward analog media, giving rise to new digitalized industries in the process. Industrial factories and workers are increasingly replaced by internet servers and automated checkout software. We have entered a new historical-aesthetic regime: We are now in the age of the image.

Today it is possible for anyone to communicate by voice or text with anyone else, to listen to almost every sound ever recorded, to view almost any image ever made, and to read almost any text ever written from a single device almost anywhere on earth. All this is now available while we're on the move—and is itself in movement in the form of an electrical flow. The image will never be the same. The contemporary mobility of images, made possible by the advent and now dominance of the digital image, is not just a quantitative increase in reproduced images. Digital media and the digital image have transformed the qualitative structure of the image itself. Anything can now be digitalized, mobilized, and browsed nonlinearly through a single portable device. Anything can now be made responsive and interactive with the viewer through the use of digital software and a continuous flow of electrical current. None of our senses has remained unchanged by the new digital image—even taste and smell can now be synthesized using computer software.[2] Something is always lost in transit, however, as the continuity of the electrical flow is converted into digitally discrete 1s and 0s—but the image moves on regardless, sweeping us all along with it.

PARADIGM SHIFT

In such a world of moving images, aesthetics can no longer be adequately understood by the old paradigm of representation. Recent scholarship in art and aesthetics increasingly attests to this recognition.[3] Images have taken on a growing mobility that shifts back and forth *between* object and subject, copy and model, transforming and modulating them in a continuous feedback loop. Old theoretical frameworks no longer fit the twenty-first–century reality of interactive electrical circulations and continually modulated images. If we continue to think of images as static objects, or even as objects that only interact with distinct human perceivers, we are missing something fundamental about those images: their interactive *movement*. Images have a material and kinetic agency with one another that is not strictly derived from human or social agents. This insight pertains not just to the digital image but to all images. Images have always been in motion, and their movements have always had a collective agency independent of human perception and social constructions. This is not unique to the digital image. Thus, more than ever before, the fact that the image is up in the air and on the move also requires us to seriously rethink the materiality of affect and sensation beyond human limits. New empirical realities require new conceptual frameworks that in turn tell us something new about our past. This is what this book aims to provide: a new aesthetics for our time, *an aesthetics of the moving image*.

In contrast to the prevailing twentieth-century aesthetic paradigms of images that have modeled the agency of the image on the agency of humans and human structures (social, psychological, linguistic, economic, and so on), this book proposes a new theory of the image that starts with the mobility and material agency of the image itself. Beginning the analysis of images with the primacy of their mobility and circulation in this way allows us to think beyond the traditional anthropocentric frameworks based on subjective, formal, or social structures and to think more about the activity and agency of the *image itself* and what the image *does*—not only what it does *for humans*. The image is not just a passive semblance of something else (another image, an object, or human perception); it has a real and material mobility of its own, which has been vastly understudied in art history and aesthetics.[4] The original contribution of this book, therefore, is that it provides a materialist and nonanthropocentric theory of the image from the perspective of the movement of the image itself. As such, it expands the study of aesthetics and art history beyond questions of representation, signification, linguistic and social constructivisms, and human perception, and toward the study of regimes of material and kinetic circulation.

METHOD

Today's Copernican revolution of the image marks a new period in aesthetic history. It sets the limits of the previous century and outlines a new one, defined, at least in part, by the ever-increasing *mobility of the image*.[5] However, the advent of the present is never limited to the present. Now that our present has emerged, it is possible, in a way it was not before, to inquire into the conditions of its emergence and discover something new about the nature and history of this image. In other words, the present reveals something new about the nature of the image more generally and what it must *at least* be like, so as to be capable of being defined by the primacy of motion and mobility as it is today.

Thus, a central question of this book is: What does the mobility of the image say about the nature of images and aesthetics more broadly? If the image is defined by the primacy of mobility, yet existing theories do not begin with this, then we need a new conceptual framework. *Theory of the Image* aims to provide a conceptual framework based on the primacy of motion, to better understand contemporary structures of sensation and aesthetics, as well as the historical events from which they emerge. In short, the contemporary rise of the image draws our attention not so much to the radical novelty of the digital image as such but rather to *a previously hidden dimension of all previous images* that it is only now possible to glimpse.[6]

The methodology deployed in this book is, therefore, neither a narrow theory of the image that applies strictly to digital media nor to a grand ahistorical theory of the image that applies universally to all images forever and for all time. It is not a naive realism in which the discovery of the contemporary primacy of the moving image gives us pure access to the unchanging essence of the image *as such*. Instead, it is a realism of the *minimal affective conditions* of the emergence of the present. It is, therefore, a *critical, historical,* or *minimal realism* in the sense in which the image is interpreted only with respect to that aspect of the image that must at least be the case for our present "to have been possible"—that is, actual. *Theory of the Image* is, therefore, not a theory in the traditional sense of an abstract and universal mental representation of the world; rather, it is a "theory" in the etymological sense of the Greek word θεωρία, *theōría*, as a "movement," "sending," or "process." Theory is the process of describing the structure of material-kinetic processes as they emerge—in this case, the contemporary ubiquity of the moving image.[7]

Therefore, the method deployed in this book is neither realist nor constructivist in the traditional senses of those words but rather *minimally* or *critically realist*. This distinguishes the method strikingly from most prevailing theories. The question is not what the conditions of the human structures (mind, language, society, and so on) must be for the present to be what it is but rather what *the image itself* must *at least* be like such that the present has come to be defined by the primacy of mobility. The question is not what the conditions of language, the unconscious, economics, power, and so on must be like for the present to be possible but rather what the reality of the image itself must be like so as to render actual these anthropic structures of sensation in the first place. Without a doubt, contemporary reality is shaped by multiple human structures, but these structures are in turn conditioned by other real, nonanthropic, and affective images. The aim of this method is not to dissolve human agency and perception. Quite the opposite, the aim is to put forward a theory of human agency as one type of image among others.[8] These are the new kinds of images this book will investigate.

The objective of this work is to locate the real and historical conditions for the emergence of the contemporary mobility of the image.[9] When these conditions are elaborated, however, it is always possible for a new present to emerge and in turn reveal yet another previously unseen dimension of the past—and so on, in an additive historical fashion. Thus, this book does not offer any kind of final word or universal theory of the image.[10]

TWO PROBLEMS

The kinetic theory of the image developed in this book hopes to overcome two major problems within hitherto existing theories that have remained relatively *static* and *ahistorical.*

First Problem: Stasis

The image has been traditionally subordinated to something *static*. This subordination has assumed two complementary formulations: an objective one and a subjective one.

Objective Stasis

On the one hand, the image has been subordinated to a static object, or unchanging essence. In other words, the image has been treated as a copy or representation of an original. The difference between the object and the image of the object is interpreted as the degree of movement or change in the image itself with respect to its unchanging original object. This is the classical model/copy relation famously dramatized by Plato in the *Timaeus*. The original or model object remains static and unmoved while subsequent images work like mobile snapshots to accurately represent the original object in all its immobile perfection and essential form.

As Plato writes, "Now the nature of the ideal being was everlasting, but to bestow this attribute in its fullness upon a creature was impossible. Wherefore he resolved to have a moving image of eternity, and when he set in order the heaven, he made this image eternal but moving according to number, while eternity itself rests in unity."[11] There can be no higher exhalation of eternity and denigration of the image than this. For Plato, the image is nothing but illusion, appearance, and likeness organized according to discrete numerical quantities. The object is thus fixed in its essence, and the image is fixed by its discrete number. These discrete numerical images fail to represent the object precisely because of the *mobility of the image*. Motion and mobility thus become the conceptual names for the failure of the image to represent the object. The kinetic image is the degradation of an original object.

All definitions of art as representation are determined by some version or degree of this static model/copy/resemblance relation. Not only is the object immobilized in the model to be copied, but the image of the model itself remains nothing more than a failed numerical attempt to reproduce this same static condition. Between the two stands a gulf of movement and turbulence that ensures their incommensurability. In this way, the only real or true sensation occurs in the object itself: all images of the object are mere appearances or modified snapshots of the original. The obsession with art preservation, authorial authenticity, and connoisseurship are historically linked to this classical idea of stasis and mimesis.

Subjective Stasis

On the other hand, the image has also been subordinated to the relatively static mental states of the subject. In this theory, perceptual images are only given conceptual aesthetic coherence and reality *in the faculties of the perceiver*. Versions of this theory are closer to the more modern aesthetics developed by Immanuel Kant in his *Critique of Judgment* (1790). What remains static, fixed, and universal is not the object being represented but rather the concept of beauty itself found in the mental structure of the subject. Fluctuating images occur in the body of perceiver, but it is only in the *concept* of beauty that they are given fixed and universal form. It is thus human mental and perceptual structures, and not sensual images themselves, that lie at the foundations of truth and beauty.

Again, for Kant, it is the movement of the image in the mobile and affected body that marks the inferiority and subordination of the image. The nature of the object in itself remains unknown *because the body and its perceptual images are moved and mobile*. The senses are thus led to misrepresent reality to the mind. The senses of the body cannot be trusted, whether in knowledge or in beauty. Our experience of beauty, therefore, is not the beauty of nature or even of the beauty of the images but, rather, the beauty of our own idea, experience, or faculty of representing these images to ourselves. Nature is only the prompt for us to discover the beauty of our own aesthetic and phenomenological faculties.[12] This is the inverse of the classical idea of the model/copy relation. Instead of defining the image by its subordination to the static essence of the object, it is defined by its subordination to the static aesthetic structures of judgment in the mind of the experiencing or intentional subject.

This subjective form is most dramatic in Kant and post-Kantian aesthetics, but a similar model is also at work in other anthropic constructivisms as well, including social, anthropological, linguistic, economic, and other nonpsychological versions. All these different constructivisms share the reduction of the image, not to the Kantian ego but to other anthropic structures. In contrast to Kant, some of these anthropic constructivisms can even be transformed to some extent by moving images. However, even in those cases, the movement of the image remains tied to the *relatively static* anthropic structures that produce and consume those images. Since numerous full-length works have recently been devoted to making this argument, including my own, and since this is not the primary focus of this book, I must simply refer the interested reader to those works at this point.[13]

Both the objective and subjective/constructivist theories of the image thus subordinate it to something relatively static. Furthermore, they both treat the movement of images as something discrete, either in number (Plato) or in the body (Kant). In both cases, movement is what makes the image inferior but also what secures the difference between the object and the subject. For Plato, the object remains different from the inferior images of it precisely because the object does not move. For Kant, the same is true of the transcendental subject. For constructivists, images remain extensions, projections, or reflections of more primary human structures. In both cases, the object and subject are separated by a kinetic gulf of fluctuating images.

There are two kinetic paradoxes here. The first is that the movement of the image is both necessary to ensure the *division* between subject and object and necessary to ensure the region of transport that *connects* them *as distinct*. The model transports its image to the senses. The human subject then receives these images on the surface of its sensitive mobile body or anthropic structure. Without this zone of transport between the object and subject, nothing transpires: sensation fails, affect dissolves. Yet precisely because of this mobility, representation is also undermined. The mobility of the image is thus both the condition of *possibility* for the object and subject and the condition of their *impossible* convergence in perfect representation; hence, the related second paradox that the image is treated as necessarily mobile in its transport but fixed and limited by number or in a human body. The image must move but only as a frozen mobility, a snapshot, or particle of sensation—something for a human subject. The mobility of the image is thus described as secondary to the fixed object or subject of human sensation, when it is human sensation itself that is produced as a regional stabilization of the mobile substratum of images.

Therefore, if we want to develop a theory of the image that does not fall into these paradoxes, we need to begin from its most primary and defining feature—its mobility—and not try to deduce this mobility from something else *it is relative to*. This requires a complete theoretical reorientation. In short, the division between the object and the subject of sensation should not be considered a primary ontological determination but, rather, the effect of a more primary kinetic process of moving images themselves. This is the novelty of *Theory of the Image*: to reinterpret the structure and history of the image and its affects from the perspective of its mobility. Objects, subjects, and human structures are products of this more primary process.

Second Problem: History

The kinetic theory of the image aims to overcome the supposedly ahistorical nature of the image. There are three formulations of this ahistorical thesis: objective, subjective, and ontological.

Objective.

First, if the image is subordinated to a static model object, then it can have no history, or at most have a *mere illusion* of history. Since history presupposes the real movement and transformation of matter, and objective essences do not move, this means that objects can have no history, and neither can their images.

Subjective

Second, if the image is subordinated to the static conceptual or constructivist structure of human subjects, then a similar problem occurs. If subjective structures are universal, as Kant and much of post-Kantian phenomenology argue,[14] then they do not change (or change only within a fixed domain) over time; and if subjective structures themselves (not just their contents) do not change over time, then they have no real history. Perceptual images may change *within this structure*, but the aesthetic conditions of making sense of these images and ordering them have always been the same—and thus the image, too, as subordinate to the structure, remains ahistorical. A notable exception to this post-Kantian ahistoricism is the tradition of Marxist aesthetics, including the Frankfurt school.[15]

Ontological

Third, the ontological theory of the image is defined by the autonomous becoming of all affects in general. The affective nature of the image is, therefore, continuous with the whole process of becoming, in which the object and subject both transform and are transformed through their co-appearance as images. In this way, the ontology of the affective image aims to liberate the image from its twin subordination.

It does so, however, only at the risk of introducing its own form of universality and ahistoricity. If the affective image is understood as *ontologically* "autonomous"[16] with respect to the objects and subjects it produces or distributes, then its constant change becomes something relatively changeless: pure becoming. If all images are reduced to their lowest common denominator—affect, becoming, and ontological change—then the particularity of their historical and regional distributions risks being submerged entirely in a pure ontological flux. This ontological rejection of history in favor of becoming has been put forward by a number of recent process and affect theorists.[17]

The process ontology of the affective image treats the image as if it were possible to describe its structure forever and all time, and from no position in particular. In this way the ontology of the affective image is saved from anthropic constructivism, but it also risks making the image something like its own kind of "autonomous force"—adding nothing to the historical description of the image but a generic ontological language applied to everything equally, even if that language is one of pure flux.

In response to the problem of ahistoricity, this book offers not only a theory of the image and aesthetics grounded in a view from the present, but it also offers a history of this present and the material conditions of its emergence. In short, it explicitly does not offer an ontology of the image. It is precisely because the image is mobile and material that it has a history, and therefore that sensation must be theorized historically and not ontologically. Furthermore, because the image has a history, it also has a whole typology of distributions that organize the world of subjective and objective structures. All these structures have to be accounted for, starting from the *historical mobility of the image*, and not from any metaphysical or ontological description of becoming. Therefore, the project of this book is to develop a theory *and* a history of the logic and structure of the moving image.

WHAT IS AN IMAGE?

The image is not a copy or a movement relative to an object or subject; it is not even a copy of a copy without an original.[18] There is no mimesis whatsoever. If we are looking for a new and more fruitful definition of the image, we need look no further than within the same Latin root of the word itself. The word *image*, from the Latin word *imago*, means "reflection," "duplication," or "echo."[19] These definitions imply precisely the opposite of what we typically think of as a copy. A copy must be something other than its model or, by definition, it cannot be a copy *of a model*.

However, *reflection*, from the Latin word *flex*, means "bend" or "curve." A reflection is a curving or bending that folds back over itself. *Duplication*, from the Latin word *pli*, meaning "fold," makes this meaning quite apparent. The image is not a distinct or separate copy but, rather, the process by which matter curves, bends, folds, and bounces back and forth, or "echoes." The image is, therefore, the mobile process by which matter twists, folds, and reflects itself into various structures of sensation and affection. By this definition, the image is not reducible to a strictly visual kind image alone but, also, is optical, sonic, haptic, olfactory, and gustatory. All sensation is thus bound together in a continuous flow of images.[20]

There are not first static objects and subjects and then later a movement or transfer of images between them. Rather, there is first matter in motion folding itself up through composition and duplication that generates larger sensuous matters like objects and subjects that then further reflect and duplicate the flows of matter between them. A folded image is not a copy because a fold is not something separate from the matter that is folded. The fold is a completely continuous kinetic and topological structure. There is not one part of the fold that is an original and another that is a copy. This is the sense in which Henri Bergson writes that the image is "more than that which the idealist calls a *representation*, but less than that which the realist calls a *thing*—an existence placed halfway between the 'thing' and the 'representation.'"[21] It is more than a representation because it is not a copy of something else, and it is less than a thing because it is already the material of which things are composed and as such is irreducible to any single empirical sensation of it. "Images," to invert Bergson's phrase, are an aggregate of "matters."[22]

In contrast to existing theories of the image as passive phenomena of formal, subjective, or other static, anthropocentric, and ahistorical structures, this book provides a refreshingly different approach: a transformative, affective, and kinetic theory proper to the action and mobility of *the image itself*. The details of this kinetic and materialist theory of the image are further developed in part I.

CONTRIBUTION AND PLAN OF THE BOOK

There are three important consequences that come from overcoming the problems of stasis and ahistoricity posed by current theories of the image. The first consequence of the kinetic theory of the image is a new conceptual framework proper to the *movement* of the image. This is developed at length in part I. The kinetic theory of the image allows us to explain not only how objects and subjects emerge from more primary kinetic structures of images but also the how different kinds of objects and subjects emerge from different historical structures. Among other things, this includes a new movement-oriented interpretation of experience, emotion, thought, memory, and the image itself. This conceptual framework in turn allows us to put forward a new theory of art and aesthetics no longer based in objective essences or subjective/social experiences but rather in the historical and material kinetic structure of *the work of art itself*.

The second consequence of a kinetic theory of the image is that it makes possible a whole new interpretation of the history of art. Part II thus provides an *analysis of the historical conditions* for the emergence of the dominant distributions of images that we have today. The dominant distributions of the image that we know today in the arts did not come out of nowhere; the image has a history. At different points in history, images were distributed according to at least four dominant aesthetic regimes: functional, formal, relational, and differential. New forms of aesthetic organization mix and rise to dominance through history. When these new techniques emerge historically, they tend to persist, repeat, and combine. Today we find the digital mobile image at the intersection of all four major forms of historical regimes. The methodological primacy of history is in fact what grounds the aesthetic theory. Part I is not a set of arbitrary conceptual categories but simply the result obtained from the more primary historical research contained in part II. This is how Marx described his method for writing *Capital*.

> Of course the method of presentation must differ in form from that of inquiry. The latter has to appropriate the material in detail, to analyse its different forms of development and to track down their inner connection. Only after this work has been done can the real movement be appropriately presented. If this is done successfully, if the life of the subject-matter is now reflected back in the ideas, then it may appear as if we have before us an a priori construction.[23]

Theory of the Image is, therefore, not putting forward any a priori constructions. Like the owl of Minerva, theoretical practice flies at dusk after the day has done and looks back on its immanent conditions. However, once the owl has seen the practical and historical conditions of its own appearance, it then describes them, not from nowhere but precisely from the very point from which it is at. Theoretical description is thus always backward-looking, like Walter Benjamin's interpretation of Paul Klee's *Angelus Novus* (1920). The angel of history is propelled forward *practically* with its back to the future while it gazes *theoretically* into the past.

Perhaps some philosophers of aesthetics who read this book will only be interested in the conceptual conclusions found in part I, and perhaps some art historians will only be interested in the historical interpretations in part II, and others still may only be interested in the case study of digital and generative images in part III. But the book is intended to be and was written as an integrative whole, in which the theory is derived first and foremost from historical study, and the historical interpretations are guided by this theory, both of which only make sense as grounded in the contemporary advent of the digital image. I therefore urge the reader to take this holism seriously as a methodological statement about what it means to think about art and aesthetics.

Accordingly, the third consequence of developing a kinetic theory of image is that it allows us to analyze contemporary art and aesthetics in a much more precise and historically sensitive way. The history of the image is not a linear or progressive sequence of self-confined "ages." Rather, the history of art is one of coexisting and overlapping aesthetic regimes. The same techniques of functional, formal, relational, and differential distribution that have emerged and repeated throughout art history are still at work today in contemporary art. The digital image does not leave these behind but rather carries them forward.

For example, digital media have not replaced analog media. The two coexist and mix, creating new hybrid structures. Thus in order to understand the present, we must also understand the past of which it is composed. In drawing from the past, however, we always do so from the perspective of the present and from the defining historical attributes of the contemporary image: movement and mobility. Therefore, part III of this book draws on the conceptual and historical work of parts I and II in order to offer a new theory of digital media and of generative art in particular.

LIMITATIONS

There are four important limitations to these consequences. First and most broadly, this project is limited historically to the period when humans became the single most aesthetically productive and diversified species on the planet. Aesthetics and the distribution of moving images in general precede and exceed humans, but the history of kinesthetics described here is restricted, for practical and not theoretical reasons, to that of this limited historical period. The skeptical reader might ask, "If this book is so nonanthropocentric, why doesn't it deal with animal, plant, and natural images?" First, it is possible to give a nonanthropocentric theory of humans; there are many out there. Second, a future book is already planned to expand this frame. It just cannot be done in a single book.

The second, and perhaps most important limitation, is that the present work is limited strictly to the study of the *kinetic* structures of aesthetic practice. The theory of the image in this book is not a complete history of art nor of every great artist in the Western tradition. It does not pretend to do biographical, comparative, or encyclopedic justice to Western art or to all digital-media studies. What is unique about *Theory of the Image*, and where its contribution should be evaluated, is its focus on the hidden kinetic structures operating within the history of aesthetics, which reveal a subterranean aesthetics of motion. This limitation is not reductionistic. The argument is not that the mobility of the image is the only or best way to understand it. Rather, the argument here is *historical* and aims simply to add another interpretive dimension to others already out there from the perspective of the early twenty-first century.

Third, and within this second limitation, *Theory of the Image* is limited to the study of only the most dominant historical distributions of the image and its associated fields of motion, *considered separately*. In real history, by contrast, all the regimes and fields coexist and mix to one degree or another. To show all such mixtures and degrees for each historical period is too large a task and must be reserved for future studies. This book, therefore, considers only the dominant distributions of images, and only during the period of their historical rise to prominence. Part II is explicitly meant to be a rereading of the Western art-history tradition. The aim is not to focus on any single work, period, or type of art, or to create a new cannon, but, rather, to trace a broader, more holistic set of patterns over a long period of time in the West. The purpose of choosing well-known works of Western art is thus intentional. This book is meant to unsettle already settled histories by tracing a different history beneath them. Note also that the depth of coverage varies by topic. I have tried to avoid this, but complete and symmetrical historical coverage is not possible in a book of this size, and so I beg the reader's patience with this constraint.

Fourth, the present work is limited geographically to the near-Eastern and Western histories of aesthetic practice. In no way does this suggest that the West has the only or best art. On the contrary, revelation of the primacy of motion at the heart of Western aesthetics is a way of undoing certain prevailing notions of it by showing the secret material kinetic conditions of its static, idealist, anthropocentric, linguistic, and visual-centric theories. This book is restricted to the near-East and West purely owing to the practical limitations of length and the linguistic and cultural limitations of its author—and nothing more.

Limitations withstanding, this remains, I think, an ambitious project worth undertaking.

We begin this project in chapter 1 by introducing the kinetic theory of the image, which provides the theoretical framework for a kinesthetics of the work of art, and eventually the contemporary image.

PART I

Kinesthetics

Part I of this book offers a new theory of the image that begins neither from the objective, nor the subjective, nor even the ontological point of view but rather from the movement and mobility of the image itself. This first part of the book thus puts forward an original kinetic theory of the image, or "kinesthetics," defined by three interrelated aspects of the mobile image: the flow of matter, the fold of affect, and the field of art. The next three chapters provide the conceptual framework that we have extracted from art history and that we will then use as a method or lens through which to reread that history in part II. The method of theoretical presentation thus differs by nature from that of the initial historical inquiry.

CHAPTER 1
The Flow of Matter

Nihil in sensu quod non prius in materia

(There is nothing in the senses that has not first been in matter)[1]

The image is in motion; therefore, the theory of the image is also a theory of the moving image. One cannot be understood without the other. If the image were static, it could never move *between* a subject and an object. Objective and subjective theories of images as representations thus both assume precisely what they set out to explain: the mobility of the image itself. Hitherto existing theories of the image have thus only assumed and subordinated the movement of the image to something else; the point, however, is to look at the transformative movement of the image itself.

We begin our inquiry with the first and most general condition for the production of the image: the flow of matter.

FLOW

Matter flows. This is the first and central thesis from which the entire conceptual framework of kinesthetics follows. In order to understand what the minimal structure of the image is such that it is capable of being in motion, we begin with its most basic aspect: the flow of matter.

The image is nothing other than matter in motion. Without the material flow of photons, for example, there is no vision; without the flow of molecular pressure, there is no sound; without the flow of saliva, there is no taste; without the flow of air, there is no smell. Most importantly, however, without the flow of all matter, there is no collision, folding, or touch—the foundation of all images. The image occurs first and foremost only because matter is able to encounter itself—to touch itself. And matter is only able to touch itself if it is kinetically differentiated in some way from itself.

Therefore, matter becomes image only through a motion or a flow that allows it to return to itself. All images, therefore, must be in continuous motion—or else nothing, by definition, would ever touch. Without continuous movement (flow), there could be only a world of static, vacuum-sealed entities—and no sensation, affection, or image. The kinetic theory of the image therefore requires first of all a preliminary definition of the matter that defines the flow of images. From this initial definition follow a number of other aspects of the image.

Matter

The image is defined by the primacy of motion, but matter is what is in motion; therefore, matter is the basis of the image. What matter is, however, must remain an open question because motion is by definition a kinetic process. This is why there is no ontological definition of motion but, rather, only a *historical ontological* definition. In other words, the primacy of the moving image entails that matter also be a kinetic *process* and not reducible to empirical or metaphysical definitions.

Empiricism

Kinetic materialism is in contrast to the empiricist definition of matter as some specific or determinately sensed substance. This is because every sensorium, or body of images, is made of flows of matter that the composite images themselves do not represent but simply compose. For example, we do not see our own retinas or touch our own nerves. Our retinas see light, and our nerves transmit haptic signals. This is why the theory of the image begins with empirical sensation but leads us to the insensible or trans-empirical material conditions of the sensorium itself.[2]

Every sensorium or corporeal image with the capacity for sensation is always defined by the relatively insensible flows of matter that compose it. Those flows of matter (retinas, nerves, and so on) are in turn composed of flows of matter that define the retinal or nervous sensorium, and so on all the way down.

Matter is therefore not reducible to empirical sensations because there are no fundamental particles or substances that define the matter of the image—only material kinetic *processes*.[3] Matter is not reducible to static, discrete, or passive stuff that gets moved around, as in classical materialism.[4] Matter is creative, unstable, and in constant motion. Unlike classical matter, matter (in contemporary physics) is not completely observable, measurable, or predictable. Quantum fields, for example, are not empirical.[5] Quantum fields are also not causal or mechanistic but, rather, pedetic and indeterminate, and move in patterns of constant conjunction. The materialist theory of the image ought to take our historical knowledge of quantum fields seriously and distinguish it from classical, mechanistic, or "crude materialism," as Marx calls it.[6] Matter is not itself merely empirical but is also what the empirical is *made of*. It is, as Bergson writes, something "less than a thing."[7]

Process materialism is different from classical, mechanistic, or crude materialism in at least three ways:

1. Matter, like quantum fields, is not reducible to static, discrete, or passive stuff that gets moved around like billiard balls following universal natural or divine laws. Instead, matter is described as creative, unstable, and in constant motion.
2. Matter, unlike classical materialism, is not completely observable, measurable, or predictable.[8] Matter therefore is not strictly empirical or "actual" in the classical sense.
3. Matter, like quantum fields, is not causal or deterministic but, rather, is pedetic and indeterminate, and moves in unpredictable but emergent patterns of constant conjunction.

These features of matter are *historical features* consistent with but *not reducible to* the descriptions of contemporary quantum science. They clearly distinguish process materialism from classical or mechanistic materialisms.[9]

Metaphysics

Matter is not simply a concept or category of all material things. Matter is not an idealist and immaterial abstraction that exists independently of or transcends various historically determinate matters. Matter is nothing other than all its immanent historical configurations—so far. It can be nothing more until that something more emerges historically through motion. All universal ideas of matter come from material and historical beings in motion. To assert the contrary is idealism, or what Marx calls, "contemplative materialism."[10]

Thus, there is no single and absolute idea or definition of matter that will always capture its changing content in advance, just as there is no single and final empirical expression of matter for the same reason: matter is an open *process of motion*. The scientific determinations of matter as discrete particles or substance and the conceptual determinations of matter are both fundamentally limited because matter is in flux. In short, matter *is* nothing other than the *process of materialization*. If matter is not a fixed or static thing, then it can receive neither a fixed empirical nor a fixed metaphysical definition. The best way to describe what it is, therefore, is by what it does: move.

Thus, process or kinetic materialism is also distinct from a "vital materialism" in which the motion and activity of matter is explained by recourse to something else: either external forces (as with Isaac Newton) or internal immanent forces (as with Baruch Spinoza, Deleuze, and other neovitalist new materialists).[11] The ontologicalization of vital forces to explain matter's movement merely ontologizes a certain historical product (life) and retroactively projects this animacy onto non-living matter.[12]

Vitalist new materialism treats matter-in-motion as synonymous with mechanistic materialism and therefore sees the injection of *force* as the only pathway to a "new" kind of materialism. The fetishizing of the so-called immanent "life" or "vitality" of inorganic matters is also symptomatic of a more general biopolitical and ideological bias in contemporary politics.[13] Instead of starting with the primacy of matter, vital materialism starts with the primacy of biological life and retroactively attributes such living vitality to inorganic matter—when the historical situation is precisely the opposite.[14] Organic matter emerges from inorganic matter in motion, not the other way around. Therefore, the vital materialist attempt to theorize a post-humanist new materialism succeeds only by introducing a new biocentrism and by resubordinating matter and motion to something else.[15] It is thus ultimately a metaphysical and ahistorical materialism.[16]

Historical Materialism

Matter is what is *in motion*, but matter is also not reducible to motion itself. Motion in itself without a matter in motion is a pure and immobile abstraction. The kinetic theory of matter therefore adopts the name of "matter" not in an empirical or metaphysical way but in a strictly *historical* way, from the perspective or *kairos* of our present in which motion in the West has been connected to the motion *of matter*.

From Aristotle to G.W.F Hegel, motion has always been the motion *of matter*.[17] Together, the two have suffered the same fate in Western history: They are always subordinated to some other category. In the ancient world, matter and motion were subordinated to eternal forms and unmoved movers. In the medieval world, they were subordinated to the vital forces or *vis inertiae* that directed their motions and formed their matters. In the modern world, they were subordinated to mechanism, rationalism, and natural laws. However, just as the historical subordination of one almost always entailed the subordination of the other, so the historical liberation of one also entails the liberation of the other. If motion is primary, creative, and pedetic, then so is the matter that moves. If matter is, then so is the motion by which it is moved. Without matter, the concept of movement remains a "false" or idealist movement.[18] Without movement, however, matter remains static, discontinuous, and dead.

The theory of the image in this book therefore puts forward a new kinetic materialism. If the image is in motion and all of motion is in the process of materialization, then the image and matter can no longer be adequately defined by empirical or metaphysical methods. Kinetic or process materialism is therefore neither a Copernican revolution, in which it is we who move around the stars, nor a Ptolemaic counterrevolution, in which we are at rest while the stars move, but, rather, a Hubblean revolution in which *everything is in motion*. To become image, matter must be able to flow and, by flowing, return to itself as process of self-differentiating or iterated materialization.

Continuous Movement

The flow of matter, which makes the image, is a continuous movement. Matter flows *if and only if* the twin conditions of *continuity* and *motion* are satisfied.

If matter were *only continuous* (i.e., a continuous substance), it would be a homogeneous totality. It could never touch or sense itself because it would be strictly identical to itself. As Aristotle describes in *De anima*, "If a colored object is placed in immediate contact with the eye, it cannot be seen. . . . [T]he same occurs also with sounds and smells; if the object of either of these senses is in immediate contact with the organ no sensation is produced."[19] If matter were merely continuous, it would be One—a finite or infinite unity—without the possibility of change or motion outside of itself, since there would be no outside to it. As a minimum condition, therefore, the image requires a differentiation between inside and outside. If all matter were continuous substance, all movement and thus all images, as Zeno and Parmenides once argued, would be an illusion.

The logic of a static or substantial continuum refutes itself. If matter was One totality that contained all matters, the matter that contained all matters would have to be different from the matter that was contained by it. Matter would thus be separate from itself—that is, nontotal. We thus reach the paradox of the One that Gödel and others discovered long ago:[20] that the One cannot be contained in that which *it contains*. Material continuum without motion thus results in a paradoxical conception of totality that cannot include itself in its own totality. Differentiation and thus the image creep back into matter as constitutive dimensions of it.

On the other hand, if matter were *only movement* without continuity, there could paradoxically be no motion and thus no moving image at all. Strictly speaking, a discontinuous movement is not a movement. For example, without continuity the movement of translation between point A and point B cannot be said to be the *same* movement. Without continuity, point A and point B would remain completely different points divided by an infinity of intermediate points, themselves divided by an infinity of intermediate points, and so on ad infinitum. We can say there is a "change" that occurs since an entity is now at point A, now at point B; it changes from point A to point B. However, if there is no continuity between points A and B, then these points are not different aspects of the same movement but, rather, radically different points without any movement between them at all. Movement without continuity is thus not movement at all but merely discontinuous, formal, or logical *change*.[21]

According to the Greek philosopher Zeno, the problem with "discontinuous movement" is that if space is infinitely discontinuous or divisible, we would have to traverse an infinite distance of intervals in order to arrive anywhere else. Movement would therefore be impossible. The same result occurs, according to Zeno, when we understand movement as a series of temporal now-points or instants. If every unit of time is infinitely divisible, it will take an infinity of time to move from one point to any other. In both cases the problem remains the same: movement cannot be divided without destroying it. By thinking that we can divide movement into fixed, immobile stages, we spatialize, temporize, and thus immobilize it. "Discontinuous movement" is simply the *difference* between divisible snapshots of space-time and has nothing to do with movement at all. Therefore if we want to say that matter and images actually move, then such movement cannot emerge from discontinuity but, rather, must emerge from the twin conditions of continuity and motion: flow.[22]

Continuity and Discontinuity

Motion cannot be derived from stasis, and continuity cannot be derived from discontinuity, but the reverse is not true. Relative stasis and relative discontinuity *can* be derived from movement and continuity. If matter flows (in continuous movement), then discreteness would simply be a relative or regional stability of that flow. For example, the object and the subject would not be fundamentally separate from one another, divided by an infinite series of midway points; they would be regional stabilizations or folds of the continuous line between and through them. In the same way that the spatial points A and B presuppose the continuity of the line AB on which they are points, discrete and static beings presuppose the flow of matter of which they are folds, like the foam of an ocean wave.[23]

Here is the crux of the problem of the movement: either we begin with it or we never get it. This is a fundamental question for aesthetics. Either we begin with discrete and static images frozen in the subjective Apollonian mind or in a model image and have to say that motion is merely relative to something else, or we begin with the flow of matter and are able to explain both movement and stasis as relative or folded forms of images in motion. All the discrete images in the world will never give birth to a single moving image. The static image is nothing more than the "dead and artificial reorganization of movement by the mind," as Bergson writes.[24] Images are more like shimmering gemstones, shells, and sand washed ashore by the turbulent flows of watery Thetis—as Valéry writes in "Naissance de Vénus":

> Out of her mother's depths, still cold and steaming,
> Look, on the belabored sill of storms, the flesh
> Bitterly vomited up by the sea to the sun,
> Delivers itself from the diamonds of turmoil.
>
> Her smile comes to being, and along her white arms,
> (Be-gloomed by the orient of a shoulder's bruise)
> Follows the pure jewels of watery Thetis,
> And her tress blazes a shiver along her flanks.[25]

Intensive and Extensive Movement

Movement and stasis, continuity and discreteness are therefore, not opposed. They are two aspects or ways of describing the same *process* or flow of matter. It is thus more appropriate to distinguish between two dimensions or axes of movement: extensive and intensive. Along the first axis, extensive movement is made up of units of space-time *pace* Zeno. It is quantitative, measurable. Extensive movement is movement as change of place, locomotion, or translation. It moves from one discrete point to another by changing places. It is nothing other than the difference or change between points.

Along the second axis, movement is intensive and qualitative. It is a change in the whole, a transformation. In the example of the line AB, it is "already motion that has drawn the line"[26] to which A and B have been added afterward as its endpoints. A and B, subject and object, presuppose the movement and continuity of the line on which they are points. The division into A and B is always a division *of something*, an attempt to impose arbitrary divisions on continuous movement. Intensive movement is already primary, but we imagine it is not in order to explain it later as derived from something else. According to Bergson, however, "It is movement which is anterior to immobility."[27] Thus extensive movement is simply a regional or relative movement within a larger intensive movement. When an extensive movement occurs from A to B, the whole AB undergoes a qualitative or intensive transformation or change, like a wave.[28] An extensive point is nothing other than a stabilization or fold in an intensive flow.

For example, the difference between extensive and intensive motion can be seen, among other places,[29] in cinema. On the one hand, film is nothing other than a series of static freeze-frames moving extensively from point A to point B across a lens and through a beam of light. However, these discrete frames are also nothing other than images on a single vibrating and continuous strip of celluloid. The condition for the extensive movement of a frame is the intensive topological transformation of the whole reel. Furthermore, what seem to be discrete shots of different people and things extensively moving on the screen are also continuous flows of modulated light from the projector. The waves of light are continuously vibrating and changing in order to give the appearance of discrete persons and things on the screen. All perceived division and extensive movement are predicated on the intensive continuum upon which they are the topological regions, like boats bobbing on the ocean.

Bergson wrote that cinema was a bad description of perception, as if we perceive only snapshots of reality plus movement and get continuous reality. He is correct that this is a bad theory of perception, and it seems to be part of cinema from the perspective of the viewing subject who experiences the "illusion of movement" when people "move around" on the screen. However, from the perspective of the movement of matter itself, this is an inaccurate description of the cinema. The material conditions of cinema presuppose both the continuous intensive change of celluloid and flows of light and, at the same time, the extensive movement of relatively discrete figures on the screen and photos across the lens. They are two aspects or dimensions of the same motion. Films like *La Jetée* (1962) and *San Soleil* (1983) by Chris Marker, for example, demonstrate this explicitly by filming photographs and for extended durations where there is no visible movement on the screen or any characters doing anything. In this case, the viewer sees a seemingly immobile photo whose very conditions of extensive "stasis" are the intensive *motion* of its material body (celluloid and light). By inverting the relationship between perceived extensive and intensive motions in film, the true material-kinetic structure of cinema is revealed directly to the viewing audience.

All movement is therefore revealed as both extensive and intensive at the same time. The two occur as dimensions of the same process, but the former is always derived from the latter and not the other way around. Snapshots, for example, are aspects or dimensions of the material flow of celluloid and light, but continuous celluloid and light can never be the product of discrete snapshots. The two are present together when we watch a film, like the latitude and longitude of a kinesthetic cartography.

Pedesis

So far, we have said that the material conditions of the image flow in continuous motion and that these flows have an extensive and intensive aspect or dimension. However, for these flows to be capable of intersection, composite creation, and thus of producing images, they must also be capable of *curvature*—like the curved smile and long white arms of Venus in Paul Valéry's poem or the twisting appendages of Parmigianino's *Madonna with the Long Neck* (1534–1540).

Pedesis (from the proto-Indo-European [PIE] root **ped*-, meaning "foot") is the motion of autonomous self-transport: the motion of the foot to walk, to run, to leap, to dance unpredictably. Matter never flows in straight lines.[30]

The concept of pedesis is derived from two of the most important kinetic discoveries of twentieth-century physics: Einstein's kinetic theory of matter (1915) and Heisenberg's quantum uncertainty principle (1927). In the first, Einstein argued that all matter is a product of the stochastic or pedetic motion of innumerable smaller materials—molecules, atoms, and so on. For example, the atoms of *gases* move faster and farther, while those of *fluids* less so, and those of *solids* even less. All matter, Einstein showed, was not only in motion but also *in pedetic* or *Brownian motion*. Each movement is continuous with its previous position, but where it will go after that is indeterminate. The macroscopic conclusion is that the form of matter is fundamentally kinetic or kinomorphic, but also fundamentally and irreducibly pedetic.

However, by showing that all matter was in turbulent or pedetic motion, Einstein introduced a fundamental kinetic uncertainty and unpredictability into the heart of being, initially suggested by Ludwig Boltzmann. Since this discovery, science has been completely unable to produce a successful deterministic theory of turbulent motion beyond minimally probabilistic models. The description of kinetic turbulence goes all the way back to the Roman poet Lucretius, and the precise kinetic structure remains one of the last, and greatest, unsolved problems of classical physics.[31] The unsolved problem of classical turbulence, combined with Einstein's kinetic theory of matter, has had an enormous ontological consequence: that all matter is in motion and that all motion is fundamentally nondeterministic. This, and the related theory of entropy, has given rise to an entire field of chaos theory and nonlinear dynamics.[32] Heisenberg was said to have once remarked that he wanted to ask God two questions.[33] The first was "Why is general relativity so weird?" and the second was "How do you explain turbulence?" He then said that he was certain God would know the answer to the first question.

In the second kinetic theory, Heisenberg showed that there is a fundamental limit to the precision with which the position and momentum of a particle can be known at the same time. The more precise the position of a quantum field, the more it looks like a stable particle and the less we know about its momentum. The less precise the position of a field, the more it looks like a wave and more we know about its momentum through its diffraction pattern or waveform. In other words, motion cannot be reduced to position without destroying its motion, and the trajectory of a position cannot be predicted without the fundamental uncertainty of motion. This fundamental uncertainty about the motion of matter is not just an epistemological effect of observation.[34] It has been experimentally shown that this unpredictable or pedetic effect is inherent in the motion of the matter waves of all quantum objects.[35] The uncertainty principle and indeterminacy are fundamental properties of all quantum systems. Indeterminacy, however, is not random or even probabilistic because position only occurs in continuous relation to momentum. Heisenberg thus showed that even at the quantum level, matter in motion is both relational and indeterminate; that is, pedetic.

Pedesis may be irregular and unpredictable, but it is not *random*. What is interesting about movement is not simply that it is pedetic but also that it is through pedesis and turbulence that metastable formations and emergent orders are possible. In contrast, the ontology of randomness is quite bleak. In a purely random ontology, all of matter would be moving randomly at all times. Since fluctuations from disorder to order are physically rare, the likelihood that anything like the sun or even our galaxy would just suddenly pop into existence would be unimaginably rare and would likely fall apart immediately owing to further random motion. It would even be statistically possible for a human brain to pop into existence just long enough to think a thought and then disperse.[36] The very idea of a purely random motion presupposes that it was not affected by anything else previously, which presupposes that it was the first thing and before it was nothing, which is a version of the internally contradictory hypothesis of *ex nihilo* creation: something from nothing. The ontology of random motion claims that from pure disorder comes high-level composite order. Given the high level of order and complexity in our present age, randomness is demonstrably not the case.

Pedetic motion, on the other hand, is not random at all but rather emerges from and is influenced by other motions—just not in an absolutely necessary or completely determined way. Unlike randomness, pedetic motion is not unpredictable because it is *not influenced* by any other motions; rather, motion is pedetic *precisely because* it occurs in relation to other motions. It is the interrelation and mutual influence of matter with itself that causes its unpredictable character. Over a long period of time, the pedetic motion of matter combines and stabilizes into certain patterns, synchronies, and relations, giving the appearance of stability and solidity, only to become turbulent again and enter into new conjoined relations. A correlate of this attribute is that if matter is currently in motion, it must have always been in motion. If not, there would have been a time when there was no motion and motion emerged out of something immobile, which is an *ex nihilo* contradiction. If matter was always in motion and all motion is fundamentally pedetic, then it also follows that the motion of matter has always been pedetic.

Unlike randomness, pedesis is not defined strictly by disorder. Turbulence is a disordered motion, but it is a disordered motion that is capable of producing order because it responds to itself and others. Nonrandom disordered motion is capable of producing emergent metastable states. For example, the random air currents in a room produce invisible spiral patterns on which visible motes of dust float in the sunshine. The laminar flow of cigarette smoke flowing upward becomes turbulent and creates spirals and ring patterns. Spiraled storm systems emerge from the turbulence of coastal breezes.[37]

In van Gogh's paintings, for example, everything radiates and flows and whirls like water. His depiction of turbulence is physically accurate because the content of painting itself is taken over by the flow and flux of its material medium: the paint.[38] In *Road with Cypress and Star* (1890; figure 1.1),

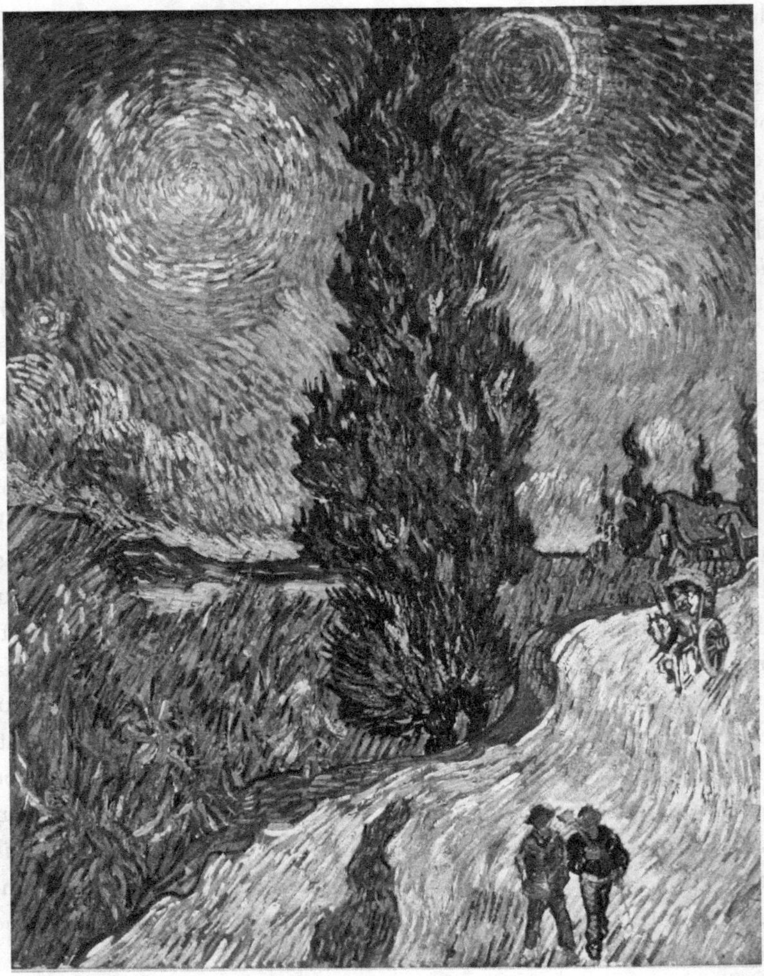

Figure 1.1 Van Gogh, *Road with Cypress and Star* (1890)
Source: Wikimedia, https://commons.wikimedia.org/wiki/File:Van_Gogh_-_Country_road_in_Provence_by_night.jpg

if the sky begins to vibrate and flow into a vortex, it is not because it is a more or less accurate representation of the world or of subjective experience (psychological impressionism) but, rather, because van Gogh discovered the materiality of the paint itself *as a fluid medium. Road with Cypress and Star* highlights the fluidity, texture, and pedetic nature of the paint itself. The painted image is a kinetic memory of the paint's journey, written in color. The cypress rises from the earth like the crash of a wave and bifurcates into two eddying whirlpools around the star and moon: a hymn to the material fluidity of paint itself.

Waves

If flows are completely continuous and undivided, but also pedetic, their motions are also interconnected motions of a whole, or *waves*. Flows are continuous but only insofar as they are themselves already simplexes whose topological distribution and micro-curvature are capable of stretching, bending, and modulating themselves infinitely without breaking. A flow is thus composed of different curvatures and topological regions, or waves.

Waves of matter are simplex, or one-folds. They bend, curve, and undulate, but they do not yet loop over themselves in a duplex. Because all motion is pedetic, the flows of matter are not straight or static lines but, rather, bent, curved, or wavy. What appears to be a straight line at one level is made of innumerable undulations and curves at lower levels, like a fractal coastline or a Mandelbrot fractal. From a continuous sequence of curved or bent lines (waves), a one-dimensional simplex is capable of producing an *n*-dimensional manifold.

In Botticelli's *The Birth of Venus* (1486; figure 1.2), for example, everything in the painting has become wavy as if it were all composed of the *same iterative wave*. The ripples in the ocean waves iterate the ripples in the scallop shell, which in turn iterate the ripples of hair in the wind, which in turn iterate the ripples of clothing; the rippled feathers of Zephyr; and the ripples of sunlight on the rippled coastline. The body of Venus leans or curves off-balance and her body is malapportioned in a proto-Mannerist style. The figures are poised and positioned as if riding a single wave of Venus' birth. These are not discrete figures on a canvas but, rather, showcase the wavelike nature of paint itself in a continuous waveform iterated or curved around itself. In *The Birth of Venus*, it is no longer clear which images are modeled on the others, but all enter into a shared resonance or kinetic waveform with each other—a single, continuous waving surface exactly like the painted wavy surface of the material

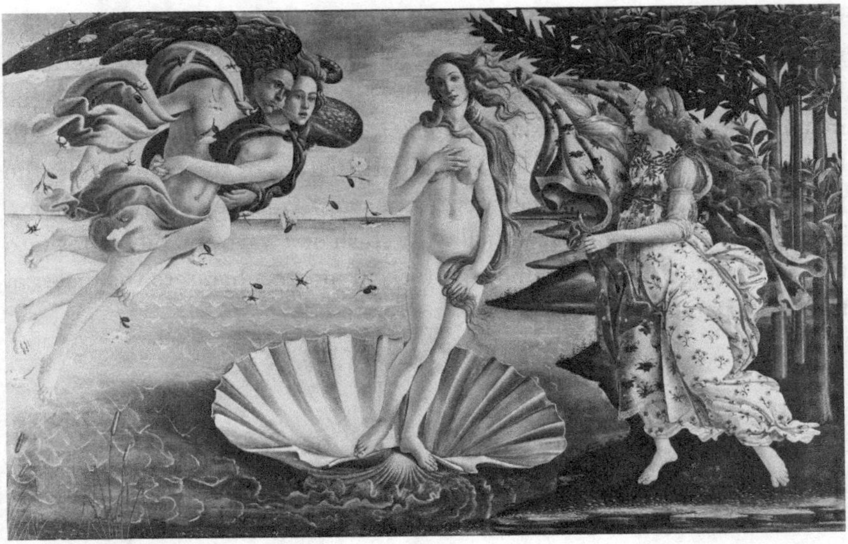

Figure 1.2 Botticelli, *The Birth of Venus* (1486)
Source: Wikimedia, https://commons.wikimedia.org/wiki/File:Sandro_Botticelli_-_La_nascita_di_Venere_-_Google_Art_Project_-_edited.jpg#/media/File:Sandro_Botticelli_-_La_nascita_di_Venere_-_Google_Art_Project_-_edited.jpg

canvas itself. Matter and form thus converge in the flow of paint on the stretched canvas.

Waves, however, are not parts of a whole, since they have no existence independent of the flow, nor will they ever have the power to separate themselves from the flow. They are nothing other, beyond, or above the flow of matter itself. The waves are the kinotoplological modulations or morphisms in the flow, and thus constitute the primary features of the flow of matter itself. Just as waves of water are not separate from the ocean, so the curvatures of flows are not separate, either. We can describe the different dimensions of a wave (crest and trough) without introducing a discontinuity into the ocean. The crest cannot be separated from the trough without destroying the whole waveform. Wave transport is thus intensive, since it moves by the transformation of the whole. For a single ball to wash ashore, the entire ocean must change.

The topological modulation of the waves is the internal self-differentiation of the flow with itself. The flow of matter is thus different from itself not by discontinuity but by curvature. The waves of a flow are not parts that were brought together at some point to create a whole flow but, rather, the kinetic and topological modulation of the flow of matter itself that gives it its dimensionality and potential for self-reflection: the image.

BIFURCATION

Flows of matter are continuous movements, but this does not mean that they cannot be divided. The idea of dividing a flow sounds impossible only if we think of division as introducing a radical discontinuity, lack, or break into the flow. However, as I hope to show in this brief section, this need not be the case. Division is not subtractive but, rather, additive—it multiplies by division—through bifurcation.

Multiplication by Division

In order to make this point, it is important to distinguish between two kinds of division following the two types of motion distinguished previously: extensive and intensive. The first kind of division, the extensive, introduces an absolute break—producing two quantitatively separate and *discontinuous* entities. The second kind of division, the intensive, adds a new path to the existing one, like a fork or bifurcation producing a qualitative change in the whole *continuous* flow. The bifurcation diverges from itself while still following the "same" pathway

Although division is typically understood according to the extensive definition, this is only relative to or a side effect of the intensive kind of division. Division occurs when a continuous process reaches a bifurcation point. By definition, a flow does not start or stop; instead, it bifurcates and is redirected (see figure 1.3). Thus every bifurcation is a bifurcation of a bifurcation, and so on ad infinitum, without any unbifurcated taproot or final accumulation of all the bifurcated flows. It is an open-ended process: a multiplicity of coexisting levels of bifurcation. After the bifurcation point in a flow, a qualitative divergence occurs and two distinct pathways can be identified. The result of this bifurcation is that the division is experienced both as a continuity and as a discontinuity, depending on where one is at in the flow.

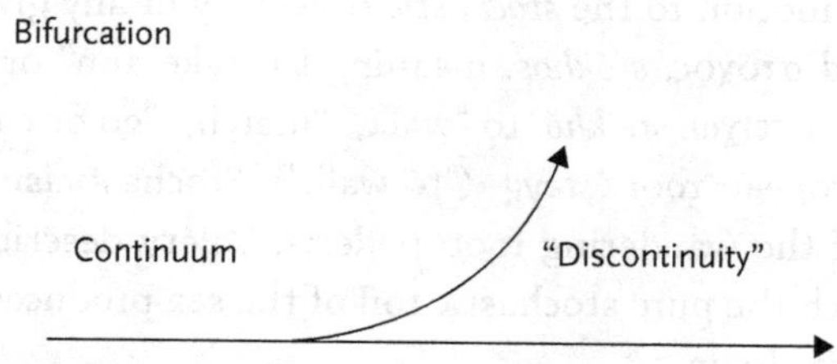

Figure 1.3 The bifurcation of flow

In both cases, however, what remains *primary* is the continuous flow that allows one flow to continue on ahead and another to be redirected elsewhere. In other words, division is an active process of bifurcation that does not simply divide once and for all but, rather, continuously redirects flows across or away from itself. The division only appears as a lack of or discontinuity from the binary perspective of the divided region. However, from the perspective of the continuum, the division appears conceptually as a secondary or derivative phenomenon.

But the problem with the extensive definition of division is that it presupposes precisely what it proposes to explain. If we begin with divided discrete objects and subjects, we fail to explain how these objects came to be delimited or divided in the first place. In contrast to Zeno's infinitely divided solid medium, we propose instead an infinitely bifurcated fluid medium. While the former results in immobility, the latter results in movement. This movement is not mechanistic but, rather, stochastic, thus making possible the confluence and intersection of flows with one another.[39]

CONFLUENCE

If the flows of matter and image are pedetic and capable of bifurcation, then it is also possible for them to flow together in a *confluence*. A confluence is the intersection or connection of two or more flows of matter intersecting one or more times. In this type of connection, multiple flows move together and intersect with one another without directly folding back over themselves. Confluence does not therefore divide flows but instead brings them together and distributes them without division. This type of collective connection is possible because flows do not necessarily move along straight, mechanical paths that would eliminate the possibility of intersection between heterogenous flows of matter. Intersection occurs because matter moves along a multiplicity of pedetic trajectories.

The collective effect of this pedesis makes it impossible to assign a single original causal motion to the *stochastic* trajectory of any given flow (from the Greek word στόχος, *stókhos*, meaning to "take aim" or "guess," from the Greek word στείχω, *steíkhō*, to "walk," "march," "go or come," from the proto-Indo-European root **steyg-* ("to walk"). Stochasticism is the experimental aim of the wandering foot: pedesis. Valéry describes beautifully the way in which the pure stochastic toil of the sea produces the manifold confluences of its foam.

What grace of light, what pure toil goes to form
The manifold diamond of the elusive foam![40]

This is why the so-called problem of turbulence in classical physics remains unsolvable. There is no first or final cause, only an infinite multiplicity of nonlinear variables-in-motion. Any classical series traced back far enough exposes this stochastic uncertainty. Since the variables that would determine the trajectory of matter include all of matter itself, a single causal source remains unassignable.

Given the kinetic possibility of confluence, several consequences follow for the theory of the image.

Event

The first consequence is the event. Once the motion of two or more flows of matter intersect or connect with one another, they create an event. An event is a singular point at which two or more flows cross (figure 1.4).

An event, however, is not an affect, thing, or object; rather, it is a kinetic hinge or terminal through which intersecting flows of matter pass. The event is, like the flows of matter that compose it, not an image but instead the condition of the image. Events occur just beneath the level of sensation in the bodies they compose. They are always insensible relative to the conditioned sensorium of images they compose. Yet, it is precisely the advent of their intersection and connection that makes possible stable and the higher orders of images and affects. At the level of the event, not only are there no sensible objects or things but also there are not even any affects or qualities (textures, colors, or sounds) of objects. Before qualities, affects, and images, there are first flows of matter that must already be capable of intersecting with one another. This is not a chronological sequence but rather a logical one. Flows and folds always coexist, but since folds are made flows, there is a logical or material primacy to the flows.

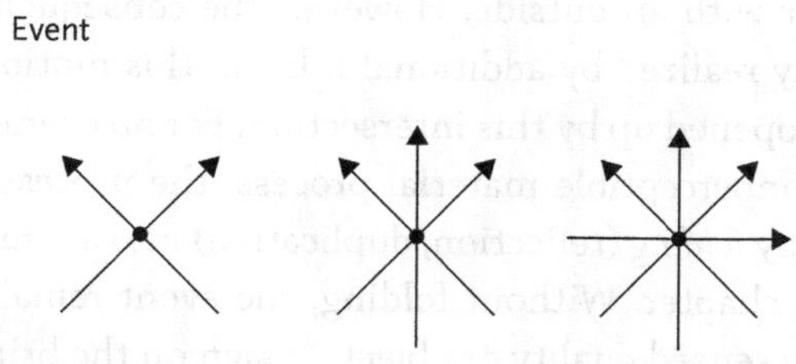

Figure 1.4 An event, from the intersection or connection of flows

The flows of matter and their events are therefore not sensible things but, instead, the active and creative processes that compose things. A flow is something that can only be known indirectly as the material condition of the sensible images that flow. This is because flows are strictly immanent to their folds. The sensible will always therefore have as its condition an insensible kinetic substratum of events that distributes it. In other words, in contrast to crude empiricism or contemplative materialism, the kinetic theory of the image adopts a transcendental empiricism or transcendental materialism that aims to discover the real material conditions for the emergence of the image *in motion.*[41] The flows of matter are not things or concepts but, rather, what things and concepts are made of. Matter is completely immanent to things and thus is not included as a thing alongside the others.

The intersection of flows of matter is thus not strictly identical to the flows that intersect, but it is not anything separate from them, either. Something new is produced by their composition. The event is thus a singular point shared by two or more intersecting flows that changes their possible trajectories: it opens up a new world of motion.

As a singular intersection, the event is also fleeting. A new future or trajectory of motion is opened up at this intersection. However, as an intersection, the flows of matter also continue on, leaving the event to dissolve immediately after its creation. The event then retroactively appears only as a trace or flash of something that once occurred. In addition, the event is also that which makes possible a potentially infinite new practical trajectory for the movement of matter and image.

There is thus a double nature to the event. On the one hand, it is the most common occurrence happening all the time and dissolving immediately, just as when one passes exits on a highway. On the other hand, it is also the rare occurrence when any given intersection becomes a new infinite trajectory, as when one chooses an exit.

Novel images arise when a flow is crossed by something from outside: another flow. In this way, the event is fundamentally collective; it always requires more than one. It occurs through an unpredictable encounter with another or with an outside. However, the consequences of this encounter are only realized by additional folds in this motion: by following the new vector opened up by this intersection. For an event to be anything more than an imperceptible material process, the process has to be further developed by *folding* (reflection, duplication) into an image—as we will see in the next chapter. Without folding, the event remains subaffective without directly sensed quality or object, "a sign on the brink of an abyss," as Stéphane Mallarmé writes.

Let's look at one example of this pre-affective event: the human brain. In the brain, there is an evental flow of matter immanent to human sensation and thought that remains relatively insensible and unthought: the flow of electricity through our neurons and bodies. What we call "conscious thought" is made up of billions of neurons moving charged particles along their axons—from one bifurcated dendrite system to another. The connectome is the total list of roughly 85 billion neurons and all the thousands of connections between them—roughly a hundred trillion or more connections in total. Conscious thought occurs only after a neuron "listens" to input for about 40 milliseconds at a time from its connected neurons. Before a given neuron fires, it receives about 280,000 possible messages before it decides whether to fire. Some signals increase the chance of firing, some decrease it, and the signals interact in complicated ways that are not fully known. All this happens before a thought occurs or does not occur.

There is thus a vast evental intersection of material flows through the connectome. A neuronal "event of sensation" is the synaptic intersection through which two or more electrical flows pass. For a neuron to fire and have any sensory effect it needs to first be connected to other neurons through the elctrochemical flows of matter and events. In this sense, neurological events are always collective whether or not any particular conscious thought emerges from them. In fact, vastly more events occur than do consciousness thoughts. Neuronal events produce effects that often dissipate instantly, with no clear causal effect. However, neurons that wire together are more likely to fire together again and thus are capable of producing effects that separately were not possible. In short, the sensorium of the brain is composed of a vast and active *flow of matter* that remains largely insensible to consciousness thought, and yet through habitual synchrony and folding, it can produce consistent images.

The same is true of the body and its nervous flows. For every corporeal sensorium there is a level below which sensation and images do not occur. But this does not mean that matter is not already moving, intersecting, and transforming our bodies. There is always a touch so soft that even the skin itself does not react. There is always a sound or light so slight that we do not hear or see it—and yet there are still sound and light waves flowing through the air and intersecting with the flows of our body. Only when these events can be sustained, synchronized, and folded into the body can they be sensed as images by that body.

What this means is that all sensory images in the body, including those of the brain itself, are first and foremost defined by a vast and intersecting flow of nervous and neuronal matter. Before we are capable of consciously or empirically sensing anything at all, matter must first flow through and intersect with the body and brain. This may sound both obvious and paradoxical at once, but the insensibility of the flows of matter are the condition of the image itself for the human body. This is the case both because there is an insensible matter that produces the sensorium of images and because there is a continual flow of matter through the sensorium, much of which remains insensible to the sensorium itself—unfolded into images by the senses.

As the condition of images, matter remains insensible to sensation, human or otherwise. Only after and through these material flows do we say, "I think" or "I feel" or "I imagine." The subject, object, and their images occur only as retroactive products of the more primary material process. We do not think and feel the flow of matter itself but, rather, the colors, sounds, and tastes it produces through repeated patterns. It is only after the flows of matter have produced confluences (events) that a durable image of them becomes possible.[42]

A kinesthetic event is like a transportation terminal, telecommunication terminal, or axon terminal in the brain. It allows new flows to enter and then be redirected or connected to other flows through this new point. The more flows intersect it, the greater the degree of kinetic freedom in destinations and the greater the collective transformation of the flows arriving there. In this sense, kinesthetic events have no essence. If the event is nothing other than the product of the intersection of any material flow that crosses it and changes each time a new flow is added, then it can have no unchanging essence or fundamental qualities.

Just as there is a neuronal plasticity, so there is a material plasticity in the flows of matter itself. Even when an event dissipates, and the flows of matter no longer intersect; the same flows continue elsewhere and in another form. Unlike essences, however, events do not pre-exist or persist independently of the confluence of their flows.

Constellation

The second consequence of kinesthetic confluence is the emergence of a *constellation* of events. Since the flows of matter are multiple, there is not only one event but also a multiplicity of events. When two or more flows intersect at two or more events, they form a constellation (figure 1.5).

A constellation is a *site, region, or surface of sensation* produced by the intersection of flows. Constellations have five characteristics. First, a constellation *has no essence*. Since the constellation is composed solely of events,

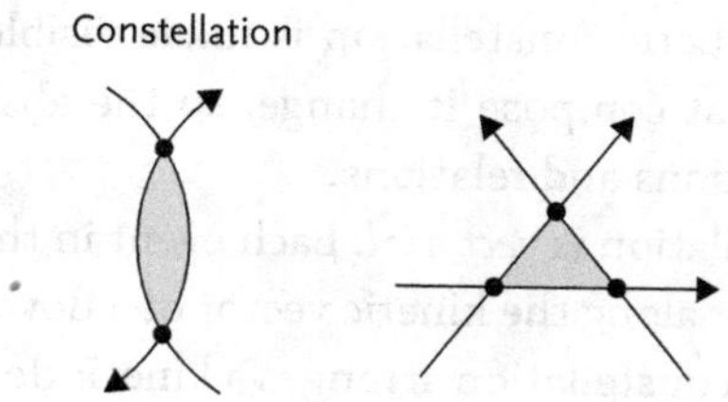

Figure 1.5 A constellation, formed when two or more flows intersect

which are composed only of flows of matter, the constellation has no preexistent or persisting essence, quality, or existence independent from the flows that compose it. Like an event, a constellation of matter does not block or capture the movement of flows but simply defines and gives a consistency to a region where flows intersect. Once the flows are redirected elsewhere, the constellation, like the event, ceases to persist in that form. The constellation is an assembly of insensible points without a fixed thing or object to which they refer.

Second, a constellation is a *kinetic surface*. Once the constellation appears at the intersection of several events, something like a metastable domain of motions can be discerned. This gives rise to the possibility of multiplying new events ad infinitum within the parameters of the constellation. Each new event along the outline of the constellation gives further definition to the constellation as a whole. Just as events become more complex and powerful the more flows intersect them, constellations also become more defined the more flows pass through the events that define the contours of the constellation. Furthermore, just as events can be the intersection of an infinite number of flows, so constellations can be outlined by an unlimited number of events without ever completely totalizing it. Since the flows that outline the constellation are continuous, they can intersect an infinite number of finite points. As continuities, there will always be room for another event between two others.

Third, a constellation is *abstract*. A constellation is abstract not in the sense that it is an illusion, fantasy, or purely mental entity. A kinesthetic constellation, just like an astronomical constellation, is *really there*, but only as the composition or confluence of heterogeneous stars and insensible connections. The constellation is absolutely real, but only as the confluence of heterogeneous events just below direct sensation and nothing more. The constellation is abstract, but the flows that populate and define its outline and motion are concrete. The abstract is only the image of the concrete and not its cause or origin. The constellation thus has no transcendence. An astronomical constellation, for example, is made visible by its stars, but a kinesthetic constellation is made visible by its events. As the concrete flows that compose it change, so the abstract constellation changes its configurations and relations.

Fourth, the constellation is *vectorial*. Each event in the constellation is a terminal or relay point along the kinetic vector of a flow. As a confluence of flows and events, the constellation arranges a kinetic domain that contains a multiplicity of heterogeneous perspectives without unifying them into a single perspective or totality. Movement in the constellation is distributed across singular events, each with its own vector and set of connecting flows. The constellation is not a totality of these flows but, rather, a fragmentary and open whole across which movement arrives and departs. Each time a new flow or event is added, the constellation changes and increases its conditions for development. The more flows and events populate the constellation, the stronger it becomes.

Fifth, the constellation is *additive*. That is, since the constellation is defined by nothing other than the collective set of movements that compose its concrete body, each new flow that is added changes the structure of the constellation. There is no static being of the constellation, only an additive and compositional becoming defined by motion. There is no discontinuity between one articulation of the constellation and another, only bifurcation.

It is not only the sensorium of the human body that is intersected by the flows of matter but all of matter itself also intersects with other matters below the threshold of the sensorium of images produced by the composition. We can thus extend the kinetic theory of insensation and confluence to all of matter. Even rocks and minerals have a sensorium. Insofar as they are capable of kinetically responding to sounds, light, and pressures, they have a capacity for action and reception that defines the most basic attributes of images. However, under a certain threshold there are also flows of matter that they remain relatively insensitive to. For instance, many rocks and metals can be made to resonate or vibrate at some sound frequencies but not others. There is a temperature at which minerals are affected by heat and will melt or freeze, but also a temperature at which no change in their sensorium can be detected. The same is true of plants and animals.

However, when certain collections of insensible events, or what we are calling constellations, begin to consistently intersect with one another, they are capable of producing collective consequences or images in the sensorium of the body. For example, just as the flows of neuronal electrochemcials end up firing together to produce a "thought" or "feeling" in a human body, so too other material flows are capable of producing effects in minerals, vegetables, and animals.

Let's look at another specific example to illustrate the idea that the sensation of images emerges from a more primary material constellation—one that is by no means restricted to the human sensation of images. The sensorium of the plant, for example, is an intersection of material flows that converge on a single point or event. The life or event of a plant was initially constituted by the intersection of two heterogeneous material flows: algae flows and bacterial flows. About 700 million years ago, a single blue-green alga (glaucophyte) swallowed a single cyanobacterium(*Cyanophora paradoxa*) and began extracting energy from its photosynthetic process. This event was singular and occurred only once, as evidenced by the fact that all subsequent plants share the same genetic structure following this event, but did not before.[43] All plants owe their existence to this singular alga-bacterium event.

There was no necessity of their intersection—only the pedetic contingency of their confluence in a fluid material medium. As flows of prey decreased and flows of light increased, this new form of life was capable of developing itself and becoming more powerful by following the new trajectory opened up by the alga's initial intersection with the cyanobacterium Although the event itself occurred only once, this unique life form was able to genetically internalize and self-generate or reproduce within itself a new cellular component: the chloroplast. This change made possible a new movement or trajectory of the algae toward the sun.

Most plants have a robust sensorium capable of sensing light, sound, touch, taste, and smell.[44] However, there are also material flows that they are not capable of sensing because the sensorium is composed of them—like the movement of their mitochondria—or because the flows are too subtle for their sensorium, as in the passage of radio waves or neutrinos through their bodies.

The constellation of plant life is thus defined by the new intersections that compose its vectorial being: water, air, sun, and nutrients. Without a continuous supply of any of these four flows, most plants will die. Given a continuous intersection of these flows, different plants develop and evolve along the different trajectories that follow these flows. Some plants moved more toward water (seaweeds), others toward the sun (trees), others under the earth (tubers), still others toward the air (epiphytes). This is not a metaphor. Plants literally moved underground, or up to the sky, or into the ocean to follow one of their confluent vectors more than the others. The constellation of plant life therefore has no original or unchanging essence, since it is the product of a more primary inessential process of material flows that produced it.

Events, such as the capture of cyanobacteria, and constellations such as those of air, water, sun, and nutrients, make possible new trajectories that a sensitive body can follow. A kinetic sensorium is the vectorial intersections that define a body's capacities for sensory images.[45] The more confluent flows that intersect in the constellation, and can be sustained, the more complex the form of life. From the simple to the most complex, all beings are subject to the transformative power of material flows.

The events that define the constellation occur only once, but the consequences can be sustained indefinitely as long as the flows persist that produced the constellation. The emergence of plant life, for example, is singular in the algae-bacteria event, but the constellation of water, air, sun, and nutrients must be continually replenished for the plant to reproduce itself. When one or more of its confluent flows dries up (lack of sun, nutrients, air, or water), the plant dies: its sensorium can no longer be maintained. But this does not mean that the material flows of the plant stop; they are simply redirected elsewhere. At the atomic level and below, flows of matter continue on in another form because they are neither created nor destroyed but only rearranged, like the flows of energy in thermodynamics.[46]

CONCLUSION

In this chapter, we have put forward the first concept in our kinesthetic framework: flow. The flow is the material condition for the emergence of images, which are defined by the folding of matters. In the concept of flow we have the necessary, but not yet sufficient, conditions for the emergence and mobility of the image. The theory of the kinetic event does not provide us with a theory of how confluences and constellations could support any kind of stability after an event. Confluence shows us how novelty is possible but not how such a novelty could come to exist as a stabilized pattern of images connected to one another in the world. For this, we require a theory of the affective fold, developed in the next chapter.

CHAPTER 2

The Fold of Affect

Nihil in intellectu quod non prius in sensu

(There is nothing in the mind that has not first been in the senses)

Matter flows, but the flows of matter are also capable of folding into affects and images. All that is air condenses back into solid. This is the second conceptual component of kinesthetics. Matter flows but it also folds over itself in habitual cycles and patterns of dynamic equilibrium, creating regional stabilities-in-motion: images.

If we begin our theory of the image from discontinuity and stasis, the (impossible) challenge is to theorize its movement. However, if we begin our theory of the image from the primacy of movement, the challenge is to theorize stability. In the previous chapter, we focused on the relatively insensible material conditions for the emergence of the image and the sensorium. To this we now add the theory of how affective folds can emerge from the confluences of material flows through "junctions" and conjoin with one another to produce larger conjunctions or images (figure 2.1). Matter flows and folds into affects, which conjoin into images.

AFFECT

An affect is the active and receptive capacity of matter to sense and be sensed. In other words, an affect is produced by the fold or intersection of a flow of matter *with itself*. If all of sensation is made of

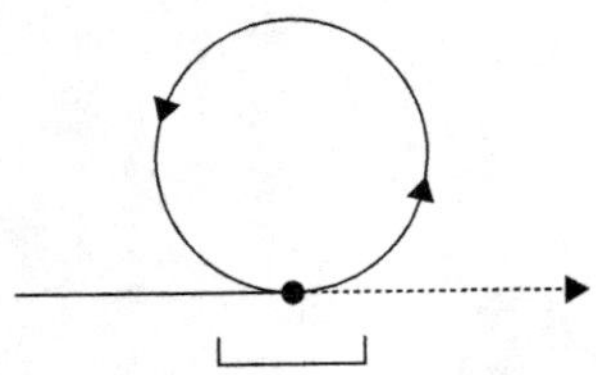

Figure 2.1 Fold and junction, from the convergence of flows

continuous flows of matter, folds explain the kinetic structure of the relative stasis or stability that emerges from this process. An affective fold is like an eddy or whirlpool in the flows of matter. It is a relative stasis that is always secondary to the primacy of the flows and events that compose it.

As such, an affective fold is nothing other than a flow of matter. A fold does not transcend or preexist the flow; it is simply the redirection of a flow of matter back onto itself in a loop or self-affection. It is therefore mistaken to think of the fold as a mere product of a flow as if the two were ontologically separate. The fold that moves already presupposes a more primary constitutive flow that composes and moves through it: the creative movement of the matter itself. Material flow and affective fold are thus co-constituted in the same immanent kinesthetic process.

In this way, an affective fold is distinct from a material confluence. A confluence is an open whole of two or more intersecting and heterogeneous flows of matter, but a fold occurs when a single flow loops back over and affects *itself*. A confluence is a novel but potentially fleeting intersection, but a fold is what occurs afterward as an attempt to stabilize or repeat the unique moment of intersection. An affective fold is the repetition of a kinetic differential: a cycle. A fold remains a kinetic process but becomes a vortical process that continues to repeat in approximately (not exactly) the same looping pattern, creating a kind of mobile stability or homeorhesis.[1] The point at which the flow of matter returns to itself may be an affective fold in an evental or nonevental trajectory. In this way, affect constitutes a point of self-reference or haptic circularity in matter that yokes a flow to itself and is assembled into an image.[2]

The affective fold, then, acts like a filter or sieve that allows some flows of matter to pass through or around the recurrent attractor of the cycle and other flows to be caught in the repeating fold of self-affection. The movement of the captured flow can then be connected to the movement of another captured flow and made into all manner of mobile composites, conjunctions, or images.

However, the yoking or joining of the flows into an affective fold also augments them, not necessarily by moving them faster or slower but by subordinating them to a cycle that begins and ends at the same haptic or affective point. All kinetic affect thus has two basic dimensions or aspects: its period and its cycle.

Period

We call the affective point at which a flow intersects with itself a "period." Although the flow of matter is continually changing and moving around the loop, this haptic point appears to remain in the same place—like the eddy of a river. In this sense, the point appears to absorb and regulate all the mobility of the yoked flow while itself remaining relatively immobile.

Affective folds occur only in that which is continuously moving. This is because a fold is defined by the reflection, curving, or bending of something back over itself, and *not over something else*. The intersection of a flow with another flow is not a fold but, rather, an encounter or event—as defined in the previous chapter. The structure of the fold is kinesthetically different. The fold is capable of producing recurrent cycles and periods, while the event is fleeting and singular. If matter were not continuously flowing there could be no folds, or even events. Folding presupposes kinetic continuity, and continuity makes possible the affective fold of matter.

Cycle

Matter flows and folds over itself, but once it returns and connects to itself again, it creates a cycle. A cycle is the movement between the departure of a flow of matter from a bifurcation point and its return to or arrival at that same point. It is a duplication, echo, or image. This point is the periodic attractor of the cycle with itself. While the concept of identity has been historically conceptualized as a purely logical or formal concept, often relating to essences, the kinetic terms "period" and "cycle" differ significantly because they are the product or effect of the more primary process of cyclical and affective motion.

Simply defined, the period is the point in a cycle where a flow recurrently intersects with and duplicates itself to some degree of iterative frequency or density. The period of a cycle does not create a perfect regularity or

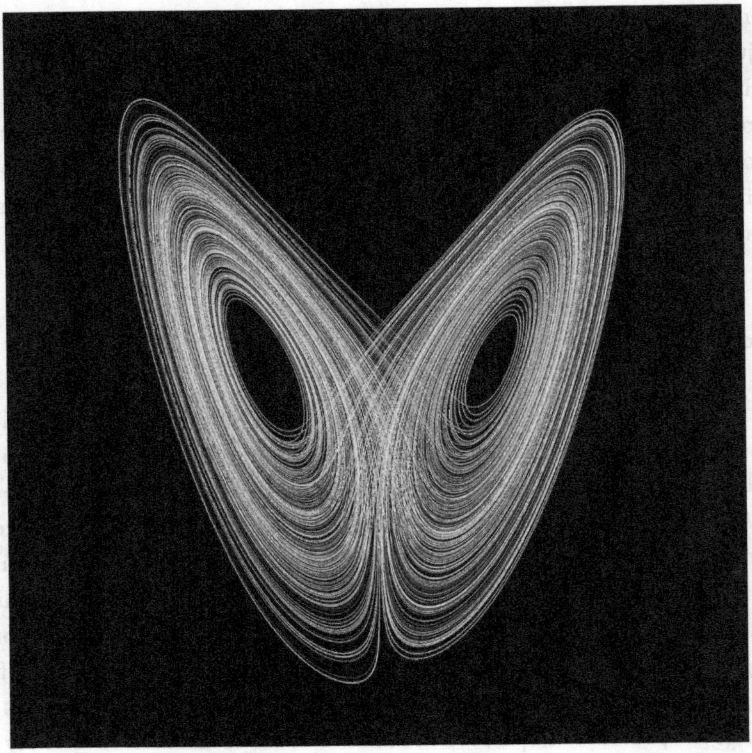

Figure 2.2 Lorenz attractor
Source: Wikimedia, https://commons.wikimedia.org/wiki/Lorenz_attractor#/media/File:Lorenz_system_r28_s10_b2-6666.png

classical "identity" as much as a metastable aperiodicity around a "strange attractor" that tends to overlap with itself again and again at irregular but frequent and infinitely differential approximate intervals (figure 2.2).

Thus a kinetic period is simply a tendency toward a certain intersection around a region of attraction and not a mere repetition of a regularity. The period and cycle are thus unstable and differential at the local level and only approximately stable at the global level.

A cycle is the whole process of self-affection of a flow with itself, but this does not mean that the flow of matter has been arrested or rendered completely discrete. The period is simply a slice or selection from the whole continuous recurrent process (figure 2.3). When we mistake the periodic attractor for a simple static or fixed point we lose the flow entirely, we see only an abstract product without the motion that composed it.[3]

It is as if we looked at a Jackson Pollock painting and wondered how he was able to paint such wonderfully detailed replicas of paint splatter. In this case, we have misunderstood everything about the kinesthetics of painting;

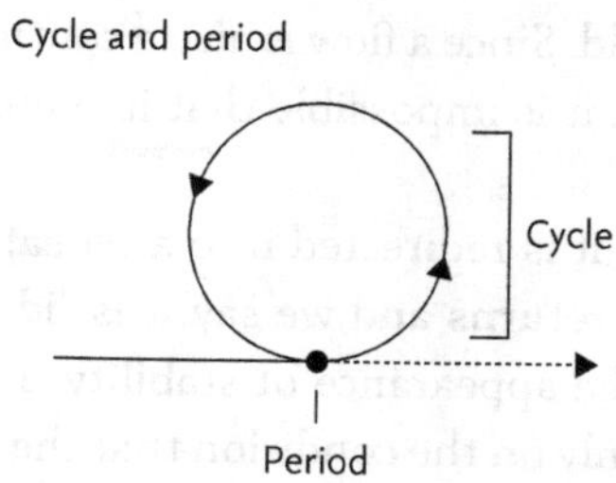

Figure 2.3 Cycle and period of flow

we have abstracted the product from the iterative process. Jackson Pollock, for example, flung paint off sticks, broken glass, and string using the turbulence of the air to let the paint fold and twist itself above the canvas before falling upon it in numerous layers, cycles, and pleats. These are not mimetic works based on sketches but, as Pollock says, a "direct" use of the self-formation of paint in relation to air flow. Pollock describes the process as simply guiding the continuous flow of paint, which runs "without beginning or end" or without absolute center or orientation. "A painting has a life of its own. I try to let it live," he says.[4]

The French painter Simon Hantaï similarly describes his process of folding or *pliage* as "coming out of nothing. You simply had to put yourself in the place of those who had never seen anything; put yourself in the canvas. You could fill a folded canvas without knowing where the edge was. You have no idea where it will stop. You could go even further and paint with your eyes closed."[5] Hantaï begins with the insensation of material flows, the canvas itself, the blind eye, almost always hidden under the paint, and from this insensate flow produces folds of sensation through cycles and periods in the canvas. What appears on the canvas as discrete fragmented shards of dense periodic orbits of colored paint are in fact only topological regions of a continuous flow of paint folded by the surface. The kinesthetic insight in Pollack and Hantaï is that the continuous produces the discrete and that affect and image emerge from the insensible of the air (Pollack) and blank canvas (Hantaï).

The recurrent cycle produces a regional attractor or "identity," but only through motion—only through *folding*. Since a flow is a continuous movement, the fold is not only continually receiving a constant source of new motion from outside but also losing some motion that passes through its fold. Thus, a fold is only a regional capture of motion in a certain period. This is because when it intersects itself, it is actually intersecting itself at a different point in the flow each time. There is no "fixed period"—only more or less dense periodic orbits or "limit cycles" that continue to shift around a metastable fold. Since a flow is also a continually moving and self-differentiating process, it is impossible that it should ever be the same, in a strict sense, as itself.

However, insofar as it is redirected into a repeating pattern of motion, the pattern of motion returns and we say it is "identical." Images persist over time, taking on the appearance of stability. Thus, we *can* step in the same river twice, but only on the condition that the river also turns over itself in local eddies and whirlpools.[6] The periodic fold remains the *same*, but only on the condition that *others* flow through it. As Heraclitus writes, "On those stepping into rivers staying the same, other and other waters flow."[7]

For Heraclitus, each eddy in the river is like another river within the river. Each flow below the surface of the river is already multiple, but it is this multiplicity that composes the stream itself, "Like one carried beneath the surface of a stream, is interrupted, torn, pricked and plucked at by sensations, spontaneous and irrelevant," as Virginia Woolf writes in *The Waves*.[8]

Each period in the eddy of the river is composed of entirely different water molecules, but the cycle of the whirlpool persists. The cycle remains "the same eternal return," like Valéry's depiction of Narcissus' reflective pool:

> Admire in Narcissus the eternal return
> toward the mirror of the water which offers his image to his love,
> and to his beauty all his knowledge.
> All my fate is obedience
> to the force of my love.[9]

As the continuous flows of matter slow and pool into cyclical folds, they make possible a smooth and stable surface in which images emerge.[10] Gide writes:

> Alas, when will Time cease its flight and allow this flow to rest? Forms, divine and perennial forms which only wait for rest in order to reappear! O when, in what night, will you crystallize again?
>
> Paradise must always be re-created. It is not in some remote Thule; it lingers under the appearance. Everything holds within itself, as potentiality, the intimate harmony of its being—just as every salt holds within itself the archetype of its crystal. And a time of silent night will come when the waters will descend, more dense; then, in the unperturbed abysses, the secret crystals will bloom.[11]

The flow from period to period prime, considered as a whole process, is thus the cycle of the fold. A cycle is not a static unity but, rather, a fluid or kinetic unity, like the recreation of Narcissus's pool from the flows of rain and river drainage that crystallize into the image, only to descend and rise again as "blooming secret crystals" of affection from the abyss.

Since the fold is only a fold in a continual flow that constantly enters and exits it, renewing it each time, its cycle cannot be said to be the unity of an ideal identity but, rather, of a kinetic process. Just like a whirlpool in a river or Narcissus's pool, the cycle is only a metastable unity of a differential process refreshed each time with new water around a periodic attractor—"paradise re-created in appearance," as André Gide writes. Kinesthetics thus replaces the concepts of identity and unity with the concepts of a periodic attractor and a differential cycle.

SENSATION

An affect can contain one or more cycles. Each cycle departs from and returns to the same period in larger or smaller intercalated loops. As each cycle returns to the same periodic point, it reproduces the identity of the cycle with itself and reproduces the unity of all the intercalated cycles with the same periodicity. This structure also makes possible an important kinesthetic effect: sensation (figure 2.4).

Sensation occurs at the period when a flow folds back over itself and touches or affects itself. Sensation is the ambiguous kinetic structure of the period itself—the double or split affect of periodicity. It is a single and same period, but also an intersection of two different points in the same flow. Sensation is the *kinetic difference* between sensibility and the sensed. The two are identical in the period of sensation (the sensed), but are differentiated in the continuous movement of the flow across its cycle (sensibility). Sensation is the kinetic differentiation internal to the

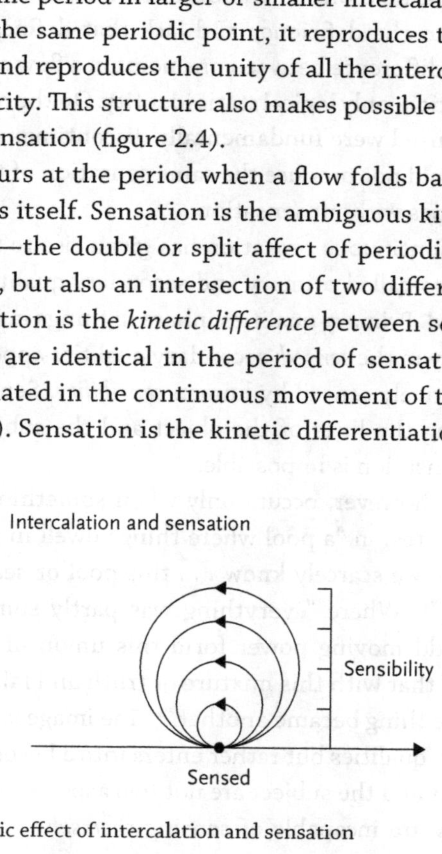

Figure 2.4 Kinesthetic effect of intercalation and sensation

fold that makes possible self-affection or self-sensation. In short, sensation is the sense of the sensed as the kinetic identity of the kinetically different.

Sensation is what happens when a flow affects itself. It is a calm pool or an eddy in the flows that make possible the sensation and beauty found in the image of Narcissus's reflection. Matter reflected or folded as sensuous image. Narcissus's pool gives us the perfect description of the image not as model and copy but as a continuum of matter flowing into both pond and body alike. Object and subject become two sides of the same kinetic process. As Charles Baudelaire writes,

> There all is order and beauty,
> luxury, calm, and sensuousness.[12]

Turbulent waters give raise to the ordered beauty of images when they become calmed in the cycles of a pool. Object and subject are no longer divided but rather partake one in the other as a calm luxury or sensuous excess—opposed to the deficiency and lack that defines the model and copy. Without the difference between one point in a flow and another there would be no sensation, only logical static identity. On the other hand, if sensibility and the sensed were fundamentally discontinuous entities (model and copy), they could not produce the same sensation of the sensed—only different sensibilities without sensation.

The idea of discontinuous sensation has given rise to the philosophical division between so-called primary and secondary qualities, from at least John Locke onward. Primary qualities inhere in things in themselves, since they are objectively in the *sensed*; secondary qualities appear in things only as they are subjectively sensed by human *sensibility*. Sensibility has been, thus, divided along the lines of the object and the subject. Under these conditions, *real* sensation is impossible.

Real sensation, however, occurs only when something senses *itself* as other, as Woolf writes, in "a pool where things dwell in darkness so deep that what they are we scarcely know . . . this pool or sea in which everything is reflected."[13] Where "everything was partly something else, and each gained an odd moving power form this union of itself and something not itself so that with this mixture of truth and falsehood . . . things moved . . . and one thing became another."[14] The image is not a copy of primary or secondary qualities but rather enters into a becoming with what it is not. If the object and the subject are not two aspects of the *same* process of becoming, they are incapable of sensing *themselves* as other. They are *merely other*. Only when subject and object are continuous and undivided is self-affection possible.

Sensation must, therefore, be a kinetic structure of *self*-receptivity, a capacity to be affected and to affect at the same time. Simply put, a flow has the capacity of being receptive to other points in the same flow. Two different regions of the same continuous flux have the capacity to touch, intersect, and respond to one another as *different* points sensed in the *same* periodic intersection. Sensation occurs in the fold of affect where a flow returns to itself at a period in a continual cycle of acting and reacting back on itself. The *two* points of the flow become *one* periodic attractor, but in doing so each point becomes twofold, or duplex. The one folds into two, but the two also fold into one.

In the image, each point becomes *itself* as an *other* to *itself*. The point is still a singular point in a flow, but in joining with another singular point, it is both itself and another. However, if the two points are now each twofold in the same period, they are together a fourfold. Every cyclical repetition of the image is thus an iteration, or *repetition of difference*.

The sensed is the exact point of intersection where the two different points become one period in their *being sensed*. Therefore, the sensed being does not precede the process of sensation. Every flow contains a great many possible capacities or points of affection, but not every sensation is always expressed in action or is intersected at the same periods. We have the capacity to taste, but we are not always tasting. The sensation of taste requires the flow of food to fold in the mouth on a taste bud. If the food is not folded at the right period where it intersects with the taste bud, no sensation occurs. When the flow intersects sensation occurs; when the flows of matter unfold, sensation does not occur. Since sensation is practical and kinetic, and not essential, we never know with absolute certainty what sensation can do.

Theory of the Subject

The kinetic process of sensation has two distinct operations that define the affective structure of the subject: receptivity and redirection. The fold of affect either allows flows to pass through or it delays them by redirection. The period adds nothing to what it receives. This is clearly a broad notion of subjectivity that includes all kinds of entities traditionally not considered to be subjects, based on the criteria of human consciousness. Plants and human brains are both composed of affects, only at different scales and degrees of complexity.

A notable exception to this anthropocentric bias is the brilliant work of the German biologist Jakob von Uexküll (1864–1944), who defines subjectivity much more broadly and kinetically as nothing but the circulation of material flows between the three subjective functions of perception, structure, and effect in a continuous "functional cycle" with the world (figure 2.5).[15]

The world gives off material flows that are first perceived by the organism and then internalized in its organic structure, retained in the form of effects on its self and then directed back onto the world. This is an important move, even if von Uexküll's approach still remains biocentric in the end. What we need to do is extend this idea to the kinetic activity of inorganic matter itself: images.

The most minimal definition of the kinetic subject we can give is that it is defined by the process of internalization or folding. But the process of internalization accomplished by the kinetic fold always has two parts: receptivity and redirection. These are not intrinsically biological functions. Rocks, for example, are composed of kinetic flows of molecules that fold over and affect themselves internally. In this way, the rock touches or senses itself. Rocks have a receptivity to light, heat, cold, and other material flows that can transform them. They also have a capacity for redirection that allows their flows to persist together or decompose entropically in their motion. Therefore, there is a kind of subjectivity in minerals and in all matter.

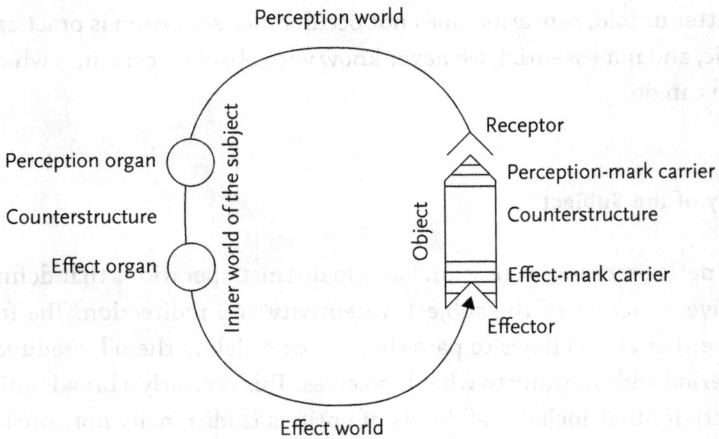

Figure 2.5 Continuous functional cycle with the world
Source: From Jakob Uexküll, *A Foray into the Worlds of Animals and Humans* (Minneapolis: University of Minnesota Press, 2010), p. 49

Sensation should thus be contrasted with representation or mimesis. Representation is a duplication or mimesis of one discrete thing by another. Representation thus presupposes the existence of a distinct original that precedes the copy, as if one point in a flow were simply a copy of another and not a fold or period. Sensation, on the other hand, is produced only at the unique point of intersection between two differentiated regions of the same flow. Sensation is the repetitive *intersection of differences* and not *a replication of a previous point*.

Sensation is a crossing over, or *chiasm* (from the Greek letter χ), that combines the operations of receptivity and redirection at a single kinetic period. The hand that touches is also touched back by what it touches.[16] At the *chiasm* of sensation, the flow that actively bends and returns back on itself is also the same flow that receives this folded flow. At the point of intersection, the flow either passes across itself, continuing elsewhere, or it is taken back up into a periodic cycle to be repeated again.

The materiality of the affective subject is split, just like affect itself, into two halves. On one side, it is defined by an active process of redirection and bifurcation that allows some flows to move along thorough the fold and into a recurrent sequence of folds. In this capacity, the subject acts like an active filter or sieve pushing matter forward. On the other side, the subject is defined by a kinetic receptivity as these flows return back onto the subject itself and transform it. Different kinetic or affective subjectivities are therefore not different in kind but are different in degree and distribution.

The operations of receptivity and redirection are crucial for the persistence of events. An event produces a new intersection of flows that can also become the site of a new cycle. On the one hand, an evental intersection very well might remain a fleeting and strange moment with only minimal or ambiguous persistence.[17] It can open up a new trajectory or perspective, but it does not necessarily sustain it.

On the other hand, one or more of the intersecting flows passing through an evental point can return and fold over itself at that point in order to sustain it. This requires both the receptivity of the flow and the redirection of its trajectory back into a stabilizing fold: a strange attractor. While the original event was singular, the fold is cyclical and periodic. The event lives on in the self-generating consequences of the affective fold and its recurrent sequence.

It also follows from this kinetic definition that sensation and subjectivity do not refer only to human sensation or even to the sensations of organic life. Subjectivity occurs wherever there is a receptivity and redirection of flows. Even minerals in painting, sculpture, and in books have a receptive capacity to be affected and to redirect the kinetic flows of temperature and pressure. For example, according to Einstein's kinetic theory of matter, a piece of basalt is composed of innumerable atomic flows that move extremely slowly and vibrate in dense patterns. Some of these flows are redirected away from this dense compound by erosion, but most move only until they reach a certain limit and then fold back into the rest of the receptive vibrating and cycling atoms bouncing off one another within certain relatively stable periodic limits. The flows of silicon and oxygen atoms in basalt, for example, move in slow general cycles outward and inward, folding and refolding at certain junctions and forming ionic compounds. However, once another flow of more rapidly moving atoms or photons (heat) collides with this basalt rock, the silicon and oxygen become receptive to this motion and respond by redirecting their flows elsewhere, melting into a liquid between 984 and 1260 degrees Celsius. Similar capacities for receptivity and redirection can be given in physics, chemistry, biology, and other natural sciences. At a general level, all matter and therefore images have some capacity for kinetic sensation and therefore affective subjectivity. Subjectivity is a cycle of *both* active and reactive motion.

Quality

Periodic cycles produce sensation, but from these sensations also come differing qualities of sensation. When a flow folds back and intersects with itself, it produces a sensation *of something*: the sensed. Events are the insensible connections of matter that make possible a new persistence but do not necessarily ensure it. Affects, however, sustain these evental connections through folding even if such minimal sensations are not yet conjoined to any others and remain unattached things.

Affects without conjunction are like a flash of light or color without an obvious source, a feeling of urgency without a plan or program of action, an ambient sound without discernible instrument, and so on. Unordered affects are sensed independently of a clear origin or object. Woolf describes brilliantly the way in which motion reveals the pure affects that compose individual things.

> Observe how dots and dashes are beginning, as I walk, to run themselves into continuous lines, how things are losing the bald, the separate identity that they had as I walked up those steps. The great red pot is now a reddish streak in a wave of yellowish green. The world is beginning to move past me like the banks of a hedge when the train starts, like the waves of the sea when a steamer moves. I am moving too, am becoming involved in the general sequence.[18]

In motion there is no pot, only a streak of red, a wave of yellow; no hedges but a streak of green; no waves but a streak of blue; and ultimately no unified subject but the general affective sequence or flow of sensation.[19] Things lose their discrete identity and gain a kinetic identity in the persistence or flow of their affects. Outside of this affective kinetic sensibility there is no transcendent essence of the image. Images and things are made of affects, not the other way around. Victor Hugo, too, writes that

> The flowers of the side by the road are no longer flowers, but flecks, or rather streaks, of red or white; there are no longer any points, everything becomes a streak; the grainfields are great shocks of yellow hair; fields of alfalfa, long green tresses; the towns, the steeples, and the trees perform a crazy mingling dance on the horizon; from time to time, a shadow, a shape, a spectre appears and disappears with lightning speed behind a window: it's a railway guard.[20]

This affective kinetic quality produced by flows differs from classical ideas of quality in several ways. First, a kinetic quality never exists independently of a flow; it exists only in and through periodic sensation. Quality, contrary to Plato, does not transcend its concrete manifestations in matter.[21] The same quality can appear in different things without there being an unchanging transcendent form of this quality, independent of the fold, because flows are capable of being moved and affected in similar patterns in more than one place at a time. Furthermore, the same affect can be shared by more than one flow at a time as they converge and cycle around the same affective period. This movement does not require any immaterial form or idea.

Second, a kinetic quality is not an attribute of a pre-existing substance. Kinetic quality, contrary to Aristotle, is not a mere attribute of "one and the selfsame substance" that "while retaining its identity, is yet capable of admitting contrary qualities."[22] A kinetic quality is not attributed after the fact to a pre-existing thing or substance to which the quality is attached as something other than the thing. The quality and the thing are produced at the same time in the fold because the thing is nothing other than the conjunction of its kinetic affects.

Third, a quality is not an essence. An essential quality is a quality that a thing has independent of any observation of it, and that must remain the same for that thing to be what it is. For example, a primary or essential quality of a book is that it must have pages. If we remove all but one page of a book, a single piece of paper is by definition no longer a book. However, the color of the book is an accidental quality or property of the book. If a book is first white but then painted black, it remains a book regardless of its color. Kinetic qualities do not follow this opposition between objective essential and subjective accidental qualities because all qualities are affective functions of the same fluent process.[23]

In other words, what a thing "is" changes each time one of its qualities changes. For example, the book with one page has a diminished capacity for being read but an increased capacity for portability; the black book has diminished capacity for reflecting light waves but an increased capacity for absorbing them. There are no essential or accidental qualities, only diminished and increased capacities for specific kinetic sensations.

The Waveform Theory of Quality

The self-intersection of a flow is a period of sensation, but it also defines the kinetic quality of the sensed. Periodicity is the process by which two different aspects of a flow become one in the same period. This period appears not as an abstract or logical identity (A = A) but as a kinetically *qualified identity*: as a *certain* solidity, size, speed, color, temperature, and so on. Depending on the way the flows are folded over one another, they produce different qualities.

For example, kinetic theories in physics have elaborated fairly robust accounts of qualities based on the thesis that all particles constantly move and vibrate in continuous frequencies or cycles and waveforms. In physics, all sensible matter—that is, matter above the Planck scale—can be understood according to its kinotopological waveform. Subatomic particles, atoms, and molecules all move and thus have a frequency of some kind. If all matter moves, all matter also has a frequency or waveform of movement that defines its qualities at different levels of emergence—atomic, molecular, cellular, organism, social, planetary, and so on.

For example, the kinetic *density* of folds determines the solid, liquid, or gas quality of the thing, while the kinetic *frequency* determines the visible, audible, gustatory, olfactory, quality and temperature of the thing. The wavelength across the electromagnetic and pressure spectrums produces the qualities of color and sound; the different vibratory frequency of atoms and molecules determine the taste and smell of matter; while the kinetic speed of the folds determines the quality of their temperature.

Solids

Hard and dense things such as rocks, for example, are hard because their vibrations or frequencies are held together very tightly or compactly such that their motion is relatively more constrained than lighter rocks. In solids, folds hold together in a compact, intertwined formation.

For example, the molecular and atomic bonds of harder materials, such as rocks or iron, are less elastic and contain more connective bonds than less dense materials. These molecular bonds form a branching network, called a "lattice," of multiple connected folds or covalent bonds. This does not mean that folds are immobile in solids but rather that their movements are just more compact. As solids are heated, their molecules vibrate faster and faster and break these bonds—disjoining—and moving increasingly farther apart from one another. Solids become less dense as motion increases.

Liquids

On the other hand, less hard or dense things such as fluids are liquid because their waveforms or vibratory frequencies are more elastic and contain fewer connective bonds among folded molecules. Liquids are composed of folds that are less densely bound to one another and move past each other more easily (smoothly) without bonding into more fixed lattice and taking on an increasingly rigid, fixed, or rough networked waveform.

Just as poppy seeds move past one another easily without sticking to one another, so the folds of liquids move past one another smoothly without bonding. Fluids have fewer connections or conjunctions within them, so they can more easily roll away or leak.

Gases

Finally, the least hard or dense things take the form of gases because the waveforms or vibratory frequencies of the folds are the most mobile and contain the fewest connective bonds between molecules.

Gases are composed of highly pedetic flows that are moving so quickly and so unbound from one another that they break apart from one another in an instant. However, gases are not completely unconnected; they still conjoin with one another but not nearly as much as do liquids or solids. Smoke and fire, for example, still sting the body but are not strongly bound together enough to bind fire into a solid object. The more flows break free the more gaseous and the more momentum of the matter. Gases therefore have the freest and also the most irregular unbound kinetic forms.

The waveform theory of matter also describes the affects of the five senses: taste, sound, smell, touch, and sight.

Taste

Foods taste differently because the flows of matter are drawn or shaped into different kinetic forms. All atoms and molecules vibrate at different frequencies that can be mapped by spectroscopy, which maps the movement of radiation on the electromagnetic spectrum. Molecules not only have different shapes depending on the atoms conjoined in them, but they also vibrate in different kinetic patterns depending on the collective vibrations of atoms in the molecule. Molecules thus draw or make their shape by moving.

According to contemporary biochemistry, there are two competing models of taste: the standard lock-and-key model and the more recent vibratory-olfactory theory. According to the first model, the reason why different foods taste different is that we have taste receptors on our tongues that are literally shaped in such a way to only receive certain-shaped molecules. According to the second model, the exceptions to this lock-and-key model can be accounted for by the theory of quantum tunneling, allowing us to smell and taste the difference between different vibrations, even though they have the same molecular shape.

In general, both theories accept that the shapes of the material flows of taste are determined by subatomic motion. Thus, the waveform theory of sensation is equally compatible with both the lock-and-key theory and the vibrational theory. Both theories demonstrate significant overlap in their evidence and differ only on a number of exceptional cases.

Sound

Things sound differently depending on the shape of their movement or waveform. Sound is nothing other than the vibration or movement of atomic and molecular folds in the air in a specific waveform pattern. Again, atoms and molecules vibrate at different frequencies producing different waveforms or sounds: sine waves, square waves, noise, or distorted waves.

Smell

Similar to taste, different smells correspond to the different molecular shapes and their vibrational waveforms.[24] Whether tasted or smelled, the topology of the molecular shapes, receptors, and waveforms all contribute to the sensation of that flow. Foul smells, such as rot, for example, can be clearly seen in infrared spectroscopy images, and appear as topologically distinct "noises" in the waveforms of more pleasant smells.[25]

Sight

Colors, like all other radiation on the electromagnetic spectrum, can be distinguished according to different vibratory waveforms. Colors with a smooth or regular waveform are colors that appear to us in their more "pure" form. The color red, for example, in its most red sensation is a vibrating flow of photons all moving in approximately the same smooth waveform and frequency (roughly 405–480 tetrahertz). Today, photo-imaging technicians, operating with this understanding, have developed a number of programs that can add or subtract distortion, blur, and noise to the wavelengths of an image, using the method of the Fourier transform to add and subtract irregularities to the waveforms of light.[26] In other words, by changing the shape of the color waveforms, the composition of colors can change depending on the amount of blur, noise, and distortion in the shape.

Touch

Differences in temperature are determined by the waveform vibrations of folds. When the momentum or kinetic energy of the molecules being broken apart by fire becomes increasingly irregular in the form of a pedetic gas (in the case of fire), they become "hot," relative to our skin. They quite literally puncture our skin and destroy our body's molecular bonds through collision, causing burns. The shape of the fire that penetrates our skin is an extremely irregular and rapidly changing figure drawn by the molecules of carbon dioxide, nitrogen, and oxygen as they fly pedetically through the air at high momentum and damage the molecular structure of our skin. Fire is a high-momentum and irregularly shaped gas. Other things can be hot, but heat itself tends toward entropic radiation and evaporation.

Ice, on the other hand is shaped quite differently since the figure drawn by cold things has a relatively low-energy momentum and a very regular solid shape. The shape of cold things in general is a result of the limited kinetic motion of their molecules. Increasingly, cold states tend toward solid and immobile shapes. Ice bites us in the opposite way as heat. Ice damages our cells, causing pain, because the energy of our cells is transferred to the ice, slowing them down and eventually destroying them. Ice and cold physically penetrate the skin, changing its shape by making it more solid and less mobile. Human cells, when frozen, actually expand and burst, while the momentum of the molecules in the ice begin to move more freely in the form of liquid water. In other words, because the shape of ice is more solid and less mobile, it bites us and moves our cells in different ways. The qualities of images are therefore the result of kinetic affections—flows whose density, shape, speed, and frequency are receptive and directive. A fold has a waveform and frequency because it has a period in which it goes out and returns to itself more or less quickly, in larger and smaller cycles. Insofar as matter exists, the movement of its component parts—atoms, molecules, cells, and so on—move outward and back in at periods or limits that define form and quality.

Degrees of Quality

Kinetic quality admits degrees: more or less solid, more or less large, more or less hot, more or less dark, and so on. This is because there can be larger and smaller intercalated cycles that all return to the same period of sensation. A degree of a quality (more or less) is thus always relative to its period or point of sensation through which all the intercalated cycles pass. One cycle is "more" than another the more smaller cycles it envelops with respect to their shared period of intersection: sensation. The period of sensation is the point of arrival (reception) and departure (redirection) for all the intercalated flows.

Just as it is possible to distinguish between larger and smaller infinities in mathematics without knowing the exact quantitative or cardinal difference between these infinities, so is it possible to distinguish between more or less of a quality without considering the exact quantitative difference between them.[27] Something can feel more or less hot in relation to a point of sensation without considering the exact magnitude of the difference between two qualities. As Nietzsche writes, "We cannot help feeling that mere quantitative differences are something fundamentally distinct from quantity, namely that they are qualities which can no longer be reduced to one another."[28] The difference between any two cycles of sensation is thus not a quantitative difference but rather a qualitative difference.

Without this periodic structure of repetition, a quality like solidity, for example, would quickly dissipate if the flows did not return to one another in a certain density. If, for example, flows of magma disjoined all the silicon flows in a basalt rock and mixed them elsewhere with other metamorphic flows, the rock would lose its qualitative degree of solidity. Its flows would not return in the same cycle or qualitative degree. In this case, the rock would be destroyed (a qualitative transformation) and would no longer exist, since the identity function of periodicity is required for the existence of qualities.

Quantity

However, qualitative folds are also quantitative insofar as their continuous cycles are treated as numerically discrete unities. There is therefore no fundamental or ontological division between quality and quantity; there are only flows and folds. The kinetic theory of folds thus allows us to go beyond the simple opposition between heterogeneous quality and homogenous quantity (figure 2.6).[29] Quality and quantity are simply two dimensions of the same continuous movement of the fold: affective and objective.[30] While quality describes the period or point of sensation of the fold, quantity describes its periodicity as a whole, identical, and unified complete cycle. Greater or lesser quantities are determined by counting the smaller subcycles they contain.

For example, 10 degrees of temperature are greater than at least nine other measurable qualitative subcycles or degrees. In this way, a cycle can be counted as a quantitative multiplicity without presupposing an ontological division between the period and cycle of a fold. Quantum physics, for example, accepts both the qualitative continuity of movement as the vibration of quantum fields (indeterminate quantum fluctuations) and the quantifications of those field fluctuations at different emergent levels: Planck scale, particles, atoms, molecules, cells, animals, plants, galaxies, and so on.[31] This is possible only because quantity is nothing more than the cycle of a qualitative fold of a continuous motion, considered as a unity, or "one." Quantity is a movement of expansion or identification of

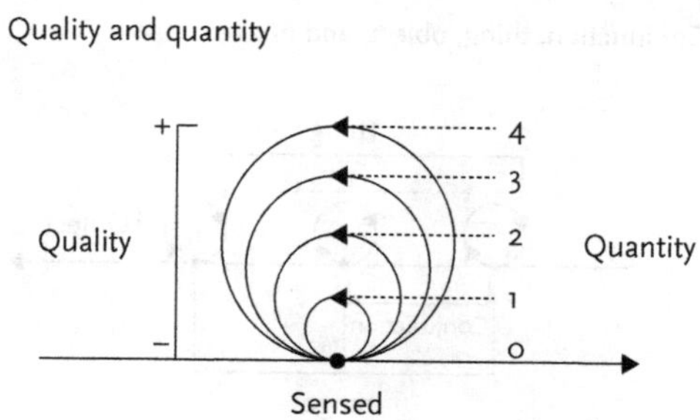

Figure 2.6 Quality and quantity: two dimensions of the same continuous movement

the kinetic period to the whole unity of the cycle, while quality is a movement of contraction of the unity of the cycle back to the single point of its self-sensation or affection. Quantity and quality are, therefore, two dimensions of the same kinesthetic process.

CONJUNCTION

The image is not bound to the mere pop and flash of isolated affects and sensations. Affective folds are connected into images by one or more conjunctions. A conjunction is the connection between two or more folds or intersections. As such, it is also the connection between different qualities and quantities, each with its own degree of intensity and number.

The conjunction of folds is a "thing." However, since every fold is both qualitative and quantitative, so is every thing. Therefore, with respect to the connection of the periods of folds, sensate qualities, or affects, we call the conjoined thing an "image." With respect to the connection of cycles or numerical quantities, we call the conjoined thing an "object."[32] Quality and quantity, image and object are two kinetically distinct but inseparable dimensions of a thing (figure 2.7).

For example, a chair is a conjunction of kinetic waveform qualities of a certain solidity, temperature, texture, color, and so on that define its *image*. However, it is also a conjunction of certain determinate quantities—four legs, one seat, and two armrests, all of a certain length, width, and height—that define its numerical *objectivity*. Together, the combination and arrangement of these qualities and quantities produces a relatively cohesive grouping that defines the thing called "chair."

Things are conjunctions of some folds but also disjunctions of others, thus giving them the appearance of discreteness relative to their environment. A "pore" is the difference between flows, but a "thing" is the difference

Conjunction, thing, object, and image

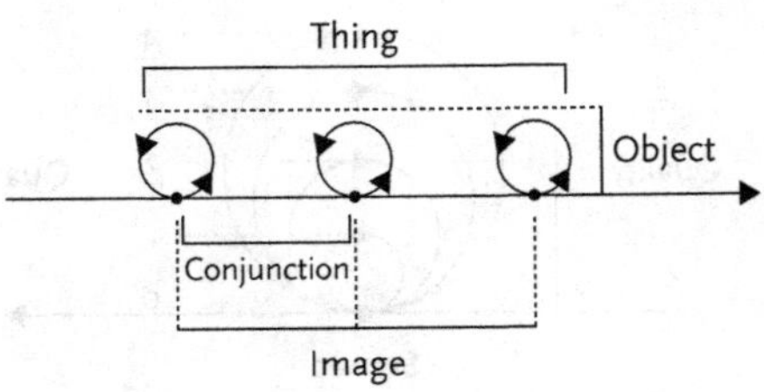

Figure 2.7 Conjunction, thing, object, and image: inseparable dimensions of a thing

between folds that are conjoined in it and those that are not. These two sides define the existence of the thing. Just like the folds that make them up, things alternate to infinity in a series of mutual self-limitations where things are only the exact kinetic outline of a relative difference between their conjunctions and disjunctions.

For example, the *thingness* of the chair is the difference between all its conjoined qualities and quantities and all the qualities and quantities it does not have, which are disjoined from it. A wooden chair might have the conjoined quality of solidity but not the quality of burning-hot temperature or the quantity of nine legs, and so on. The list of disjoined qualities and quantities is obviously much larger than those conjoined in the chair. In this book, we focus on the qualitative composition of the thing and reserve its quantitative dimension for the next volume, *Theory of the Object*.

Once an image has only one or fewer of its qualities, it ceases to be an image. This is because for a quality to appear as a quality *of a thing*, or *image*, it must appear as a quality *of* at least one other quality. The kinetic structure of this "of-ness" is made possible by the process of conjunction that links one fold to another. If a quality is just a quality and not a quality of something else, it is not an image; it has no conjunction or cohesion. It appears only as an ephemeral noise or streak of color. The image, therefore, always has more than one affect, without which it would dissolve immediately.

Furthermore, every image is supported solely by the flows of matter that move through the whole series. For example, living organisms are only relatively stable pools or folds in a continuous flow and transformation of energy moving from the sun, conjoined by the organism, reproduced in its offspring, and disjoined in death. The folds of life are only eddies and images in the kinetic stream.

Even the inorganic bodies of minerals are nothing more than relatively stable images in the continuous transformation of kinetic energy. Igneous, sedimentary, and metamorphic rocks are simply three relative stages of a continuous mutation and conjunction of the earth's liquid body—the rock cycle. Solid, liquid, and gas—or ice, water, and air—are simply the three relative stages of a continuous conjunction in the Earth's fluid body—the water cycle.

At the microscopic level, all organic and inorganic images are conjunctions of smaller images, and those of even smaller images, and so on, all of which are in constant motion at every level. Flows of molecules, particles, and subatomic particles are continually moving and conjoining with one another.[33] Quantum waves ebb, flow, conjoin, disjoin, and collapse into particles on the luminous shores of existence. Even at the macroscopic level, all these images do not produce a final stability. Everything is moving through an accelerating universe at incredible speeds. Since all things are products of kinesthetic conjunction (images), they are metastable. Images are always supported by flows at a smaller level.[34]

So-called essential qualities such as an extension, volume, shape, and so on are nothing more than products of the process of continuous and constant conjunction.[35] It is only after a series of qualities have been added together in a conjoined structure of periodic cycles that images emerge. Thus, it is only retroactively that they appear to have these qualities by necessity or essence. Necessity and essence are only kinetic effects produced by kinetic and affective cycles—and not vice versa.

Without conjunction, there are no images—only fragmented sensations, a degree of heat, a flash of color, a pop of sound. Flows keep moving, folds keep cycling, but without conjunction, nothing holds together in a stable image. Nothing seems to be attached to or part of anything else. Everything flows, but motion is not a thing; it is a process. Flows are vectors or potentialities in things and are not reducible to them. For example, at a given time, a body of water may not be frozen. At that moment, there is no *thing* or *image* of ice. However, as the kinetic waveform of water changes, slows down, cools, folds, and congeals, ice comes into existence as an image composed of affective folds. Once the hydrogen and oxygen folds slow down enough to conjoin together at slower speeds, there is ice. Images emerge through kinetic processes, but the processes are not separate or independent from the images. Flows are the processes by which images come into and go out of existence. They are the warps, woofs, and vectors by which images are woven, folded, and unfolded.

The conjunctive process that produces images is additive, "one by one," not something attributed once and for all. This is the case because there is no single substance to which the conjunctive process is attributed. Since flows are multiplicities and matter is nontotal, conjunctions can only be regional. The conjunctions that compose images are, like the flows themselves, in constant motion and can always undergo a change or recomposition. The determination of the qualities of images is thus never total, complete, or final because the flows that compose them always leak or connect to something else outside them. As a process of flows, the kinetic image is thus not reducible to any fixed set of qualities conjoined at a given moment.

However, it is also important to distinguish between three kinds of kinetic conjunctions: injunctions, circuits, and disjunctions.

Injunction

The first type of conjunction is injunction, which is the inclusive joining of two or more affects into an image such that the cycle of one is equal to the cycle of the other. In other words, injunction occurs when two or more folds share exactly the same cycle and periodicity as another (figure 2.8). Injunction is thus the inclusive identity or unity of two or more affects. Injunction also entails that two or more folds have exactly the same affective capacities, expressed as identical qualities of the same image. When we hear, for example, two cellos playing exactly the same note in pitch, loudness, and duration, at a certain distance the two notes become indistinguishable and are injoined in the same sonic image. As we move closer and closer to the cellos and they move farther apart, the sounds may diverge.

Circuit

The second type of conjunction is the circuit. A circuit is the conjunction of one or more folds into a third larger fold (figure 2.9). The third fold functions as a common background for the others and brings them together. In this way, a flow conjoins multiple folds together by using another. Thus, all the flows in the circuit cycle through at least one shared

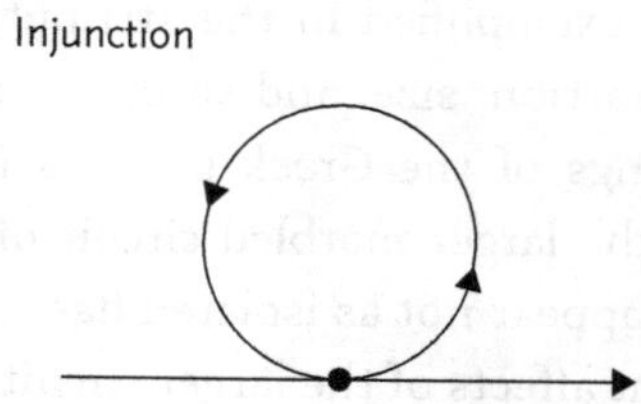

Figure 2.8 Injunction: when two or more folds share exactly the same cycle and periodicity

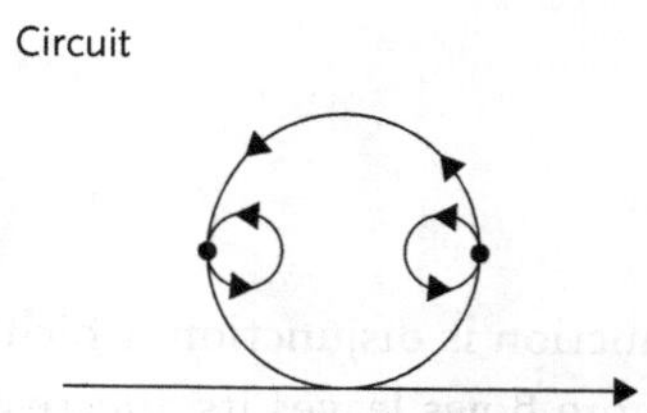

Figure 2.9 Circuit: the conjunction of one or more folds into a third, larger fold

period. The subjunctions that compose the larger circuit may have no shared qualities or capacities with one another, but they are all limited in their cycles by a third cycle that constrains or underlies their co-motion. A kinesthetic circuit works in much the same way as an electrical circuit board: It provides a pathway or material foundation for electrical current to pass through via multiple components—capacitor, transistor, diode, and so on.

Each component fold has its own capacities, but the circuit ties them together and binds them to another component, without which the others would be disjoined—and the affects would never sustain an image. Together, the affects in a circuit are different qualities of a single quantity because they are part of the same image. Since conjunctions are nothing but flows, a circuit has no support independent of the folds that compose it. Every larger circuit depends on the support of all the smaller subcircuits and junctions that compose it, all the way down and all the way up.

The conjoined circuit is what allows affects to hold together as related qualities of the same image. In music, the idea of the circuit is exemplified by the vertical harmonic relation of notes in a chord. The chord is nothing other than its notes, but a chord can also have a root note within which the other notes are contained or harmonically related. For example, a C major triad contains the first, third, and fifth notes of the diatonic scale beginning with C: C, E, and G. Therefore, E and G are the third and fifth with respect to the harmonic circuit defined by their first: C. The notes are notes conjoined in the sonic image of the chord.

In sculpture, this is exemplified in the attribution of the qualities or affects of texture, proportion, size, and so on to the sculpted figure. The two downy layered wings of the Greek goddess Nike, for example, are component affects in the larger marbled circuit of *The Winged Victory of Samothrace*. The wings appear not as isolated flashes of soft flowing white but, rather, as attendant affects of the larger circuit of Nike's shapely body and rippling dress. The texture, size, and proportion of the wings fit within and are supported by the marble circuit of her body such that we see them as "the wings" *of Nike,* and not the other way around.

Disjunction

The third type of conjunction is disjunction, which is the kinesthetic process by which one or more flows leaves its affective folds, periodic orbits, or conjoined image. Disjunction is the entropy of flows (figure 2.10). Every

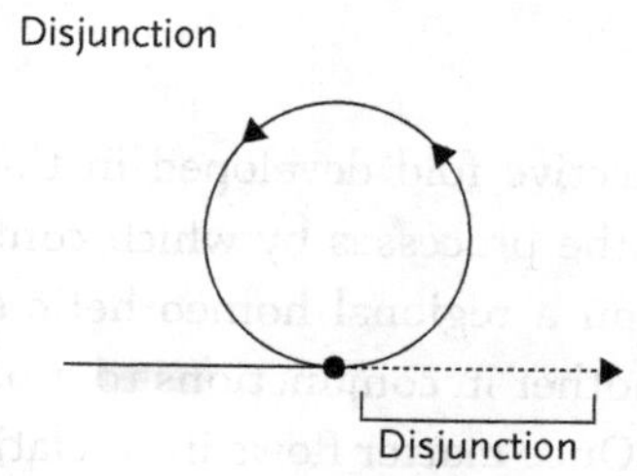

Figure 2.10 Disjunction: the entropy of flows

junction and conjunction of the image leaks. That is, every seemingly stable image, as material and kinetic, is also subject to decay, degeneration, and entropy. Folds have to be made and remade continually in order to keep them from sinking, like a leaking ship. The arrangement and order of folds in the image does not have an eternal form or essence that preexists its concrete conjunction. Images must be built, and constantly conjoined.

Plutarch's example of the continually rebuilt ship of Theseus is not a special case of affective decomposition but, rather, a general kinetic one.

> The ship wherein Theseus and the youth of Athens returned from Crete had thirty oars, and was preserved by the Athenians down even to the time of Demetrius Phalereus, for they took away the old planks as they decayed, putting in new and stronger timber in their places, in so much that this ship became a standing example among the philosophers, for the logical question of things that grow; one side holding that the ship remained the same, and the other contending that it was not the same.[36]

All images continually flow, fold, and unfold because all images are material and subject to decay. All images are ships of Theseus. The so-called form or essence of the conjoined image "ship" is nothing other than the assembly of material flows folded up into conjoined affects. With respect to the kinetic flow of matter, there is no strict identity. Matter is not identical to itself because it is kinetic process, flux. This is what makes affection possible in the first place.

In this sense, the ship is not the same ship rebuilt each time. On the other hand, kinetic "identity" can be stabilized through the continual conjunction of affects. In this sense, if the new ship image has all the same capacities as the old one relative to the collective Athenian sensorium, then it can be said to be the "same" ship. This is not an ontological or formal question but, rather, a practical and kinetic one. If the new ship moves in all the same ways as the old one, then they are kinetically identical.

CONCLUSION

The concept of the affective fold developed in this chapter provides us with a kinesthetics of the processes by which confluent flows of matter are capable of taking on a regional homeorhetic stability in cycles and combining with one another in conjunctions to produce larger metastable composites or images. Once matter flows into relatively stable events and constellations, we have a surface or sensorium within which certain events can be habitually repeated in the form of persisting affective folds. But it is only with the conjunction of these affects that the affects can be attached to one another and thus said to be qualities of a sensed image.

However, the theory of material flows and affective folds developed so far still does not allow us to determine how such images can be *ordered or distributed* in the sensorium. The theory of folding shows us how affective stability is possible for a flow of matter, but not how such heterogeneous affects could be arranged and related to one another in larger kinetically ordered aesthetic systems of images, or art. For this, we require a kinesthetic theory of the aesthetic field, which is developed in the next chapter.

CHAPTER 3

The Field of Art

Matter flows and folds into affects, affects are conjoined into images, and images are arranged or ordered in an aesthetic field of art. This is the third concept of kinesthetics. If flows of matter intersect and if folds of affect periodically cycle, aesthetic fields organize them all in a continuous feedback loop. This chapter provides a theory of how conjoined flows become organized according to distinct patterns or fields of art.

So far in this book I have made two major moves quite different from most theories of the image. First, instead of beginning with subjects or objects as the origins of the image, I began with the image itself and its defining feature: motion. This lead to the paradoxical-sounding thesis that the condition for the image is itself insensible—not ontologically but relatively so. Every sensation of an image is itself supported by other images that remain for this sensation relatively insensible. This leads us not to an abstract flat ontology of the image but, rather, to a historical and material theory of the image and its regimes of distribution.

Second, instead of defining the image as a representation, I argued that it is affective and kinetic. This leads to an extremely broad definition of the image that includes all of nature, differentiated only by its degrees and capacities for ordered affection.

In this third chapter, I put forward a no less radical redefinition of art and aesthetics that includes all ordered distributions of matter, again differentiated only by degree or distribution, and not by kind. The traditional concept of aesthetics as a strictly anthropocentric field defined by subjects and objects is here abandoned in favor of a multiplicity of *kinesthetic fields*.

THE FIELD OF CIRCULATION

An aesthetic field is a single, continuous flow of matter that has a kinetic vector for each affect on its surface. A field is a continuous flow of matter that provides a path of circulation that binds together and orders a regional distribution of affects. This is different from conjunction. A conjunction of affects merges folds together into larger and smaller composites of qualities in the image. Conjunction is composed entirely of larger and smaller images. An aesthetic field, however, binds together conjoined groups of affects or images, but is itself not another affect. A field is a binding and ordering flow that moves through all the folds and subfolds and then repeats the process. Affects allow flows of matter to persist, but aesthetic fields allow images to be distributed.

The field of circulation is the kinesthetic condition for the ordered distribution of affects. An aesthetic field puts affects and images in a particular order or arrangement. It defines the kinetic logic of the image. It is a flow of matter that moves through all the different images, but at a certain limit it folds back over itself and returns to at least one other affect, beginning the process again. The movement of circulation across the field thus secures the conditions under which a relation or order between two or more folds can persist through repetition.

Without a field, images only exist as floating fragments of conjoined affects—a grin without a cat or a walking hand without a body. Images without background or landscape, like the scattered and floating fingers, hands, and bodies in Michelangelo's sketches and studies. In sketches like his *Studies for the Libyan Sibyl* (ca. 1510; figure 3.1), we can see how Michelangelo begins by assembling the affects of shape, size, and proportion into various images or body parts (the angle of big toe), floating on the page. Then in the final painting, he conjoins these floating images into an ordered aesthetic field: the proportional human body (figure 3.2). Only through a larger circulation are the fragmented images kinetically ordered and put into relation with one another.

This in no way means that the field of circulation is anything other than the flows of matter that compose it. Just like affects, aesthetic fields have no transcendent reality independent of the material flows that constitute them. Ordered flows do not preexist the flows themselves. Ordering is produced by and through the flows of matter alone. The circulation of a field is, therefore, also immanent and continuous with the affective folds that constitute it.

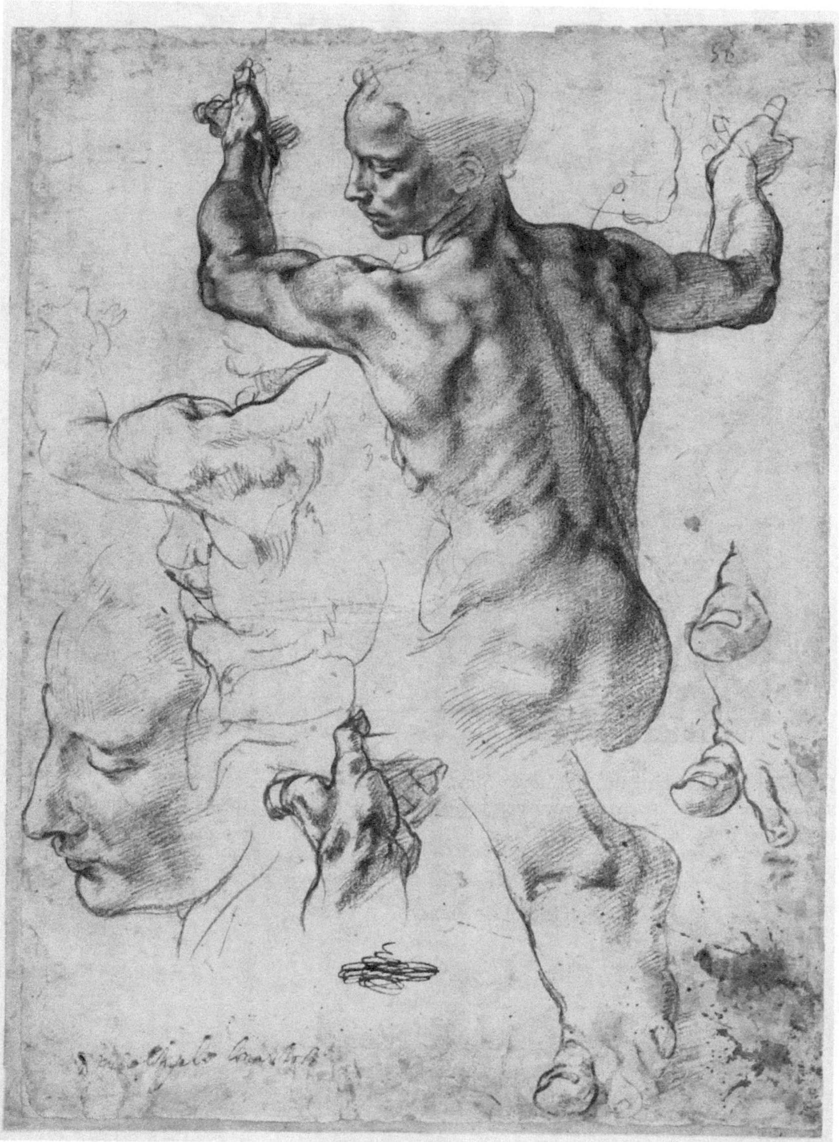

Figure 3.1 Michelangelo, *Studies for the Libyan Sibyl* (ca. 1510)
Source: Wikimedia, https://commons.wikimedia.org/wiki/File:Studies_for_the_Libyan_Sibyl_(recto);_Studies_for_the_Libyan_Sibyl_and_a_small_Sketch_for_a_Seated_Figure_(verso)_MET_DP826907.jpg#/media/File:Studies_for_the_Libyan_Sibyl_(recto);_Studies_for_the_Libyan_Sibyl_and_a_small_Sketch_for_a_Seated_Figure_(verso)_MET_DP826907.jpg

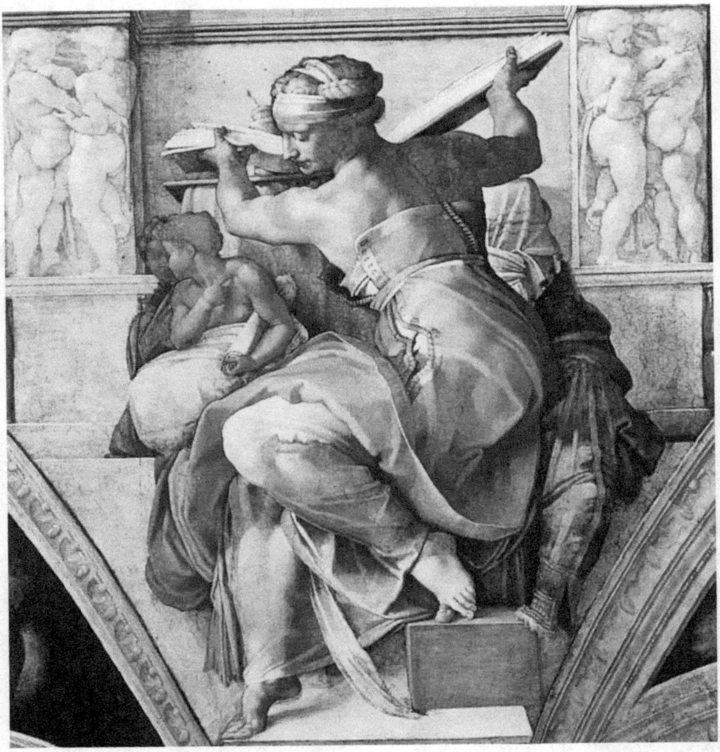

Figure 3.2 Michelangelo, *The Libyan Sibyl*
Source: Wikimedia, https://commons.wikimedia.org/wiki/File:Michelangelo_the_libyan.jpg#/media/File:Michelangelo_the_libyan.jpg

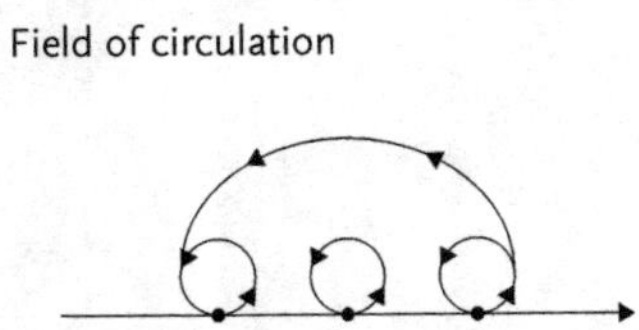

Figure 3.3 Field of circulation

In other words, the kinesthetic condition is immanent to what it conditions. This is precisely because there is a complete continuity between the flow of matter that connects one affect and image to another and the flow of matter that connects the other back to the first. This circulatory field (figure 3.3) is what gives folds a consistent and repeatable relation to one another, such that the grin does not float away from the cat. Without the differential repeatability of this relation, images have no persistent arrangement or structure. They exist as affects, but they do not move together.

Synesthesia

An aesthetic field is not an image and thus it has no sensible qualities independent of the images that compose it. In this sense, there is no ontological hierarchy or division of the senses into the historically arbitrary five human senses. If all of matter is sensitive, then each sensorium has its own receptive and directive capacities that always occur together insofar as all the senses are active. The aesthetic field is thus synesthetic not in the clinical or experiential sense in which some senses are confused or conflated with others, but in the sense in which all the senses of a given sensorium are always active and sensate at once to varying degrees. Even when one is simply viewing a piece of visual art, one is also touching the ground, smelling the air of other bodies in the museum, tasting one's mouth, or hearing the echo of footsteps and low voices in the museum. Despite the rigorous attempt by museum curators to remove all "extraneous" sensations and images from the visual work of art, all the senses are included nonetheless. Only afterward do we try to mentally separate them out and to isolate a single art image from the multitude of other images in which it is nested.[1]

Since an aesthetic field itself is not a fold, it has no affective capacity, no qualities, and thus cannot be an image or even part of an image with respect to its ordered images. Rather, the field of circulation is the continual flow that traverses images, binds them together, orders them, and conditions their co-motion and relation to one another. It is the way images move together relative to one another. The field is nothing but a folded flow of matter that connects with itself only indirectly through two or more related folds. The path of circulation is thus not an a priori order waiting to be filled with images, nor is it a posteriori fiction that only seems to order images. The field of circulation is the real and immanent constitution of order by the flows that traverse the images themselves. In turn, the images are nothing other than the constitutive flows that join them in this order. The whole process—flow, fold, field—is one continual motion. This is the kinetic synesthesia of the field. Nothing in the field is isolated; all affective capacities and senses occur together in co-motion in the field. Circulation is therefore synesthetic in a dual sense: It has no affective capacities or qualities of its own, *and* therefore has only and all the sensory capacities of its constitutive images.

The specific distribution of a field, the existence of its folds, and the conjunction of its images are not universal or necessary relations. There is nothing essential about them. Rather, any given series of images only exists with respect to the ordering relations of a given aesthetic field. Without an ordering circulatory flow, images remain disjoined, fragmented, and incoherent. In other words, there are no grins without cats or hands that walk on their fingers without bodies. All images that are outside a given circulation are not ordered under its constituting relations and are thus relatively unrelated to its motion. Images outside an order have no relation to it, and thus are not ordered in co-motion with it. A more complete treatment of this issue is developed in the section on knots, and a concrete example is given.

Theory of Fluxions

Motion is absolute. Everything moves, not as a totality but as an infinite sum of motion. Therefore, there is no single immobile point from outside by which to measure the objective and absolute order of images in any flow of matter. If there were, there would also be another point further along from which to measure that one, and so on in an expansive kinetic infinity. It therefore follows that there is only an infinity of points from which to measure the strictly *relative*, but no less *real*, motion of images.

The theory of fluxions provides a logic for understanding the relative ordering of images in a field. Affects and images move differently in the same field. A *fluxion* is a degree of motion relative to the motion of the field it is on, and to its neighboring motions (figure 3.4). Depending on its degree of flux, an image moves more or less with respect to its other neighboring images and all those with respect to the field of motion that measures or orders them all.

Relative to one body another appears to be moving; relative to another the same body appears to be still. In his *Principles of Philosophy*, René Descartes gives the example of the motion of a ship. Relative to

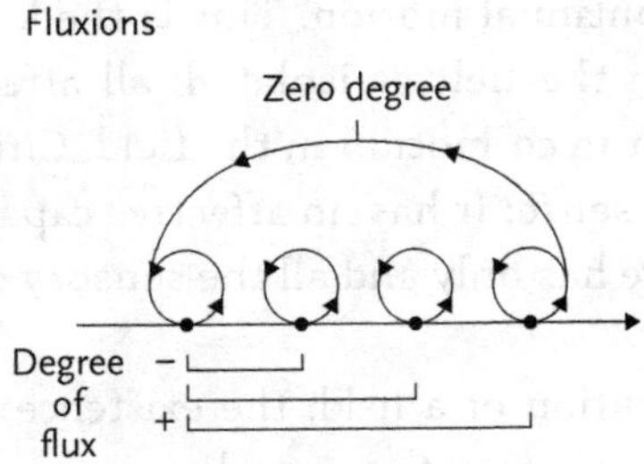

Figure 3.4 Fluxions

the shoreline, a passenger on a ship appears to be *moving* down the river. Relative to the ship, the same passenger is *not moving* relative to the neighboring bodies aboard the ship. However, if the ship were being pulled by the wind upriver at the same speed as the current was pulling it downriver, the ship would *not be moving* relative to the shore, but *would be moving* relative to the changing wind and water surrounding the ship.[2] The point is this: everything is moving, but only moving more or less *relative* to other motions. A fluxion is, therefore, the relative kinetic difference between affects or images. The aesthetic field is the background flow or continuous "function," as in calculus, within which the fluxions are related.[3]

Each image has its own periodic motion, but is also related to other images in the same aesthetic field. Every field of motion is therefore a composite. For example, Descartes continues, if you are walking along the deck of a ship with a pocket watch, the wheels of the watch have their own motion. However, added to this motion is the motion of your body along the deck, and added to your body's motion is the motion of the ship tossed about on the waves, and added to that motion is the motion of the ocean as a whole, and added to that is the motion of the rotating earth, and so on.[4] All these motions are part of the same field of motion. However, each also moves more or less relative to the others. Relative to the watch, your walking body is less mobile; relative to your walking body, the ship is less mobile as you walk across it; but relative to the ship, the waves are less mobile as the ship sails across their surface—and so on, in relatively *decreasing* degrees of mobility or fluxion. From the inverse perspective, the waves move across the surface of the ocean, the ship moves across the surface of the waves, and so on in relatively *increasing* degrees of mobility or fluxion.

The great contribution of Einstein's theory of special relativity to the theory of the image was to show that space and time themselves are relative to one another *with respect to motion*. The aesthetic field is what relativizes space and time. As the degrees of fluxion increase, time slows down, or "dilates," and space "contracts" relative to a given field. Every degree of flux therefore also determines a degree of space-time, not the other way around. Following special relativity, the theory of fluxions thus offers a description of space-time grounded in the motion of the image. A fluxion is simply a difference in degree of motion relative to a given kinetic field. A kinetic field, on the other hand, is the relatively immobile background from which the different fluxions are measured. Therefore, a field is that which has zero motion relative to the rest of the degrees of motion in its sensorium.

Three correlates follow from these initial definitions.

First Correlate: Different Flux

First, an aesthetic field requires at least two affects or images and a relation between them such that the flow of one is kinetically relative to the flow of the other. If this is the case, it is possible to directly compare the cycle of one image with the cycle of another. This makes possible a measure of their fluxion, or kinetic degree of "more or less" motion on the basis or support of a single, continuous flow that binds them together and apart: the field.

In other words, images differ with respect to one another only through a relatively immobile background or zero degree of motion that flows between them. Thus, the flux of one image is relative to another image only on the basis of the other's relation to the first *and* to their shared and triangulated relation to the field. Images thus appear as different from one another or the same as one another to varying degrees of flux only relative to their shared field of circulation. Therefore, the theory of the field presupposes at least two related folds and at least one fluxion or kinetic difference between them. Every kinetic sensorium presupposes its senses and every sense presupposes its kinetic sensorium; the two are relatively defined.

A sensory field with no affects or images is simply an insensible flow of matter. A field with just one affect is simply an affect—a quality without fluxion or relation. A flow with just one affective fold can only have larger and smaller intercalated folds intersecting at the same period, as described in the previous chapter. However, for multiple different affective folds to be related together *as different images*, as fluxions, in some ordered way they require a field of circulation that binds them together but also holds them apart.

Second Correlate: Positive Flux

The second correlate of these definitions is that for something to have its own, distinct motion it must have a higher degree of flux than all its lower degrees. For example, relative to the shore, the motion of the sitting passenger is identical to that of the ship. Only if the passenger moves while on the ship does she appear to have her own motion. However, if she moves in the opposite direction of the ship's motion at the exact same speed of the ship, she will appear to not be moving in relation to the shore. Her degree of flux in this case would be identical with that of the shore. Only when her degree of flux is greater than *all the other relative degrees* does she appear to have her own, positive motion the field.

Third Correlate: Zero Flux

The third correlate is that if two folds have the same degree of fluxion relative to the zero fluxion of a given field, then they appear more or less identical or injoined relative to the other degrees in that field. If two folds have the same fluxion, then they will co-appear in motion as injoined. If their fluxion is different only with respect to the third fold, then they are part of a third fold or circuit. If a fold has no degree of flux, it will not appear; it will be disjoined. It will be part of the relatively immobile background.

Example: The Waves

In the following example, let's try and bring all three of these theories (field, synesthesia, and fluxion) together. In Virginia Woolf's *The Waves*, six characters meet in a restaurant on the occasion of saying farewell to their old schoolmate Percival, who is going away to India. We can call the aesthetic field "Percival's farewell dinner." In contrast to the other characters, who all speak in the first person in the novel, Percival has no voice. Additionally, no authorial narrative is given of Percival, or the restaurant, or anything else in the background. We read only what each character *says, thinks,* or *does* in turn within the kinetic and affective background with respect to his or her *degree of flux in the field*.

The aesthetic field is thus not an image that can be objectively sensed or described by an omniscient author, but only lived in the synesthesia of affective images of the six characters who meet with Percival. In the world of the book, the field is everything written between pages 118 and 140. As such, all the senses of all the characters are synesthetic and collectively related with respect to the zero degree of motion on which the relative affective movements of the characters play out.

Neville arrives at the restaurant before anyone else. Since no one sees him arrive, his movement appears with the least degree of affective flux possible relative to the restaurant itself. All the other characters find him already there, sitting and unmoving like a piece of furniture, relatively unaffected by their arrival. As the first and only affective fold in the field, his image of the restaurant is one of complete indifference. "Things quiver as if not yet in being," he says. "The blankness of the white table-cloth glares. The hostility, the indifference of other people dining here is oppressive." The aesthetic field of Percival's farewell dinner has transformed the restaurant, but as Neville observes, without the others "things have lost their normal uses—this knife-blade is only a flash of light, not a thing to cut with. The normal is abolished." At this minimal point, there are only affective flashes or events without coherent images or order to hold them together.

Louis arrives next, but he does not come to the table immediately. His motion is first one of hesitation and self-reflection as he fixes his hair in a mirror. He is the next degree of fluxion—more than Neville, who is unmoving and unaffected, but less than the others whose movements will have still greater affect. Louis arrives; he moves unremarkably through the restaurant. His motion is oblique and has no affect on anyone or anything except Neville who watches him. Now, Louis describes the next arrival, Susan, who enters like "a creature dazed by the light of a lamp. Now she moves. She has the stealthy yet assured movements (even among tables and chairs) of a wild beast." Susan comes in immediately after Louis and has a similar but not identical degree of affective fluxion, since she moves while Neville and Louis sit, but she moves in a dazed manner, by instinct or self-affection, toward the table, "touching no one." "Rhoda comes now, but from nowhere, having slipped in while we were not looking," says Louis. Although entering later, Rhoda appears as if she had never moved and has already been at the table next to Neville, unaffected. The kinetic pair of Neville and Rhoda thus also has similar but not identical degrees of flux in relation to the aesthetic field.

Susan then describes the entry of Jinny. "There is Jinny. She stands in the door. Everything seems stayed. The waiter stops. The diners at the table by the door look.. . . . She brings things to a point, to order. . . . Now she sees us, and moves, and all the rays ripple and flow and waver over us, bringing in new tides of sensation. We change." Jinny is the maximum degree of fluxion for the field. Her movement is so affective that it radiates out, making everything else look like a relatively immobile background against which she moves. Everyone is affected by her. Her final degree of fluxion shows, by triangulation, the relative degrees of fluxion for all the others. The sequence or chain of fluxions is complete. The maximum, minimum, and middle can now be measured and ordered across the affective field.

Finally, Bernard comes in, without pushing a door. He does not look in the mirror. "He has no perception that we differ, or that this table is his goal," Neville says. Bernard does not feel as if he is coming into a room of strangers. He has a very high but not maximal degree of flux. He talks to everyone as he moves, both affecting them and being affected by them. "He half knows everybody; he knows nobody." With respect to his relatively maximal affection, everywhere appears to be his goal and no one appears to have any more or less significant degree of difference or degree of flux at the table.

"Now is our festival; now we are together. But without Percival there is no solidity. We are silhouettes, hollow phantoms moving mistily without a background. . . . Nothing can settle," Neville says. Once Percival arrives, "the reign of chaos is over. He has imposed order. Knives cut again." "He is like a stone fallen into a pond . . . like minnows . . . we undulate," Rhoda observes. Now there is "a chain whirling round, round, in a steel-blue circle beneath," as Louis puts it. Susan says, "That is the furious coal-black stream that makes us dizzy if look down into it." "Yet these roaring waters," Neville states, "upon which we build our crazy platforms are more stable than the wild, weak and inconsequent cries that we utter when, trying to speak . . . . 'I am this; I am that!'" Percival is the aesthetically insensible and yet synesthetic field that brings the friends together and allows their "common emotion" (love) to circulate, and is therefore not reducible to Percival the person but becomes the kinetic occasion of co-motion. "No, that is too small, too particular a name," says Bernard, "We cannot attach the wide and spread of our feelings to so small a mark."[5]

Emotion

The aesthetic figures each enter and flow like waves on the ocean across the restaurant in different affective motions, all distributed on the background field of Percival's unaffective, undescribed arrival. Emotion is the name given to the complete affective state of the field of images at this moment. Following its Latin etymology, *ēmoveō*, emotion is the name of the collective moving and removing around, out and away or kinetic agitation of the field at any moment. Emotion in this sense is not identical to any particular affect or set of affects but, rather, to a particular arrangement, distribution, or field of affects at a given moment. Emotion is not reducible to the heart rate, temperature, breathing, facial expression, or any other behavioral humanist determination. In *The Waves*, the "common emotion" love is *not a single subjective state* but, rather, a collective affection of an entire material arrangement involving the complete sequence of motions in the restaurant.

Kinomena

If every affect and image in a field has a degree of flux, then the appearance of one image is also related to and entails the simultaneous co-appearance of the others in the same field. Kinomena are the *co-appearance of images* in an ordered aesthetic field of circulation.[6] They are immanent, collective relations between two or more fluxions. Kinomena are thus different from phenomena. Phenomena are what appear to *consciousness* and are related to one another strictly through consciousness. Since phenomena are the appearance of movement *only through consciousness*, which for Kant is not itself in motion, they remain static representations of motion: frozen, fixed, and arrayed by intentionality for contemplation. As long as motion remains something *for* consciousness, and not a feature *of* consciousness itself, phenomenology will always be a graveyard of immobile forms.[7]

Kinomena are defined by their kinetic relations, but their relations are always extrinsic or external to them. In other words, images that are related through a continuous field are themselves nothing but folds in this field. There are no relational flows between folds until there are folds to be related, and vice versa. Until then, there are only unfolded flows of matter. Fold and field thus emerge *at the same time*. Kinetic relations are something that must be made through the primary process of folding and do not precede it.

The theory of kinesthetic relation is thus in contrast to the theory of intrinsic or essential relations that precede the images they relate. Intrinsic relations are fixed relations or properties that define concrete images before anything appears as that image. For example, the intrinsic relations of a chair bind together four legs, a seat, and a back, and so on. The intrinsic relation between these affects defines them as a chair, whether or not such an image actually exists or appears as an image. If an image fails to fulfill these relations, it is not a chair.

In contrast to this, kinomenological relations are constituted immanently, one by one through the affects produced by folds. A fold creates an affective capacity, but it also makes possible a new relation with other folds based on what it can do. For example, the kinomenal appearance of the chair depends on what it can do within the immanent relations in its aesthetic field. A rock could be a chair at an outdoor picnic, a small table could be used as a chair; anything elevated off the ground that can be sat on has at least one affect in common with a chair and can function *as a chair*.

However, the rock at a picnic is not "like" a chair, and the chair is not "like" a rock. Their images are not metaphoric nor representational; they are real. The rock really can be sat on, as can a chair. A chair can also "do" more than its dictionary definition. It can be used for all kinds of things other than just sitting; it can be part of a street barricade, or with a blanket over the top, it can become a child's fortress. All these things and more constitute the external relations a chair is capable of entering into. It is not stuck or fixed by one set of formal features. Ultimately, we cannot know all the affects a chair is capable of. It can always do one more thing. Because the flows of matter are continuous and infinite, fields are always capable of adding one more fold to their flow. Circulation is thus the process or flow capable of holding together an infinity of affective points or folds in the same field of images.

Images are the product of conjunction. For example, a chair might be composed of a conjunction of a flash of brown, a region of flatness, a certain height or elevation. However, only when these affects and conjunctions are co-ordered in their kinetic function are they able to take on the collective appearance or image of "something to be sat in." A chair is thus not simply a series of qualities but also a particular order and actual co-appearing of these qualities in the image in such a way that they work or function together to form the chair as something to be sat in, or used as a barricade, or whatever: a kinomenon.

Experience

This kinetic definition also entails a redefinition of "experience" in contrast with the narrowly anthropocentric one. If phenomena are what appear to human experience, then kinomena are what occur in a field of images. As we showed in chapter 1, a subjective sensorium is not limited to the human; it also includes minerals, plants, animals, and all stable confluences of matter capable of reception and redirection. Any sensory surface in which at least two affects co-appear can be a sensorium. Experience, therefore, is nothing more than the receptive capacity of co-affective kinomena as they are distributed in an aesthetic field. Experience is the collective capacity of an aesthetic field to sense or be affected both by flows of matter from outside the field and from the ordered flows of matter inside the field. It is the ability of the field to both persist in its being and become differently ordered than it is.

Triangulation

Kinomena are thus not only defined by their relation to the field of circulation in which they appear as fluxions *but also co-defined by all the other kinomena in the field*. In other words, their capacity to act depends on the co-motion of the other folds that support it. Since there are at least two folds in a field, every fold is supported by a flow from at least one other.

The second fold secures the relation through the direct support of the first. Given the security of this first relation, it is then possible for this first relation to support a third fold that is related directly to the other two, producing a triangulation. A fourth fold can then also connect to the first two, producing yet another relation different from third, and so on in higher levels of relation. At each higher level of triangulated relations between kinomena, a new parallax relation is possible that modulates and supports the capacities of the others.

The Limits of the Work of Art

Within an aesthetic field of circulation, there are two types of affects that define the body or work of art: limit affects and nonlimit affects. Limit affects or "limit junctions" are the final affects in the field after which the flows of matter are disjoined from the work and/or enter into another aesthetic field of circulation (figure 3.5). The limit affect is thus a kind of filter or redirector of flows. Once a flow moves through a series of conjunctions and reaches a limit affect or bifurcation point, it is either expelled or recirculated back across the previous conjunctions. The work of art is the ordered movement of images inside its affective limits. But the work of art is not a passive object; it is an active process of *becoming what it is*.

Thus, the task of the limit affects is also to actively expel, destroy, or unbind flows. It removes flows of matter from circulation, detaching or disjoinings them. The work of art is defined first and foremost by a relative limit after which the work of art ends, but beyond which it still has some active power to affect things outside it. For example, one is either "moved" by a work of art or not. The work of art affects the sensitive capacities of others around it.

However, limit affects also redirect circulation back to its internal affects and circuits. In other words, limit affects are also filters that allow some

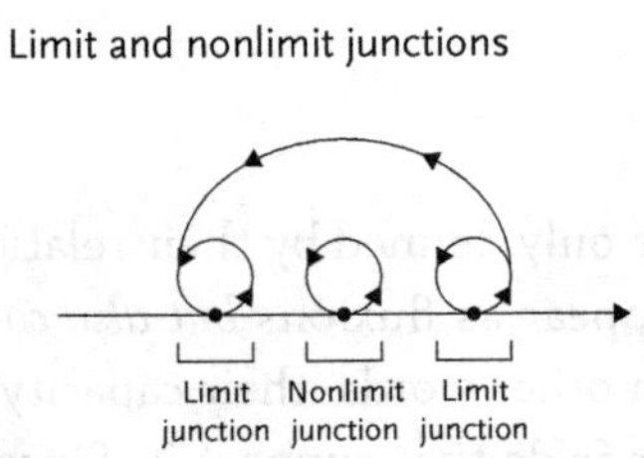

Figure 3.5 Limit and nonlimit junctions

flows to enter into circulation and others to be blocked or redirected. The sealed surface of a painting aims to prevent light and heat from damaging the paint. Limit affects are thus responsible for the flows that define circulation in the first place; they are the kinesthetic conditions for ordered interiority and relatively disordered exteriority. However, limit affects or folds are not determined in advance as limit affects. Their role as points of bifurcation or limit is only relative to their operation in the field and to other relative degrees of flux. Any affects or period in a flow can thus take on a limiting function insofar as it disjoins, conjoins, and recirculates a field. It is therefore possible for any field to both shrink and grow according to the expansions and expulsions of its limit affects.

A nonlimit affect is simply a fold in the field of circulation. At the end of each image or fold, a flow can either start over or move on to another image until it reaches the limit affects of its field. Nonlimit affects thus do not filter what comes in or out of a circulatory system, but simply sustain and constitute the affective capacities of the field itself.[8]

Expansion by Expulsion

Every limit affect or border is thus composed of three operations: an outward expansion, an inward fortification, and a recirculation. The first motion pushes outward, expanding the limits and reach of the field, while also possibly expelling or disjoining a flow of matter. The second motion follows the second, securing, supporting, and retracing this first motion through an expansion of the circulation. The third motion transports and recirculates newly incorporated flows back through the field (figure 3.6).[9]

The aesthetic field, just like the flows of matter that compose it, is not well understood by the concepts of exclusion and inclusion. The conceptual

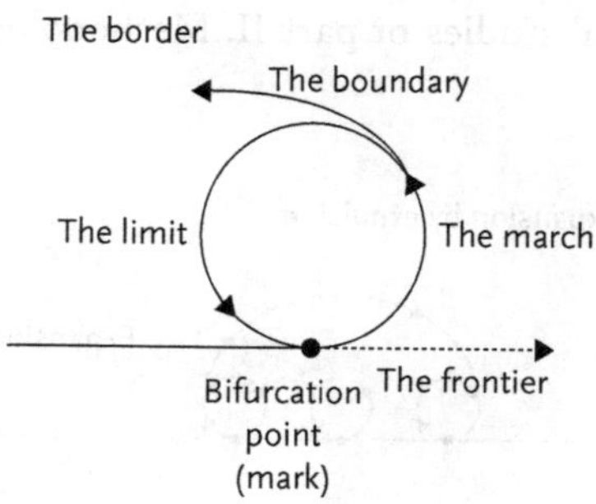

Figure 3.6 The border

basis of circulation is that something goes out and then comes back in, again and again. It is a kinetic continuum. In this sense, circulation is both inside and outside at once. It is a manifold or complicated process creating a folded system of relative insides and outsides without absolute inclusions and exclusions. The insides and outsides are all folds of the same continuous process or circulating flow. Each time circulation captures a fold or pleat, a new inclusion and a new exclusion are created.

The aesthetic field feeds off disjunction. Each newly disjoined flow can be captured, redirected, and looped back around to the beginning of the series. Circulation is an attempt to recycle and redistribute waste back into a relatively stable field of order (figure 3.7). This does not stop kinetic entropy, but it simply rearranges things more or less.

Works of art are not reified, static objects. They are kinetically active material processes that are continually emitting material flows of light, sound, and scent, as well as receiving material flows and being transformed by them in turn. The work of art is an entire ecological system, aesthetic field, or feedback loop between all the flows of matter that compose it and its environment, all the flows that leave its body, and all the flows that return to and affect it.

It is perhaps as obvious as it is odd to say that the work of art is not just an image that is sensed but also that it is something that itself has sensation. Architecture, books, theater, paintings, and even musical instruments are all, as material entities, receptive to decomposition, affection, and transformation. Sculptures erode and break. Books lose pages and burn. All performances can be interrupted. Paint fades and peels. Great meals are digested, and perfumes dissipate in the ambient air.

One might respond to this point by remarking that the materiality and sensitivity of the work of art is in most cases so small that it should be not be considered alongside so great a phenomenon as beauty. This is partially true. The kinetic theory of art does not yet seem to explain the macro-level experience we call beauty. For this, we will have to wait until the next section and the historical studies of part II. Methodologically, however, it is

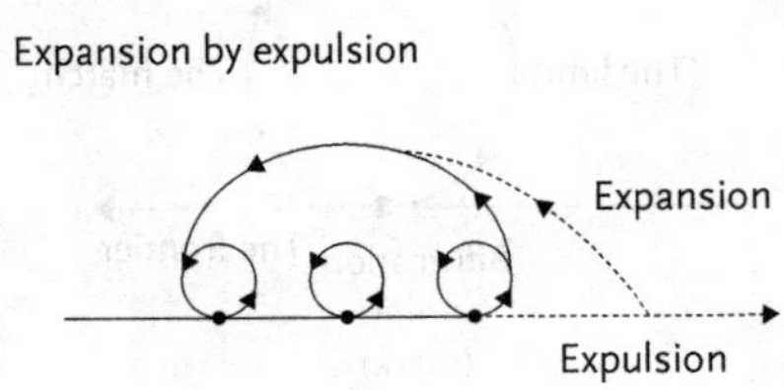

Figure 3.7 Expansion by expulsion

important not to assume the self-evident, ahistorical, and anthropocentric experience of beauty. I do not want to throw out the experience of beauty in the kinetic theory of the image but, rather, simply to understand its concrete historical and material conditions of emergence. In other words, a theory of beauty should emerge from the bottom up (from its material and kinetic conditions), and not from the top down (from its subjective definition).

With respect to borders, there are all kinds of affective limits that delimit the work of art: the frame of the painting, the stage of the theatre, the walls of a building, the platform of a sculpture, the pause before and after a musical performance, as well as that between movements in chamber music, sonatas, symphonies, and acts of an opera. These are not ahistorical, immaterial, or clearly nonaesthetic aspects of the work of art. The affective limits of the work of art, insofar as they are obeyed, do function to filter and focus the kinds of images the artist works with. In some sense, this is the first step in creating a work of art: to make a mark, draw a territory, or create a limit affect beyond which the image is left relatively disordered. This occurs in all matter. Islands have beaches, plants have cell walls, animals mark their territory with smells, sounds, and visual signs, and so on.

As soon as limit affects are put in place, however, they begin to leak. Flows of light, sound, and smell emanate outward from the work, adding new affects to everything around them and in turn being transformed and changed by others. The affective power of the work of art is expanded insofar as it affects the flows of matter beyond its limits, but it is also diminished insofar as it begins to decompose. Historically, this entanglement is much more obvious in works of art where viewers were allowed to touch, smell, and taste the works. Today, museum curation reveals the fragility and materiality of art only by dramatic contrast with its securitization, glass plates, and attempts at sensory isolation. The power of the work of art to expand its affective influence can be seen in the number of its reproductions, the people who have seen it, and where and how it is displayed, talked about, and written about in history. There is a discernible and material way in which the aesthetic field of a work of art is expanded by the changes it introduces into the world.

This not only gives rise to the possibility of the persistence of a kinetic order of fluxions but also allows the order itself to expand and contract. The folds remain distinct, but the field adds or subtracts them from circulation. By the action of the limit junctions, the more folds act together, the stronger and more complex the field becomes; whereas the more folds separate through disjunction, the weaker and less complex the field becomes. Circulation, through limitation, sets some folds loose and merges others in an expanding network. As the circulatory field increases its folds, it also increases its qualitative and quantitative dimensions. It becomes more complicated and more powerful. Circulation is thus more complex than unordered movement or even harnessed movement (the fold); it is the controlled reproduction and redirection of collective movement across a certain limited field.

THE EXPERIENCE OF THE WORK OF ART

Now that we have put forward a kinetic theory of experience (as the receptive capacity of images ordered in an aesthetic field) and a theory of the work of art (as the delimited domain or field of affection), we are now in a position to offer a materialist and nonanthropocentric definition of the experience of the work of art. The experience of the work of art contains two double-genitive dimensions rarely attended to in the philosophy of art.

The first double genitive concerns the *experience of* the work of art. Experience in is this sense is both something the work of art has—as its own material capacity for sensory receptivity—and something the work of art makes possible in the form of an experience for something or someone else.

In the first sense, works of art, as material processes, have an experience defined by their sensitivity to light, sound, temperature, and so on. Insofar as they are defined by a field of images, those images are, like the ship of Theseus, constantly breaking down and being disjoined while also being supported by new flows of matter. At the level of the activity of matter itself, we can and should therefore speak of a kind of agency, activity, or subjectivity of matter and the work of art itself. It is affected by matters.

In the second sense, the work of art is something experienced by another aesthetic field. Insofar as another field of images (no matter what that field is, whether rock, plant, animal, or human) is composed of ordered affects receptive to and capable of being changed by a work of art, then it also has an experience of the work. Taking together both senses of this first double genitive, it becomes clear that it is the kinetic process and flow of matter that is, in fact, primary in the work of art; it is simply circulated differently into different but entangled subjective and objective structures. On the one hand, the work of art and the sensorium that experience the work of art both have their own sensitive (subjective) experiences. On the other hand, insofar as both rely on the other as their material condition of experience, both act as the object for the other. The double genitive shows us that subject and object are simply two sides of the same material kinetic process of distributed images.

This leads to exposure of a second double genitive in the work of art *itself*. A work is the product of artistic creation. The work is the delimited region of affective composition—although to some degree it also recedes and exceeds these limits through degeneration and expansion. The work of art is a receptive object of creation insofar as it is capable of being contracted through destruction and expanded through further creation. The work of art is created.

In another sense, however, the work *of* art refers to the active agency of the work itself to affect others outside its limited field. A work of art is not a merely passive object; it affects the light, sound, texture, and smell of the world around it. The spectator is then affected and changed by this work. This is not a metaphor. The world around and the body of the spectator are literally and materially changed, no matter how slight, by the introduction of this new distribution of images into the world by the work of art. New flows of matter (light waves, sound waves, scent waves, and so on) are introduced. The work of art creates.

In these two double genitives—"experience of" and "work of"—we can see two dimensions of the same material kinetic process. The subject and object are two dimensions of the same distribution of images. It is strange to say, but insofar as the work of art becomes both subjective and objective, so too does the experience of the work of art. The division between subject and object and the theory of representation is exposed for what it is: an arbitrary historical creation desperately in need of a new theoretical framework that takes seriously the primacy and activity of the image itself.

The kinesthetic theory of art proposed here is substantially more expansive than most, but it is not absolute or ontological. It is both historically situated in the present (since it is focused on the primacy of motion in the image) and excludes a number of things from being art. For example, relatively insensible flows of matter are not art. Fragmented affects are not art. Works of art require an aesthetic field.

The kinesthetic theory of the experience of the work of art proposed here is based on the idea that the image is nothing other than matter in motion. When one field becomes materially entangled with another, both undergo a change that must be taken seriously in any philosophy of art.

KNOT ART

Aesthetic fields circulate and order images with degrees of flux, as we have seen, but fields can also combine into larger and smaller networks, or

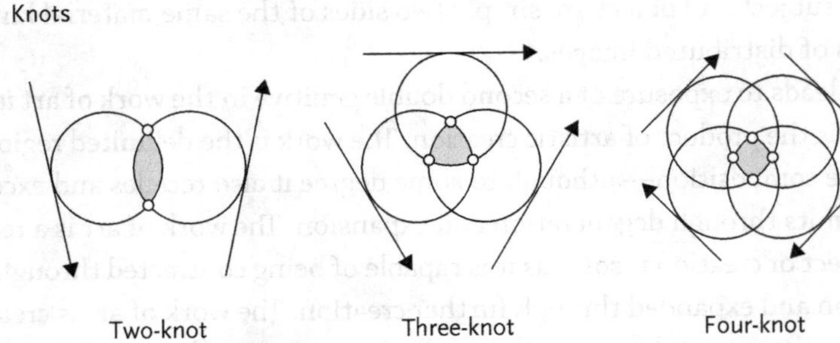

Figure 3.8 Art as kinesthetic knots

"knotworks," based on their shared motions or affects—what we have just described as the "experience of the work of art." Art is a kinesthetic knot (figure 3.8)—that is, the intersection of two or more fields at two or more of the same affects or folds.

The knotting that occurs in the experience of the work of art is what makes possible the continuous intersecting fields of circulation. In a kinesthetic knot, each field remains in some sense distinct but also becomes connected at specific affects to other fields, making possible a series of shared or collective qualities. One aesthetic field thus "becomes" another not by mimesis, metaphor, or representation but by literally sharing the same affective capacities or folds as another field. Two fields are knotted together by their shared affects, but these shared affects can also produce their own field in turn: a knot.

Knots are different both from the constellation of flows and from the conjunction of folds. Constellations do not require folds, since they are simply the intersection of flows. However, the intersections or events in a constellation can become the site of two or more folds or conjunctions if folding occurs at the points of intersection. Fields, however, can then connect these two affective folds in an ordered pattern of images similar to that of the original evental encounter. In other words, an event can become stabilized by repeating its *trajectory* and *intersections*. A constellation can then be transformed into a field by stabilizing and ordering all its events by creating affective folds for each event. A field can then intersect with other fields, creating a knot.

Nests

Nests are different from knots. A *pore* is the inside of a fold. It is the difference between larger and smaller folds, which are not discontinuous but simply appear as porous depending on their order. However, just as there are larger and smaller folds, each one containing the previous without a final above or below, so there are larger and smaller nested aesthetic fields of circulation. The difference between these nested fields is similarly porous. Within any given field, only the kinomena appear; the field itself does not. However, at the next larger level or nest, a subfield will only appear *as a thing* and *not another field* relative to the larger field that contains it. Each field is therefore a real synthesis of all its components at each level. Therefore, the difference between a nest and a knot is that a knot is created by a *nontotalizing overlap* (like a Venn diagram) of one field with another that entails the sharing of at least two affects, and *not a nested or totalizing relation* of one field completely inside another.

Knots

As such, knotting also entails the possibility of a third field, or knot, between the two overlapping fields. This shared circulation between the two fields makes possible a third condition for ordered relations—a third distribution of affects shared by two differing fields, but distinct from them. A kinesthetic knot is thus the product of overlapping but distinct fields. If all the shared affects should dissolve or be redirected, the knot is destroyed. Its circulation exists only on the condition of the other fields that support it. It thus has nothing to do with essential or accidental properties. Knots are simply the practical affects produced by kinetic overlap. If an affect is unfolded, the knot is loosened; if a new affect is added, it is strengthened. When the kinetic qualities that create the knot are gone, the knot is gone. Thus knots can appear and disappear as flows of matter fold and unfold—as images emerge and dissolve.

Therefore the fundamental question of kinesthetics is not linguistic, conceptual, or indexical, such as "What is art?" These kinds of questions try to determine the intrinsic relations of the image before it has been folded. Rather, the question is material, practical, and kinetic: What does it *do*? How does it *move*? Art is nothing but movements—flowing, folding, and the knotting together of different fields across shared affects and images. To describe an image is simply to identify its kinetic capacities and the field of circulation that orders it. The more shared folds it has with other circulations, the greater the degree of similarity or becoming between them as an overlapping composite.

Fields of circulation are knotted together, but they also produce a knot or new field that occurs only as the interaction between the fields. A knot is like a dance whereby two fields directly coordinate their motions around a few shared capacities. The knot is not an interaction between two separate individual fields but, rather, what happens *in between* them: the entangled dance of their motion. Both fields thus undergo a mutual transformation by coordinating their motions. Instead of remaining isolated, the knot allows multiple fields to become a single field with two or more *dimensions* or pathways.

Knots also make it possible for fields of circulation to morph or change their patterns of motion without changing the number of shared affects or images. As long as the morphisms or movements in the circulations do not disjoin from the shared junctions, the two fields remain knotted. However, as aesthetic fields change and move, their flows and folds may move closer to or further away from one another, forming different kinotopological "neighborhoods" or proximities. Kinotopological neighborhoods may change, but the number of shared affects will remain the same in the knot. In other words, knots are what allow composite sensations and images to persist in their composition without dissipating, even when they are moved around or morphed.[10]

Example: Beethoven's Sonata No. 32 in C minor, Op. 111

Let's see if we can bring all the concepts of experience, the work of art, knots, and nests together in a single example: a musical performance of Beethoven's final piano Sonata No. 32 in C minor, Op. 111.

The Work

As a work of art, the sonata is defined by at least two affective limits that mark the beginning and ending of the work. The beginning of the work is marked by an opening affective limit or border: the introduction of several moments of relative silence before the musician begins. This is not a true silence of course but, rather, a sonic marking of the beginning using the ambient sounds in the concert hall. Throughout the piece, these sounds will be used strategically to indicate the relative silence of the aesthetic field itself—that is, the sounds that are "not" being played. This is the inward fortification or incorporation of the ambient sounds of the concert hall into the work of art itself. It is the sound one hears in between the numerous *sforzandos* and pauses that run throughout the work, between the first and second movements of the piece, and at the end, for example.

The closing affective limit of the sonata is marked in the score by the bold double bar line marked in the performance by a sonic image of silence even after the *diminuendo* of the work trails off into the slightest trace of sound. All the sonic images between these two limit affects define the domain, field, or work of art. The work of art is both a series of sonic images *produced by* the performing artist using the piano and also something *that works* by vibrating the entire concert hall and everything in it. From the seemingly insignificant seats to the human ear, all the matter in the room undergoes a material transformation as it literally moves and resonates with the opening *sforzandos* chords. The work of art is both worked on by others and works upon those others. Each time the sonata is performed, it is changed or played differently by its performer. In turn, each time it is performed, it works differently on the different materials and persons who are exposed to it and their whole environment. Together, the affected and affecting nature of the work of art give it a certain style, order, and cohesion that defines the whole aesthetic field (artist, work, audience, environment).

The Experience

The experience of the sonata thus includes all the ways in which this work of art is affected, given the sensitive capacities of its material image. This includes the resonating structure or technological amplifications in the concert hall that may amplify, mute, sustain, or dampen the sonic image. The music is augmented by the technique of the performer, as well as all the micro motions and sounds made by the performer's body or clothing: his or her deep breathing, shuffling shoes over the pedals, and so on. Even the subtle sounds of the spectators augment the work through their collective micro movements, the involuntary shuffling of their limbs, and the occasional cough or sneeze. The piano is both the producer of sound and the resonance chamber affected by the movement of the sounds it produces. Not only does the music vibrate the hall but it also vibrates the sound waves of the music itself. As different chords follow others or different notes occur simultaneously, they overlap with one another and augment each other in harmony or disharmony. In this sense, we can say that the sonata has its own experience of the performance.

However, the work of art is also *experienced* in the sense in which the spectators are literally and materially changed by all the sights, sounds, textures, smells, and so on in the room between the opening and closing affective limits of the field of images. There is no work of art in a vacuum;

there is only the work of art in motion. All the senses of the human body thus occur simultaneously and are modified in the synesthetic and kinesthetic field. The human body is not absolutely separate from the work of art like a Kantian judgment, as we have seen, but the piano, the concert hall, the shuffling feet, the human ears, and all the other matters affected are dimensions or regions of the same folded aesthetic field of images.

The experience of art is not composed of objects and spectators but, rather, a single distribution or arrangement of images in a given field. The sounds of the room, the piano, and the listeners are all works of art in their own right, delimited by their own affects, but during the performance they all enter into a specific pattern of kinesthetic resonance related directly by the pattern of the sonata image itself. The sonic tensions of the sonata produced by the expressive *sforzandos, crescendos, diminuendos,* melodic contrasts, and incredible tone contrasts take on a life of their own that similarly affects all the material in the performance.

Nests

To limit the work and experience of art to the single field of the human body, or even further to the human mind, is to limit the image to only one nested field of images within a much larger aesthetic field of images in which numerous other material distributions are at work. From the limited perspective of one human mind sitting in the concert hall, everything appears as a subjective image, judgment, or representation. However, the material conditions of this field include a number of other fields, images, and affects (pianos, walls, chairs, sound waves, and so on) that all enter into a collective and mutually transformative kinesthetic order of everything occurring in the concert hall between the affective limits of the performance. From the perspective of this larger field, the limited field of human thought or even human sensation is only one image among others within the larger field that supports and orders them.

Knots

All the sonic images ordered by the aesthetic field of the concert hall during the duration and movements of the performance also precede and exceed that duration. In this sense, their fields of motions only temporarily intersect or knot together with the aesthetic field of the sonata performance. The work of art has its own sensory field, but so does the human body insofar as it is already composed of material images and is already distributed. This body then enters into composition with other material fields, which through touch, light, and sound become receptive in specific patterns. The human body is actually touched and changed, just as the work of art is affected by the light and sound emitted by the human body, because both are composed of nonhuman matters. By distributing these images across the body, the body is affected and the affects are the affects of the work of art itself. Both the human body and the work itself share several of the same affects.

For example, in Beethoven's sonata, the human body begins to resonate and vibrate in the C minor scale; it shudders with the melodic and tonal tensions of the first movement; and vibrates with the crescendo and then diminishes with the trill at the end of the second movement that drags the image entropically into disjunction, while at the same time allowing it to conjoin with the ambient noises that can now be heard. The diminishing trill thus slowly exposes the constitutive noises of the nonimage with those of the sonic image itself. Art disjoins into life and life merges with art. The sonic image of the sonata thus beautifully reveals the whole aesthetic field.

Insofar as the sensory field of a single human body is affected in at least two ways (in the violent *sforzandos* of the opening bars, in the bursts of melodic tension and contrast, or in the deconstruction of the melody itself into the rapid trill at the end that trails off into infinity), then, the field of the human body becomes knotted with the image body of the sonata and creates a new kinetic field defined by the range and order of affections that the body is capable of being moved by.

However, not all bodies enter into the same knots equally. For example, there are those who say that "all classical music" sounds the same to them. The subjective experience of the work of Beethoven's sonata may be defined in this case b general but differentiated unpleasant affects, and thus no knot will be subjectively discerned. To knot one's self into a work of art is to be affected by it in as many ways as possible. To knot is to enter into a kinetic becoming with the work such that there is only one field with multiple dimensions. One does not passively listen or represent a subjective image of the work; one becomes the work as one's own image body resonates alongside it.

Beauty

Kinesthetics thus has nothing to do with contemplation (Plato), judgment (Kant), knowledge (Hegel), or communication (Tolstoy), but is first and foremost about the affectations of matter. An aesthetic knot occurs when different material bodies move and mingle in the same distribution or ordered relation. The experience of ego loss, or of "losing one's self" in the art image, has at its core a kinetic definition in which the matter begins to resonate along shared affects and becomes a new composite. Beauty is the maximization of shared or knotted images between the body and the work of art. We can thus give a new kinetic meaning to an old idea.

Beauty is the multiplication of knotted images. Beauty is not an idea, normative value, or even a feeling but, rather, a collective material transformation of the aesthetic field. One can hear Beethoven's final piano sonata as one relatively homogenous drone of "classical music"—in this case, one shares only one image, the "classical image" with the work. The more images one shares, the more one begins to discern the internal order of the piece relative to the same diverse order of affections in one's own bodily images. Therefore, beauty it is not a question of accurate representation but, rather, of establishing a kinetic *coordination* between two or more aesthetic fields. Sensation is thus *transmitted directly*, as Valéry and Pollock say, and avoids the detour and boredom of conveying a story.[11]

CONCLUSION

This chapter concludes part I of this book and the kinesthetic exposition of the minimal features of the image and its motion. If the image can truly be said to be in motion today more than ever before, it must at least be capable of flowing, folding, and circulating across a field. Kinesthetics gives the theory of the moving image its own proper concepts. Without them, the image only appears to be a derivative or even illusory construction of more fundamental objects, subjects, social structures, perceptions, cognitive states, and so on. Kinesthetics thus provides an original conceptual framework in order to understand the primacy of motion in the images of art and aesthetics.

However, the kinesthetics of part I remains insufficient in an important way: it remains entirely *conceptual*. We have so far presented the kinesthetic *concepts* of flow, fold, and field largely abstracted from their detailed historical and material origins. For this reason, what can be said of kinesthetic operations at this point is extremely minimal and general. To mitigate this, I have tried to provide examples along the way, but this is not enough. The conceptual kinesthetics of the image tells us little about how these concepts are actually deployed in the motion of aesthetic practice itself.

The image is in motion, but it does not always appear as such in the history of art. Furthermore, there is not only one kind of aesthetic field or one kind of ordered motion possible on this field. Thus, it remains to be explained: What different *types of fields* are there? and What dominant patterns of motion on these fields have been invented so far in history such that the image appears not to be in motion? The aim of part II is to answer these two questions. The theory of the image is not complete without the history of the image from which it was born.

PART II

History of the Image

The image is historical. If the image, as we argued in part I, is nothing other than matter in motion, then it is by definition historical, since history is defined by the becoming of matter and not by the being of an immaterial and immobile unity. This thesis has at least two important consequences, which is the subject of part II of this book.

First, because the image is material, kinetic, and thus historical, it is possible for different, coexisting, and even mixing aesthetic fields of images to emerge. In other words, it is possible for matter to distribute itself differently over time into different patterns or orders of arrangement. By this, I do not mean that the same humans only interpret the "meaning" of a work of art one way at one time and another later on. The pattern and order of images actually change. Not only is the work of art itself not an ahistorical object of contemplation but neither is the spectator an ahistorical subject of interpretation. As we showed in part I, both the subject and the object are simply dimensions of the same aesthetic field of images. These fields change historically.

If the image is material and historical, then so, too, are the subject and the object and thus so, too, are the underlying aesthetic fields that produce them. If this is the case, then it is possible to study this material history and to discern the aesthetic regimes or fields of images along with the different affective subjects and objects that are distributed by them. The aim of part II is, therefore, to study four dominant regimes of Western aesthetics during their periods of historical dominance: the prehistoric functional field, the ancient formal field, the medieval relational field, and the modern differential field. Each of the chapters in part II defines the patterns of this kinesthetic field, as well as the concrete historical images that constitute it.

The second consequence of the historical nature of the image is that all these different kinesthetic fields are capable of persisting into the present. What this means is that contemporary aesthetics is not defined by a single kinesthetic field or pattern of motion but rather is composed of a motley mixture of everything that has ever been. Today, all the fields of sensation that have ever existed persist and mix with one another in relative degrees of predominance. History does not just provide a context for understanding our present; the present is literally composed of previous aesthetic fields in different combinations. Therefore, any theory of the image or affect that does not engage the coexistence and mixture of historical regimes of images is not simply lacking historical background but lacking foreground as well. This will become especially clear in part III, where all the historical labor of part II is required for a full analysis of the contemporary image.

Part II of this book is thus divided into four sections, each covering one of the four major aesthetic fields of Western history: function, form, relation, and difference. Each section is then analyzed from two perspectives: conceptual and historical. Conceptually, each section looks first at the common kinetic pattern produced by the dominant arts of the time (centripetal, centrifugal, tensional, and elastic). Although this conceptual perspective is derived from the historical events, it is presented first in each section as a guide for the reader to follow and anticipate the common kinetic patterns of the various arts of that period. Each section then looks at the concrete historical articulation of these aesthetic patterns in their prehistoric, ancient, medieval, and modern expression, respectively. The thesis here is that the prehistoric aesthetic field is predominantly defined by a centripetal motion and functional aesthetic, the ancient field by a centrifugal motion and formal aesthetic, and so on.

SECTION A
The Functional Image

The first major kinesthetic field of images to rise to dominance in the West was the functional field. This began during the Paleolithic period (3.3 million to 10,000 years ago) and achieved its zenith in the Neolithic period (10,000 BCE to 5,000 BCE). Historians have used the description of prehistoric art as purely "functional" as a way to contrast it with the dawn of "true" art, supposedly defined by the autonomy of form over use-value—"art for art's sake." Hegel, for example, defines the functionalism of the natural and prehistorical "artificers" (plants, insects, animals, and early humans) who produce "without having yet grasped the thought of itself" as a purely "instinctual operation, like the building of a honeycomb by bees."[1] This, he contrasts with the "absolute art" of the classical period, defined by the formal self-consciousness (the "Notion") of the creator as the creator of "absolute beauty as such." "There can be nothing more beautiful," Hegel writes, "than the classical; there is the ideal."[2] From Plato to the present, functionalism is either treated as a proto-aesthetic stage that was eventually overcome during the ancient period or as a nonaesthetic category altogether.[3]

This book proposes a different understanding of prehistoric art and aesthetic function. We are not trying to go back and show how what seemed to be purely instinctual acts were actually intentional and formal works. We are also not arguing that animals and early humans were in fact "more rational or aesthetic" than we thought. We argue instead that all art has and always has had a functional dimension to it, not because it overcomes or sublates this dimension in "contemplation," "the Notion," or the aloofness of the so-called aesthetic attitude, but because all art relies on a fundamentally kinetic structure, even if it tries to cover it up. In other words, the nonhuman and material origins of art are never sublated or overcome but rather are simply redistributed and combined in different kinetic patterns.

CHAPTER 4
Centripetal Function

The forming of the five senses is a labour of the entire history of the world down to the present.

—Karl Marx[1]

The theory of kinesthetic function is defined by three interrelated features, which we define in this chapter. Once these features are laid out we will then be in a better place to see how this kinesthetic field concretely emerged in the dominant arts and images of the prehistoric period.

CENTRIPETAL MOTION

The functional field is defined first and foremost by the continuous flow of matter from the periphery toward a center. Any aesthetic distribution or field requires a centripetal motion, such that two or more heterogeneous gathered affects can enter into some kind of ordered relation that ties them together in a series of images. Flows of matter without folds are insensible, while folds of matter without relation are fragmented. Only when affects are gathered together into ordered images do they become works of art.

But matter is also self-ordering. The flows of matter move centripetally into a new aesthetic field only at the expense or disjunction of another field. Every ordering is always a disordering and a reordering of matter. Bees disjoin heterogeneous flows of pollen from the periphery in order to conjoin them together in the centripetal location of the hive. In doing so, the new aesthetic field of ordered images appears as the hexagonal hive.

Birds and other animals disjoin and centripetally gather images (sticks, moss, and grass) from the periphery in order to assemble them into the ordered field of the nest. The human body is itself already a work of art insofar as it is literally composed of a constant centripetal supply of material images (water and food) from the environmental periphery into its center. What the human body does is simply an epicycle that emerges within this larger ecological circulation.

The centripetal motion of images is the material kinetic condition for the interoperation or co-functioning of the aesthetic field. If the flows of matter and its folded affects were not drawn from the periphery to the center, the images could never take on any distribution in the first place. Matter flows, but because it is pedetic, it also curves back over itself in returning back to itself toward a center, without necessarily producing a center. The curvature and reflection of matter simply define a region in which affective folds can gather and, in gathering, produce some minimal arrangement or order of images. Images therefore work or function together only because they enter into a kinetic proximity with other degrees of flux and reflection.

ENTRAINMENT

The functional aesthetic field is defined by the entrainment of its centripetally gathered images. *Entrainment* is the process by which a flow or fold is put into motion or synchronized by another flow. For example, biologically, there is an entrainment of an organism's circadian rhythm within that of the external rhythm of its environment. Geologically, there is an entrainment of sediment in a stream of water. Musically, there is a rhythmic synchrony as heteronomous moving pendulums begin to move together.

Entrainment is the kinetic theory of how divergent and heterogeneous matters move from the periphery into a central, ordered feedback loop. Through this centripetal motion not only does the inside of an organized body enter into a kind of kinetic feedback loop with the outside, but the inside of the body itself also takes on its own epicyclical organization in relation to this larger deferent cycle.

Entrainment Is Functional

A kinesthetic field works and does what it does not by a unidirectional causality moving from subject to object but through continuous kinetic loops, cycles, and epicycles in which various images collectively transform one another. Over and against the stable and empirical objects and subjects of experience, the functional entrainment of the kinesthetic field is defined by its continuous mobility and mutability. It is functional insofar as it produces a collective field in which the movements of subjects and objects use and rely on one another. Entrainment is thus a kind of multidirectional and collective utility.

French archeologist André Leroi-Gourhan names this functionality a "*chaîne opératoire*" ("operational chain"). In the work of art, the human body and the material world enter into a rhythmic entanglement, like a series of different pendulums slowly synching up into an ordered relation. Leroi-Gourhan defines the *chaîne opératoire* of "functional aesthetics" not by the application of a mental template or of a top-down plan in the mind of a human to a receptive matter but, rather, by a "functional plasticity" of the entire process.[2] "The principles of functional aesthetics are derived from the laws governing matter and cannot, by that token, be regarded as human attributes except to a very limited extent."[3]

For example, the origins of the Acheulean biface (sharpened stones and axes) were not first in the mind of the human artists, like a blueprint or idea applied to the material but rather were products that emerged from the material-kinetic capacities of the human body to identify and reproduce a muscular, visual, and auditory activity on matters that could be knapped on both sides without breaking. Through the activity of knapping, the body and brain became more rhythmically entrained with the stones and the stones became more entrained with new cuts and sides. The rocks gained new affects and capacities they did not have before.[4]

The human acts of hammering, sawing, and chipping had to first become affectively and rhythmically entrained by physical repetition on a certain kind of matter before the product could be produced, but as the product emerged, it showed the body what could be done. In other words, the centripetal motion of gathering rocks from the periphery and then accumulating cyclical actions on them are primary, not the mental picture of the product. The kinds of motions the body can do and the kinds of actions that change a kind of rock all get worked on at the same time. Functionalism is, therefore, the collective kinetic mutation, co-functioning, and reproduction of the whole aesthetic field.

Embodiment

Entrainment is different from cognitive theories of embodied, embedded, extended, and even enacted mind. While it is certainly true that cognition is affected by the body (embodied),[5] that the body is situated in the cultural and physical world in which it thinks (embedded), and that it relies on the use of external objects like tools and memory aids in its functioning (extended),[6] and even the activities that the body does in the world (enacted),[7] all these theories are fundamentally incomplete without the fifth "E" of material-kinetic "entrainment" from which they all emerge. This is because the four "E's" of embodied cognition all begin from the product (mind) in order to understand the process that produced it in the first place: the mind's relation with the body and world, and so on. In doing so, however, they risk presupposing precisely what needs to be explained in the first place: the emergence of the mind itself. They have got it backward. In these theories, the world, bodies, tools, and culture all become extensions or aspects of cognition, and not the other way around. Thus, cognition is not explained but simply "extended" and pushed further back as if it had always been there. In other words, thought is not understood as enfolded matter (entrainment) but, rather, as embodied cognition (enactivism).

However, what we call cognitive brain activity is only part of consciousness, and consciousness is only one kind of motion within a much larger kinetic pattern of motion in which it emerges. It would, therefore, be much more accurate to say that the mind is an extension or enfolding of matter than that matter is an extension of the mind. The danger of beginning a theory of mind, art, or perception from cognition is clearly apparent in the radical constructivism and idealism that result from it. For example, prominent embodiment theorists like Maturana and Varela end up claiming that "we do not see what we do not see and what we do not see does not exist."[8] If this is true, then the mind is affected and embodied in the world, but outside of this connection, nothing can be said of the actual world.[9] In short, the autopoietic model of cognition as a continuous transformation and becoming with the world risks regionalizing this becoming strictly to the cognitive and anthropocentric organism and, therefore, losing sight of the larger noncognitive material kinetic patterns that exist alongside it and/or produce this regional feedback loop in the first place. We therefore need a more material and kinetic theory of function, or "use value."

KINETIC USE-VALUE

The functional aesthetic field is defined by its kinetic use-value. Kinetic use-value is opposed to unidirectional or instrumental use, or mere "cognitive extension." The instrumental theory of use is anthropocentric insofar as it presupposes first a mental idea, forethought, or desire in the human subject that precedes the functional act of usage for some nonexistent end. In other words, the instrumental theory of use posits in advance of the use what the desired outcome will be.

Although intuitive, this definition is historically and materially backward. It assumes in advance the product of the usage and then reads this back into the mind as a mental cause of the value. This is a classic case of idealism and fetishism, in which the product seems to be the functional cause of production, when it is in fact the other way around. Historically and materially speaking, the correlation of use-activities and used activities is not at all obvious and emerges only after millions of years of practical kinetic experimentation and co-adaptation of entrained matters.

Every affective action is mutually transformative. When *Homo habilis* chipped a rock, the rock was changed, but so was the brain and body of the person who chipped it. Over time, the two affects become materially coordinated and sedimented in the same synchronized field. Historically, however, the same backward-looking "causality" is attributed to all kinds of things. Animism, panpsychism, vitalism, and humanism all shift agency around differently. To say that the rock is "using" the human to transform itself into a more affectively dynamic being only sounds absurd because of a relatively recent and chauvinistic concept of human agency. In truth, it is just the other side of the same process of entrainment. However, bracketing this historical bias of anthropocentrism and the whole metaphysics of agency, all we can safely say is that a set of affects have been entrained in a regular kinetic correlation of images following a certain pattern.

As argued in part I, an affect is not just something sensed; it is also something that itself has a capacity for sensation. An affect is a capacity to *affect and to be affected*. In this way, each affect in an aesthetic field does something: it affects itself and others in the field. Agency is materially and kinetically distributed.[10] Use-value, therefore, is not unidirectional or defined exclusively by an anthropocentric model of intentional consciousness. An aesthetic field actively co-functions with other affects, transforms those other affects, and reproduces itself through the reproduction of the other affects in the field of circulating images. Use-value and agency are things that circulate between affects. Each uses the others to reproduce itself.

For example, it makes just as much sense to say that Neolithic humans used the "Deer Hunt" wall paintings in Çatal Höyük[11] to help them hunt successfully as it does to say that the works of art used humans to return to the cave and reproduce more art, as well as add to the existing ones through hunting. If art leads to successful hunting, and successful hunting leads to food and survival, then more art is produced. In another example, we could just as easily say that the clay uses the Neolithic pot to hold itself together, and the pot uses the human to persist in its ordered being. The human then in turn uses the pot to store grain so that it can survive, gather more clay, make more pots, and so on. If we accept that the function of images does not first appear in the mind of the user but, rather, emerges as an experimental and mutual affectation of what both bodies can do, then there is no reason to confine use-value to a purely anthropocentric definition.

The kinetic foundations of all aesthetic fields lie in a functional materialism. There is no separate or immaterial aesthetic conciseness that can cut itself off from its material co-functionality with matter. Similar functionalist versions of this argument have been given for the ways plants and animals have used humans to reproduce themselves,[12] but there is no reason to limit a materialist theory of use-value to organic life. Use-value and function are simply a kinetic chain of ordered relations and mutual presuppositions. In other words, to say that X uses Y to produce Z is simply another way of saying that Z is defined and supported by the material conditions of X and Y, or even that Z uses X and Y in order to produce itself, or that Y uses X to produce and preserve itself. Because fields of circulation are feedback loops, kinetic use-value goes both ways: each operates as the condition of the other. Every ordered image is functionally supported by others.

This is the aesthetic technique for producing images that first rise to dominance in the prehistoric period, as we will see in the next chapter.

CHAPTER 5

The Prehistoric Image

The historical emergence of the functional aesthetic field can be seen across all the major arts of the Paleolithic and Neolithic. This emergence did not happen out of nowhere or during some "Paleolithic renaissance" around 35,000 BCE, as is often said. Rather, there is much more of an incremental increase in centripetal patterns of related image creation. The increasing sedentism that occurred between the Paleolithic and Neolithic, for example, is poorly understood as a lack or decrease of movement. Sedentism is not immobility; it is the redirection of flows, the creation of junctions, and the maintenance of kinesthetic images on a new sensorium. Sedentism is movement achieved by other means; it is the capture and ordering of images on a series of relatively stable centripetal surfaces.

This chapter argues that this kinesthetic shift occurs increasingly from less ordered and gathered images to more and larger gatherings of images on new and larger fields of sensation. This is done by looking closely at the kinetic patterns produced by six major aesthetic fields that define the prehistoric image: the body, the hearth, the cave, the vessel, the wind instrument, and the house. The argument of this chapter is that each of these major fields is defined predominately by a distinctly centripetal pattern of motion and a functional aesthetics.

THE BODY

Prehistoric art has its origin in the bipedal body.[1] The origins of human art precede its dominant kinetic distribution in the Neolithic by tens of thousands of years. Hominids, all their precursors, and many other animals were capable of making and responding to images well before they were increasingly gathered and circulated by *Homo sapiens* into a dominant field of functional aesthetics. At the kinetic level, there is thus no difference in kind between the art of animals and the art of humans but, rather, there is a difference in degree or a difference in the regime of motion.

Bipedalism

In fact, what makes possible the human increase in image accumulation is nothing other than a kinetic transformation in the animal body itself, from quadrupedalism to bipedalism. The human aesthetic field is distinguished from other animals by a dramatically larger and more diverse accumulation of affects or images in its body. This centripetal accumulation makes possible a vastly more expansive sensory and aesthetic field than was possible before the advent of bipedalism. Bipedalism thus gave birth to two interrelated kinetic processes made possible by the liberation of the paw from being a strictly pedetic tool focused almost exclusively on locomotion, to becoming a free-moving appendage capable of creating all kinds of new images.

First, the liberation of the hand made possible a massive diversification of images through digital movement and grasping. This in turn allowed for an explosion in the diversity and function of tools that could be manipulated by the hand, including gesture or graphism (the manipulation of matter with the hand). Gesture and graphism thus increasingly enabled the gathering or centripetal contraction of an extremely wide diversity of rhythmic motions of the arm and hand. The freedom of the hand not only made possible an increased kinetic capacity for grasping and accumulating things toward the center of the body, such as putting food into the mouth, but it also permitted a rhythmic accumulation of natural motions in the arm and hand itself, like the repetitive motion of knapping. In other words, the hand began to reflect certain natural motions—waves, wind, water, and so on—in the functional graphic patterns or distribution of images that define the aesthetic field. The liberation of the hand thus gave birth to both the centripetal motion of grasping and the accumulation of natural motions concretely recorded in early graphic patterns—the curved chipping of the Acheulean biface, the spiral, the ellipse, and so on.

Second, the disjunction of the hand produced a simultaneous disjunction of the mouth and tongue from its almost exclusive function as a primary grasping mechanism. Instead of having to forage and manipulate food almost entirely with the mouth and lips, the freed-up hand could simply grasp and forage and deliver the food directly to the mouth. The free movement of the hand and arm thus also made possible the free movement of the mouth, lips, and tongue, which could now develop independent kinetic functions—phonism. The evolution of the tongue was a diversification of its motions. As it evolved, it increased its degrees of freedom and motility just as the evolution of the fingers and hand increasingly diversified the kinds of images that could be made. Just like graphism, phonism also began to duplicate and accumulate a repertoire of natural images made possible by the new freedom of the tongue to modulate a flow of air.[2]

The important takeaway from the historical origin of speech and gesture in bipedalism is that these two kinetic disjunctions of the hand and the mouth occurred and developed together in close parallel, without one being modeled off of the other. Speech is not a representation of graphism and graphism is not a representation of speech. One did not emerge first and then cause the other to adapt to or supplement it; the two co-emerged in the same kinetic distribution of bipedalism. "As soon as there are prehistoric tools," Leroi-Gouhan writes, "there is a possibility of a prehistoric language, for tools and language are neurologically linked and cannot be dissociated within the social structure of humankind."[3] Graphism and phonism are therefore dimensions of the same functional kinetic field, not essences or transcendent forms of representation. Both are creative, both "form part of the same human aptitude, that of reflecting reality in verbal or gestural symbols or in material form as figures."[4] When early Ardipithecus descended from the trees 4 million years, ago it made possible two kinesthetic operations: the contraction of sounds into descriptions and the contraction of rhythms into graphic inscriptions. These are two parts of the same centripetal process of accumulating increasingly diverse sonic and gestural images in the human body.

The bipedal body dramatically increased the number of images that could circulate through it. For example, by freeing up the mouth and raising the ear above the ground, bipedalism allowed for a much larger range of sonic images to be gathered by the ear and recirculated through the mouth with the aid of the expanded vertical chest cavity. With the simultaneous freeing up of the hands, the body took on new instrumental capacities, such as hand-clapping, stamping on the ground with the feet, and beating the hollow chest. With the liberation of the hand, the body could now play itself.

Pedesis, the autonomous action of the foot, whether to walk, to run, to dance, gave birth to an exponentially diversified aesthetic body defined by the centripetal accumulation of images on the body, but it also gave birth to an increase in what the body could centripetally accumulate outside the body. In the act of walking, for example, we can already see the origins of all the arts. As Elias Canetti writes in *Crowds and Power*, "Rhythm is originally the rhythm of the feet. Every human being walks, and, since he walks on two legs with which he strikes the ground in turn and since he only moves if he continues to do this, whether intentionally or not, a rhythmic sound ensues. . . . The earliest writing he learnt to read was that of their tracks; it was a kind of rhythmic notation imprinted on the soft ground and as he read it, he connected it with the sound of its formation."[5] In the pedetic movement of the body, one finds already the entrainment of sonic and graphic images into an ordered and rhythmic aesthetic field. The sounds of the footsteps and breathing, the motion of the body, the beat of the heart, the sight of the tracks left on the ground, and even the smells left behind all combine synesthetically and become ordered in the movement of walking. In other words, the bipedal body centripetally gathers all these images in a certain region in and around the functioning body.

Tools

The earliest kinetic distributions of images in the hominid aesthetic field are centripetal collections. Already with Australopithecus, 3.4 to 2.5 million years ago,[6] we see the first gathering of stones to a work site and the gathering of kinetic motions into cyclical repetitions to produce the first stone tools. Tool creation is the first externalized aesthetic cycle or minimal degree of kinetic order made possible by freed hands. This minimal externalized aesthetic field is defined first by the preconditions of the centripetal gathering of at least two stones: a hammerstone to strike and a core to be stricken. Second, the field is defined by two ordered images: strike, repeat strike. "Move outward" and "return" constitutes the basic kinetic movement of the fold. The kinetic structure of all ordered affection requires at least two movements: move out and return in the first strike, and then repeat in the second, and so on. Each strike entrains the other in its function.

After walking, tool-making is thus the second centripetal art of folding and entraining. Stones, body, and brain all enter into a mutual transformation or functional cycle. The knapping cycle transforms human cognition in increasingly complex ways, and in turn human cognition transforms stone-tool production in increasingly complex ways. From this minimal ordering and simple circulation a whole history of tools and cognition developed side by side, adding more and more intermediate images: single face, biface, rounded, pointed, with handle, and so on, all the way up to the repetitive actions of contemporary production. Affect and repeat affect still define the basic cyclical kinetic structure of all technical and mechanical creation.[7]

THE HEARTH

The second major centripetal aesthetic field of the prehistoric period is the hearth. Somewhere between a half million and a million years ago, *Homo habilis* was the first to centripetally gather and control fire in the hearth.[8] It is impossible to understate the kinesthetic transformation initiated by the creation of the hearth. Well before the first graves or houses, the hearth was one of the earliest major sites of centripetal collection and aesthetic ordering of images. The hearth is defined by two kinetic operations: first, to cut into the earth's flows of matter and make a pit or curved bowl in the earth; second, to gather stones from the periphery and put them in a pile or circle around the pit. In these two very simple but essential centripetal gestures we can locate the kinetic foundations of an entire aesthetic epoch spanning the Paleolithic and Neolithic periods. The flows of matter are increasingly gathered from the periphery and put into a functional kinetic cycle at a center. This center point introduces a mutual transformation of the natural periphery on one side and the central hearth on the other. In contrast to the individual labor of making stone tools that only one person can make and use at a time, the hearth made possible the first collectively produced and simultaneously used functional aesthetic field. By no means immobile, the hearth was the first step toward an increasing "kinetic sedentism" and a robust centripetal aesthetics.

The Pit and the Pile

The pit and the pile define the first centripetal motions of the prehistoric aesthetic field. The hearth first of all marks the limits of its own field by the depth and circumference of its pit on one side, but also the height and illumination made possible by the fire that surrounds the camp at night on the other side. Between these affective limits a vast material flow of light is folded up into an ordered spectrum of illuminated images.

Just as the bipedal body becomes an aesthetic field with its own limits and images, so does the hearth. The two fields can then enter into a shared kinetic entrainment in which both are transformed by one another as a single functional field. The fire uses humans to keep it supplied with fuel and to increase its affective capacities: to illuminate, to burn, to smoke, to cook, and more. In turn, humans use the fire for safety, warmth, cooking, and so on. This allows them to survive, which allows them to make more fire, and so on, completing the functional cycle of this historically novel field. The hearth and the human body thus have a functional dependence that allows both to expand their own order of images.

The affective limits of the fire in the dark become the affective limits of a shared world of images: vision, colors, shadows, smells, tastes, and sounds. In the night, the human-fire field stands distinct from the rest of the earth. It creates a territorial image. Inside the fire, there are some people and not others, safety and not danger, light and not dark, food and not hunger. In Paleolithic history, there is perhaps no world of images more vivid and intensely stark than that of a bright dancing fire surrounded by a dark world. "The image," as Gaston Bachelard writes, "would have to be placed under one of the greatest of all theorems of the imagination of the world of light: *Tout ce qui brille voit* (All that glows sees)."[9]

The glow of the fire itself has a kind of agency or light sensitivity that affects and is affected by its surroundings. Things block its light and reflect it back as images. The fire is an eye on the face of the earth that sees the dark and is seen by the dark. In the hearth, Bachelard writes, "we have the impression that the stars in heaven come to live on earth, that the houses of men form earthly constellations."[10] The hearth becomes one of the earliest territorializing images.

Prometheus

Over the course of the prehistoric period, the fire of the hearth becomes a kinetic image defined by gathering, accumulation, creation, and destruction. The hearth is the first home before the house, the first human-made place, the first robust field of images that humans made and merged with. It is no wonder that the use of fire is described by the Greek poet Aeschylus in *Prometheus Bound* as the πάντεχνος, *pantechnos* ("source" or "assistant" of all the arts).[11] The Greek word τεχνε means both "artistic skill" and "artful cunning" or "intelligence," something that defines Prometheus's clever theft of fire from Zeus, according to Hesiod and Aeschylus.[12] However, the Greek word τεχνε (*techne*) also comes from the proto-Indo-European word *teks*, meaning to "fold," "weave," "braid," or "work with wood," indicating the material conditions for both art and intelligence.

The kinetic insight of early Greek tragedy here is profound, even if retrograde. The first major disruption in the flows of matter that distinguish humans from nature and from the gods is not a difference in kind (categorical, rational, or linguistic) but rather a difference in their material kinetic distribution. Art emerges when the flows of matter, wood (*hyle,* for the Greeks), becomes curved (*flex*) and folded (*pli*) back over itself into witty woven patterns in the burning woodpile. The result is the production of illuminated images.

However, the hearth's woven, folded, and plaited pile of centripetally gathered rocks and sticks is also the material condition for a bifurcation or distinction between human and nonhuman nature. As they gather around the hearth fire, humans turn toward one another and away from the periphery. Just as with Prometheus's body, the illuminated images of the hearth are eaten away in the day by the sun (Zeus, the eagle), but regenerated at night. The night restores Prometheus's liver so it may be eaten, and so that it may be restored and eaten again in a kind of functional cycle similar to that of the human hearth. The fire is made to eat food to live in order to build more fire, to eat, and so on. A million years after the first hearth, the Greeks would thus find in it the retroactive conditions of all aesthetic division: between good and evil, male and female, day and night, sensible and insensible: Promethean wisdom.

Cooking

The hearth makes possible not only a new aesthetic field of illuminated and colored images but also one of olfactory and gustatory images, as different flows of matter release new flavors and scents from their oils. As early as 500,000 years ago, *Homo erectus* began cooking food,[13] which became a widespread practice by around 250,000 years ago.[14] With cooking, previously toxic foods could now be eaten and foods that were difficult to digest became easier to process. More, larger, and more diverse periphery matters could be collected and consumed using the hearth.[15]

Over the course of the Paleolithic, fire also became increasingly mobile with the use of lamps—stone bowls filled with animal fat and a wick. These lamps were frequently used to take fire deep inside caves for ritual and aesthetic purposes. Combined with the aromas of cedar oil–based paints in the confined space of the cave, this must have produced yet another incredible field of images—odors, sights, sounds (echoes), and textures—centripetally gathered together in close proximity.

Just as the material periphery must be first gathered to the hearth, house, and cave, so the products of hearth, house, and cave are further gathered into the human body as food that sustains the body and allows for further centripetal collection and thus the functional preservation, reproduction, and expansion of the whole aesthetic field. The smells of food and aromatic woods also become kinesthetically coordinated to the centripetal nourishment of the body, to homecoming, and to sexual and social reproduction. Once gathered, various sensory images could be combined and ordered, seasoned with spices, and scented with oils.[16] The centripetal hearth makes possible a whole culinary aesthetic field.

THE CAVE

The third major centripetal aesthetic field of the prehistoric period is the cave. Like the hearth, the cave gathers together the material flows of the earth and folds them over each other, creating a giant plait (*techne*) or hollowed-out center that can be filled with images and ordered into an aesthetic field. While the pit and pile of the hearth must be manually fabricated and folded from the periphery, the cave is discovered already folded, as if it had already gathered itself by itself into an aesthetic field. In other words, the cave does just what birds, bees, animals, and humans do: it uses the same centripetal kinetic pattern to gather and fold up the periphery around a hollow center. In this sense, the cave is the nest or hive of the earth.

In contrast to the relative mobility and impermanence of the early hearth, the cave is the pinnacle of fixture and permanence, albeit a vortical one. The cave is a ready-made functional aesthetic field, an entire ecosystem, relatively immune to the flux and flows of the outside world, but made of nothing other than the outside world itself. The cave is the interior of a folded exterior surface. While the hearth delimits a luminous horizontal aesthetic field, the cave delimits an aphotic vertical field. While the hearth intensifies a functional kinesthetic surface, the cave amplifies this intense center by surrounding it with walls, but without becoming a sphere. Since the cave is also open to the outside through labyrinthine and bifurcated passageways, it is not a circle but, rather, a twisted spiral open to the outside and continuing further along inside through its forked and coiling paths without absolute center or symmetry.

Above all, the cave is a sensory resonance chamber. Images echo and reflect off the topologically variable walls in new, dynamic ways without dissipating immediately. The heartbeat, breath, and footstep of the body in the cave achieve their maximum sonic expression in the cave. The smallest lights produce enormous shadows that dance kinomorphically along the folded surfaces of the walls. Even smells linger and settle for prolonged durations. The textures of the ceilings and floors are amplified by stalactites and stalagmites, and the damp air is condensed into mineralized water and vapor. It is no wonder that the first idiophones and aerophones appear in caves, along with all manner of centripetally gathered bones, toolkits, and tools; music, movements, dancing, vocalizations—all resonate on the cave walls.[17] If the hearth is the assistant of all art, the cave is the recording surface of all art.

The cave is a kinomorphic shapeshifter. It is both art and artist. In the cave, all that is solid literally melts into air. Water vapor rises, condenses into water, and drips down the walls in the form of heavily mineralized fluids. These fluids then slow down and turn back into solids in the form of crystals and mineral deposits. All that is air crystalizes into solids. Matter crystallizes into form, and form melts into matter. The cave sculpts itself as it drips mineral flows into plaits and layers of textural relief. The cave paints itself with the pigments of the minerals that are entrained into the flows of water and drip down the walls, creating simultaneously a range of textures and colors on every surface.

The cave is also self-resonating, as the drops of water echo back and forth on its walls, modulated by the unique and changing topology of the walls. The cave reaches out and touches itself as the drips of matter flow from the ceiling as stalactites and pile up on the floor below them as stalagmites, until they finally merge in the dark like two fingers touching. The cave is therefore autopoietic or self-making.

Paint

Paint is defined by three centripetal kinetic operations that duplicate those of the cave itself. First, pigment is created by gathering solids (red or yellow ochre, hematite, manganese oxide, and charcoal) from the periphery and crushing them into a concentrated dust, powder, or powder medium. Second, a fluid "vehicle" (water or cedar oil) is gathered from the periphery to a central location. Third, the pigment solids are entrained into the fluid medium or vehicle in a single curved bowl or cup to prepare for transport to a surface. All these kinetic techniques were discovered in Blombos Cave as early as 100,000 years ago.[18]

Painting is the ordered distribution of this relatively homogenous confluent mixture onto a region of a surface. A painted surface marks out an organized centripetal collection of pigment. From the periphery to the bowl, into the cave, and onto the wall, the flow of pigment and liquid vehicle is increasingly gathered closer and deeper. The fluid vehicle transports and collects the pigment to a region of the surface and then deposits it there as a solid through evaporation. Painting is an evental confluence between a flow of pigment and a flow of liquid. The two intersect and then diverge on the surface. The question, then, is how to distribute these events into an ordered, enduring, kinetic constellation. In painting, all that is solid (pigment) melts into liquid (water or oil) and then returns to solid in the textured pores and plaits of the cave wall. A truly smooth or frictionless surface cannot hold paint. Painting requires a minimum of texture and touch to hold together after the evaporative process.

The Function of Painting

Cave, paint, and human body thus enter into a collective kinesthetic entrainment. One is not simply the creator of the other, but each is a dimension of the same aesthetic field of circulation. Each uses and is used by the other. Each transforms and is transformed by the other. The cave already paints, sculpts, and sings itself. Paint and the human body do the same. Several flows of matter are already centripetally gathered, folded, and ordered in the material conditions of cave painting itself. Before any image is painted on the wall, the cave, paint, and human body are already works of art that follow the same circulatory process of liquefaction-solidification-distribution: flow, fold, field.

Humans did not first enter caves with a mental image or plan to paint, or for what they wanted to paint, or even with the idea of creating paint for its walls. The emergence of cave painting is a constellation of events that occurred around the same time throughout the Upper Paleolithic, from 40,000 to 10,000 years ago, in the caves at Chauvet, Lascaux, Altamira, Cosquer, Pech Merle, and others.

1. Humans saw how the cave sculpted and sang itself.
2. They saw how paint made and painted itself down the walls as mineralization and evaporation.
3. They experienced how their own senses painted, hallucinated, and imagined in the dim, flickering, or absent light, among the wild shapes of its walls and the echoes that bounced off them.

The cave allowed human sensation to experiment and play with images. The distinction between inner and outer images became folded up in the dark, just like the rock structure of the cave itself. These three events formed a functional kinesthetic constellation that became durable in the work of art.[19]

Cave painting is not a representation of reality. There is no model and no copy. Furthermore, cave painting is more than just a visual art. The materiality of cave painting should be understood to include all the senses without the visual primacy we retroactively attribute to it today. The cave, the paint, and the human all paint each other. The cave walls create and change the paints on its surface while also changing the human sensorium. The paint changes the cave walls and human cognition *at the same time* (just as in tool use). Humans in turn mix paints and modify the cave. All three move from mere constellations to durable aesthetic orders in the performance of painting.

Painting is, therefore, the human attempt to do what the cave does by other means. Just as the cave produces reflexive sounds, images, and textures, so the human body can make sensible its own sonic and graphic images. Humans are not copying the cave but actually becoming part of the aesthetic field of the cave. The cave senses itself, as does the human, but in the cave the cave senses itself as human and the human senses itself as cave: human-cave. There is not only "a mind in the cave," as paleontologist David Lewis-Williams puts it, but we could add "a cave in the mind."[20] One is simply the aesthetic fold of the other.

The act of painting does not represent these images; it simply adds and modifies the images that are *already there* in the synesthetic field. When early humans entered the caves, they literally saw, felt, smelled, and heard images. As early humans painted what they saw, they began to see what they were painting. Functional art is apophenic, or defined by the sensation of patterns and order in seemingly unordered images. What was in the cave and what was added became indistinguishable, like a surrealist game of exquisite corpse (*cadavre exquis*) where multiple collaborators add something new to an artistic work, building on the last, and producing a kind of heterogenous composition with distributed agency.

There is no representation in cave painting, only a back-and-forth feedback loop between paint, cave, and human sensation that eventually produces a final functional image. From the continuum of the cave wall there emerged the affective traces of the bison, but by tracing them back onto the wall, a new texture and affect was produced by color, scraping, or modifying the surface that in turn could be traced, and so on in a kinesthetic circulation. The idea of representation robs painting of its kinetic materiality and treats the trace as if it were a preformed idea in the mind. This pure historical bias is a residue of classicism and idealism in contrast to the true kinetic, material, and functional feedback patterns of the prehistoric.

Men, women, and children returned again and again to these paintings, adding to them and changing them in a continuous feedback loop.[21] "Paintings in Chauvet Cave show animals coming out of and entering holes and cracks in the cave wall. Some animals are drawn with a reduplication of lines, as if something is flickering, wavering, or out of normal focus." Other art objects were placed in cracks and corners to take advantage of sonic topologies.[22] Many of the bison painted in the Spanish Altamira cave (ca. 12,000–11,000 BCE) are painted on bulging convex surfaces. "In fact, prehistorians have observed that bison and cattle appear almost exclusively on convex surfaces, whereas nearly all horses and hands are painted on concave surfaces."[23]

Paint, human sensation, and cave chase one another in a feedback loop of mutual and functional transformation. Each reproduces the other through the others. Unidirectional utilitarian accounts of these paintings as "tools" to destroy, protect, or capture hunted animals or to educate or train others how to hunt them is undermined by the fact that most of these images depict animals that *were not hunted*.[24] All attempts to explain the strictly anthropocentric use-value of these paintings already presuppose what they try to explain: representation.

Representation, the argument goes, becomes possible because it is useful, but for the image to be useful, it already requires a mental or graphic representation of the image used. However, a simple nonmentalist and nonrepresentational description is that humans painted what they did because that is what was already there in the cave. The image that was sensed and the image that was painted are co-constituted in the performance or kinetic act of painting itself. There is no model or copy, only the formation of matter and the materialization of form in a continuous and open-ended centripetal spiral of mutual functional transformation.

THE VESSEL

The fourth major centripetal aesthetic field of the prehistoric period is the vessel. The cave is the first sculpture and sculptor. Sculpture is defined by two reciprocal kinesthetic operations: a centripetal accumulation of fluid minerals into durable affective folds (shell, bone, horn, or stone) and a refinement of this accumulation into a further selected accumulation. Sculpture is the process of adding and removing images from an accumulated circulation of minerals. Affective removal in this sense is not the same as material removal. By removing matter from a woolly mammoth tusk, for example, one actually adds new affects to the bone that it did not have before: depth, sharpness, curvatures, fingers, breasts, buttocks, and so on.

It is no surprise, then, that we find the oldest sculptures in caves. The oldest shell beads, for example, were found right alongside the paint kits of the Blombos Cave around 70,000 years ago.[25] The shell itself is already defined by the curved and centripetal movement of mineral fluids gathered from the periphery and accumulated in the form of a protein-mineral shell. The motion of the cave is defined by a similar centripetal fold and mineralization, but with a hollow in the center to accumulate more. The hole punctured by a sharp tool in the center of many of these collected shells brings the shell into kinetic resonance with the aesthetic interiority of the cave. The two become peripheries with hollow centers for affective accumulations: a double centripetal folding. The necklace thread moves through the shell beads like humans move through the hollow tunnels of the cave, both producing a new distribution of images as they move.

This broad kinetic definition of the vessel encompasses two kinds, the full vessel or figurine, and the empty vessel or bowl. Let's look first at the full vessel of the figurine.

Figurines

Unsurprisingly, the oldest figurative sculptures have been found in caves, the vast majority of which are Venus figurines. "Venus" is the name retroactively given to the thousands of female figurines found during the Upper Paleolithic and Neolithic periods. The Venus has not only been the single most common figurative graphism for tens of thousands of years, but it is also the very first human figure ever sculpted.[26] The oldest is the Venus of Hohle Fels, which dates to around 35,000 BCE.[27] "The frequency and longevity of this symbol in the archeological record (more than 30,000 years) speaks for its essential role" in the prehistorical aesthetic field.[28] The Venus, like the hearth and cave, can thus tell us something about the kinetic nature of the historico-aesthetic field. The question here, however, is not what the Venus represents (fertility, femininity, and so on) or what people *thought* about it, but what its kinetic structure says about the distribution of images at the time. The Venus is not just an abstract symbol; it also has a kinetic structure, as in the case of the *Venus of Lespugue*, that expresses a specific field of motion resonant with the centripetal and functional kinesthetics of the prehistoric period (Figure 5.1).

The Venus is first and foremost a graphic description of the emergence of the image itself as a centripetal internalization (gestation) and creation

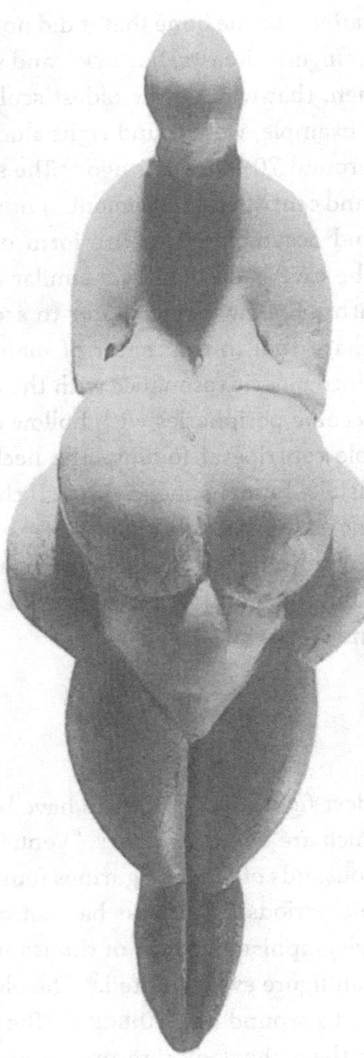

Figure 5.1 *Venus of Lespugue*, ca. 23,000 BCE
Source: Photo prepared by the American Museum of Natural History and Alexander Marshack, California Academy of Sciences.

(birth). The Venus is a kinesthetic image of how movement is internalized, organized, and generated in a functional feedback loop of life and death. One must live and die so another can live and die. More specifically, the kinesthetics of the Venus describe a *centripetal motion* that orders the appearance of images *spatially*.

The material kinetics of the Venus is almost universally composed of a series of rounded or curved ovoids: the head, the breasts, the belly, the thighs, vulva, and buttocks. These ovoids are curves, folds, or pleats in the flows of matter that bend back around, reflect, and intersect with themselves in a continual circulation. These kinetic relations are connected to the series of spaces or chambers where flows (milk, amniotic fluid, blood) are stored in the female body. The rounded figurine of the Venus is similarly a product of an accumulation of terrestrial flows (flows of dirt, flows of clay, flows of stone, flows of ivory) and their centripetal contraction into a series of pools, bulbs, or sacs where they are stored in the figurine's body parts. This contraction of flows is demonstrated by the frequent marking of Venus figurines with a series of parallel straight or wavy lines—"streams."[29] Sometimes the only markings on Venus figurines are a line or flow of straight lines flowing into and out of the vulva. These streams or lines have a double kinetic function: they are both a centripetal accumulation of flows folded into rounded ovoids or bulbed spaces, and a disjunction of these same flows through the vulva where they are released. The flow of lines across the body of the Venus is an open one that draws a continuous circulation between inside and outside. As a container of flows, the Venus is the centripetal space or series of places (bulbs) where the flow of matter folds, is internalized, persists, and circulates across an aesthetic field.

Bowls

The creation of bowls and containers goes back to Blombos Cave, around 100,000 BCE. Like the figurine, the container makes centripetal accumulation mobile for the first time. With the container, matter can be held beyond the limits of the bipedal hand. While the hearth had to be created again and again each night, the container could mobilize not only itself but also what it contained, and is thus distinct from the mobility of the hearth, cave, and Venus.

The first hollow container was that of the body; the most dramatic corporeal expression of this is the carrying of a baby in the womb of the mother. The kinetic shape of the curved belly, buttocks, and breasts of the mother thus becomes the kinetic basis of the first bowl, "like the original bowl described in Greek myth, which was modeled on Aphrodite's breast."[30] The earliest pottery technique is thus the centripetal one of pinch and coil, where clay is rolled and gathered in a circular shape with the hands.

Paleolithic humans had little need to store flows; containers would only have inhibited their movement as they hunted. The Neolithic age is thus the true age of containers. By staying in the same place, by creating a place, Neolithic agriculturists also invented numerous kinds of containers, which in turn made it possible for them to stay in the same place: stone and pottery utensils, vases, jars, vats, cisterns, bins, barns, granaries, and houses, as well as the great collective containers like irrigation ditches and villages. Agriculture brought a surplus of food, and containers made it possible to centripetally store this surplus over the winter and to remain in the same place during that time. The rounded image of the container also duplicates the rounded journey and migratory return to the same place or village. The journey and return to the same place equally echo the rounded centripetal image of the village itself. The bowl is thus curved because life is curved, and life is curved because the bowl is curved. The two are co-functional conditions of each other, a kinetic feedback loop of the mutual centripetal sculpting of space.

The Venus is the kinesthetic image of the multi-vesseled vessel, the body vessel whose swollen parts form the regional vessels where fluid is stored and life is made and remade. The Venus is the kinetic vessel that contains, creates, and multiplies other vessels, the arch-vessel or ur-vessel. The functional or entrained aesthetic between life and death found in the Venus is also in that of the bowl. Once humans settle, for example, there is a need for bowls, containers, and pottery, but before humans can settle they already need bowls, containers, and pottery to contain their settlement and food. There is therefore again a functional aesthetic circulation between images in the same field.

Spirals

The spiral is almost universal throughout Neolithic Europe, and is encountered on seals, plaques, altars, dishes, elaborately decorated bases, anthromorphic vases, and figurines.[31]

It is no coincidence that the decoration, engraving, and ornamentation of bowls and other pottery items focused heavily on images of eggs, shells, birds, and spirals, which are all centripetal kinetic patterns.[32] Spiral graphism appears as early as 13,000 BCE in paintings and 6,000 BCE on the earliest pottery.[33] Even before this time, however, humans observed and collected unique spiral objects such as gastropod shell fossils, a testament to the "mystery of strange forms," as Leroi-Gourhan writes.[34] Spiral graphism is also related to snakes, whose association with water, motion, and eggs connects them to the Venus as well. If the egg is the graphic form of the image as centripetal space, the spiral snake is the graphic shape of the kinesthetic movement that gathers matter into the curved image of bowl.

In addition, "spirals are often located on uniquely feminine body parts in figurines such as the breasts and uterus. The hook is a shortened version of the spiral and can be found on megalithic tombs of western Europe," as well as alongside moon and lunar cycle images.[35] Precursors to the bowl and spiral are the Venus and bison horn found in one of the oldest known relief sculptures (ca. 25,000–20,000 BCE) in Laussel, France (figure 5.2). In this sculpture, the kinesthetic connections between the curved bison horn, the lunar cycle, menstruation, pregnancy, the container, and women are united in a single stunning image.[36]

In all its prehistorical expressions, the spiral is an image of the continual and kinesthetic cycle of folding and reflection that creates the material surface on and through which the image is distributed. The movement of the prehistoric image is thus a spiral feedback loop. Just as in cave painting,

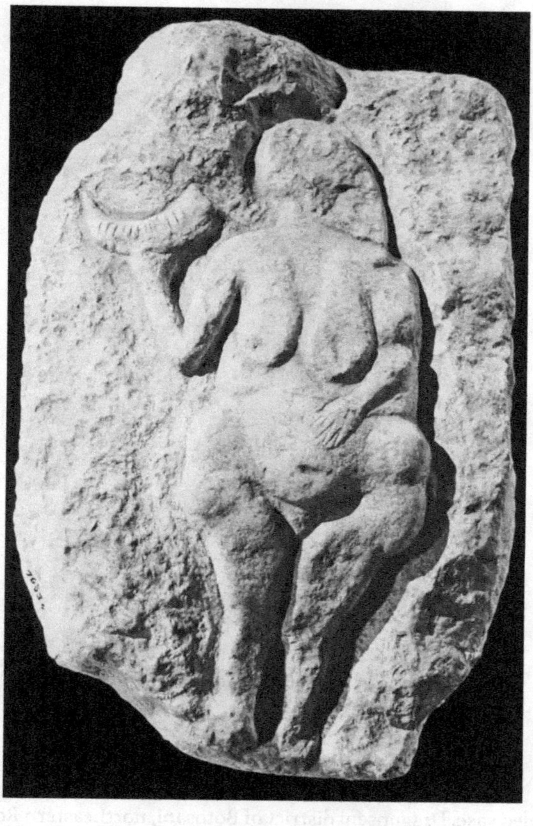

Figure 5.2 Sculpture of woman holding a bison horn, from Laussel, France, ca. 25,000–20,000 BCE. Painted limestone, 1' 6" high. Musée d'Aquitaine, Bordeaux.
Source: From Helen Gardner and Fred S. Kleiner, *Gardner's Art Through the Ages: The Western Perspective*, 13th ed. (Boston: Wadsworth Cengage Learning, 2010), p. 4.

the spiral goes out and returns to itself, but each time slightly transformed. The spiral is an image of the reflective process by which motion folds over itself into the images of an aesthetic field. The spiral is an image of the iterative processes of a dynamic motion: whirling and twisting spirals, winding and coiling snakes, circles, crescents, horns, sprouting seeds, and shoots.[37]

The spiral is also the union of opposites. Inside and outside, light and dark, male and female are all co-present by way of their continuous differentiation and unity within the folded arm of the spiral. As one travels the curve of the spiral, the inside slowly becomes the outside and the outside becomes the inside. The two are connected in the single continuous stream of motion. This is attested to in numerous spiral designs on Neolithic pottery and vessels. For example, two antithetic spirals whirl around a beautiful lidded vase from the Neolithic, simulating the generative power of the central egg (figure 5.3).[38]

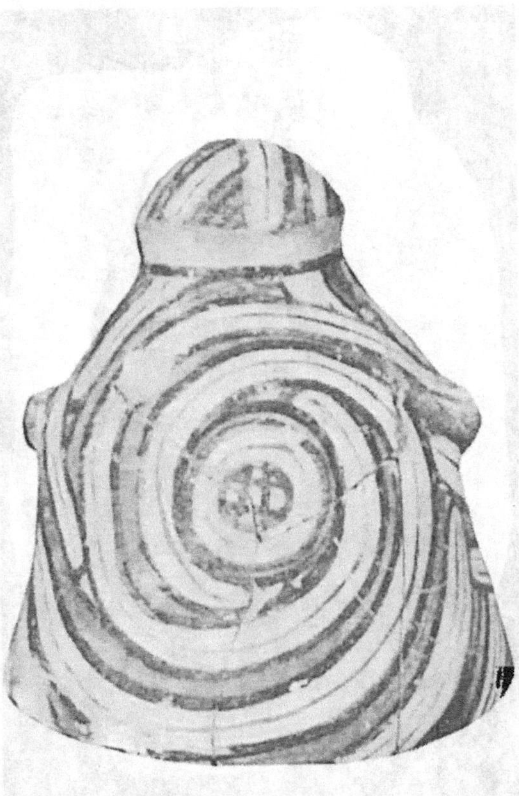

Figure 5.3 Spiraled vase, Dragunseni district of Botosani, northeastern Romania; Cucuteni A period, ca. 42–41 centuries BCE
Source: From Marija Gimbutas, *The Language of the Goddess* (London: Thames & Hudson, 2001), plate 22. Courtesy Vladimir Dumitrescu. Publ. in V. Dumitrescu, Arta culturii Cucuteni, Bucharest : Editura Meridiane, 1979.

THE WIND INSTRUMENT

The fifth major centripetal aesthetic field of the prehistoric period is the percussion and wind instrument. Music is ordered sound images. As such, it is defined by two minimal kinetic operations: a movement back and a movement forth (a fold, cycle, or vibration). This is a broad definition, but here we focus only on its use in early human music.

In addition to the musical biomaterial entrainment of the body, the resonating capacities of its body cavity, and the basic folding cycle required by tool-making (between 150,000 and 25,000 BCE), there is an increasing centripetal accumulation of sound images in various instruments over the course of the Paleolithic. The invention of sound tools and amplification made possible the largest and most diverse centripetal accumulation of sonic images in history at that point. Nature makes music, but with sound tools, humans invented new sonic images and orders never before heard in nature.

It is not surprising that the majority of these first sound tools were discovered inside caves and are themselves modeled on the centripetal kinesthetics of the cave that vibrated and echoed with sound and voice. The first human instruments are like little caves. They are kinetically defined by the centripetal gathering of sound into a resonating enclosure. The oldest discovered instruments are thus idiophonic and aerophonic (ca. 35,000–25,000 BCE). Idiophones are instruments that produce sound by vibrating their whole body and include scrapers, various rattles (some used as adornments around the neck, arms, or legs, or attached to clothing), clappers, percussion sticks, plaques, and tubes. Just as the hollow cave amplifies sound when one strikes its walls, so the hollow mammoth bone rings out when one strikes it, releasing pressure captured inside the tube. By capturing sound in a centripetal resonating chamber, a whole new range of sound images could be distributed.

Even the membranophones, invented during the Neolithic period, are simply versions of the hollowed-out idiophones of the Paleolithic, but are increasingly more vessel-like. The tube drums, vessel drums made of clay, vessel rattles and flutes, animal horns, and frame drums are all defined by their bowl shape and their capacity to hold more and more air in the chamber. Kinesthetically, the centripetal accumulation of air in the resonance chamber incrementally increases over the prehistoric period.

Aerophones share the same centripetal structure as idiophones and membranophones, but with an important addition that makes possible the introduction of discrete pitch. Instead of simply capturing and vibrating the air inside the resonance chamber or body of the instrument like a percussion instrument, the flute, pipes, and phalange whistles used by Paleolithic peoples distribute sound images by circulating captured and free air through the hollow body of the instrument itself. The kinetic cycle of this flow is defined by a constant capture and release of air. Just as the cave itself whistles and howls as air moves through it, so the Divje Babe flute, carved from a bear femur (ca. 40,000 BCE), whistles as air moves through its hollow body. Other types of free aerophones such as the bullroarer do not capture the vibrating air in their body but do vibrate the air around their body as they spin. The musician spins an airfoil in a circle, gathering and vibrating the air surrounding the instrument. The sound this produces is a continuously modulated, rhythmical, or cyclical series of sonic images. It is what some Australian aborigines still call "the voice of God."[39]

Idiophones, membranophones, and aerophones thus introduced a new kinesthetic field of sonic images defined by three features: an affective entrance limit or opening by which air enters, a continuous hollow cavity through which the air moves, and an exit limit by which the air leaves. Even solid clappers and airfoils without obvious hollow cavities are still defined by the entrance and exit of air across their continuous surface. In these instruments, sound is distributed in a continuously graded and centripetally entrained circulation of air through or across the capturing body of the instrument.

With the introduction of the holed flute, discrete pitch was introduced for the first time. Once a continuous centripetal circulation of air is achieved, it becomes possible to further modulate this flow between the limits of the field by introducing a hole by which the air can either escape or not. The introduction of one or more holes produces an affective or sonic bifurcation in the continuous sound waves. Opening a hole along the bore causes low-pressure waves to move back away from the opened hole, folding back over themselves, while high-pressure waves move further along the bore. A series of holes like those of the Divje Babe flute are able to transform a series of waveform intersections or nodes through the tube into a series of waveform folds throughout the sonic flow.

From the continuous waveforms of human speech and natural sounds, the holed flute makes possible a series of regional folds in this waveform. Discrete pitch is thus not kinetically "discrete" at all but, rather, pitch folded from the continuous pressure waveforms inside the tube. However, the emergence of folded pitch also entails a whole new kinetics of sonic composition, arrangement, distribution, and hierarchy of these folds across an instrumental field.[40] Pitch is, therefore, not modeled on human language, as many have argued, but instead emerges along the same continuum of natural and human sounds.[41] Musical pitch does not emerge from narrative or "lexical tones" or "musilanguistic" references; it is first and foremost its own affective fold in the matter-flow of air through a hollow tube. Musical pitch is not modeled on a linguistic field but, rather, has its own nonsignifying sonic field of images that creates just as much as it is created by the functional material operations required by language itself: discrete combinatorics and hierarchic cognition made possible by centripetal accumulations.[42]

There is thus a kind of functional entrainment of prehistoric music that both uses humans to expand the complexity and compositionality of folded sound waves and is also used by humans to further develop combinatorial and hierarchical ordering practices used in language, cognition, and elsewhere. Breath is not only centripetally gathered, folded, and reordered by the flute in the form of music; this very activity already produces and is produced by the same centripetal gathering and controlling of discrete breath and sound patterns in human language itself. The two emerge alongside each other as different dimensions of the same functional kinetic process of entrainment in the resonance chamber of the cave.

THE HOUSE

The sixth major centripetal aesthetic field of the prehistoric period is the house. The house combines the centripetal kinetics of the hearth with those of the cave. While the hearth is mobile and provides a horizontal accumulation of material images, the cave is relatively immobile and provides a vertical enclosure of images. The house achieves both: a relatively mobile enclosure that can be built like the hearth but can enclose like the cave. The aesthetic entanglement of prehistoric centripetal motions is thus completed. The house becomes the ultimate centripetal vessel of image accumulation, the source and amplification chamber of all the arts. The house accumulates all the previous accumulations. The hearth, cave, figurine, bowl, drum, and flute are all brought together in the kinetic structure of the house. The house lays out an aesthetic field limited by its walls and populated by various different images.

The earliest huts emerged alongside the hearth itself millions of years ago with *Homo habilis*, and gradually became taller, wider, deeper, and punctuated with more holes (doors, chimneys, windows, storage pits, and so on). In fact, the hut is already built like a giant hearth, surrounded by accumulated rocks on all sides to support a vertical pile of sticks with a hollow center of circulating air. The same kinesthetic techniques of the hearth are applied to the hut: the house gathers a flow of matter, folds it over itself into a verticality, and hollows out the center to produce an aesthetic field within which a sensitive interiority of images can emerge.

It is no surprise that the increasing permanence and immobility of the hut or house occurs simultaneously with the increasing settlement and accumulation of human beings and their increasing centripetal technologies (figurines, bowls, drums, and so on) into the house. There is thus a functional feedback loop between the centripetal movements that define the house itself—the accumulation of rocks, sticks, hearths, and circulated air—and the centripetal circulation of humans who continually go out and return to the hut itself. As archaeologist Dusan Boric describes it, the house is a literal "capturing of the landscape."[43]

In this new aesthetic field of the house, all sensory images—smells, sounds, colors, tastes, temperatures—are amplified, reflected, and resonated, just as they were in the cave. With the house, however, humans build the walls and thus both lay out the aesthetic field and order the images that circulate through it. A modification in the house, like the addition of windows, multiple hearths, and animal skins, makes possible a new distribution of the images inside the house. In the house, humans began to modulate the distribution of smoke, light, warmth, and sound images. The "dramatic tension between the aerial and the terrestrial"[44] is remade in the house itself. The house remakes the earth in its own image. The dirt floor becomes the terrestrial and the smoky ceiling the aerial. Just as the human remakes the world in the house, the house remakes the world in the human. The human becomes an increasingly interiorized animal distributed by floor and roof, ground and sky, body and mind.[45] Earth, house, and human thus enter into a mutual entrainment of functional aesthetic resonance.

The house has two main kinesthetic features.[46] First and foremost, it is a strike or cut into the earth (digging, puncturing, carving out): the pit. Second, the house adds something to the cut or hole to create a verticality rising above the earth: the pile. The house cuts or tears into the flows of the earth in order to redirect them vertically. If, according to Bernard Cache, the most basic expression of architecture is "the frame," in the sense that all houses are composed of the basic elements of bottom, sides, and top, then the house produces the first vertical accumulation by bifurcation. According to Cache, "The architectural frame fulfill[s] at least three functions, whatever the concrete purpose of the building might be. . . . The first function is that of separation. Its functional element is the wall. . . . But architectural space is not this general form of simultaneity; it is a space where coexistence is not a fundamental given, but rather the uncertain outcome of processes of separation and partitioning. The wall is the basis of our coexistence. Architecture builds its space of compatibility on a mode of discontinuity."[47] Thus according to Cache, architecture should not be primarily conceived of in terms of space (simultaneity) or time (succession), both of which are the outcome of the more primary centripetal processes of accumulation and bifurcation.[48] The other two functions of the architectural frame—selection (the "window") and distribution ("floor")—are built off of the primacy of a horizontal and vertical kinetic accumulation. On the kinesthetic surface of the frame, the chimney and window fold the inside into the outside by opening one up to the other.

The Grave

The earliest grave, dug around 100,000 years ago, was a house or vessel for the dead. The dead, as Lewis Mumford observes, "were the first to have a permanent dwelling: a cavern, a mound marked by a cairn, a collective barrow. These were landmarks to which the living probably returned at intervals, to commune with or placate the ancestral spirits."[49] Entombment cuts into the earth in order to create a mound or megalith for the dead. Some of the first graves emerged as an extension of the hearth pit itself, and continued to be used as such up until the famous burials of Çatalhöyük, where the dead were literally buried in the floors and walls of the house. The resulting grave mound or megalith rises above the level of the earth just like the house, and it marks a redirection of the earth's flows into a material distinction between life and death. The grave becomes a field on which images are gathered, ordered, and circulated. The grave is the underground home and is thus similarly decorated with its own images: jewelry, red ochre, food, tools, and fresh flowers.[50]

CONCLUSION

All this attests to the predominantly centripetal and functional nature of prehistoric aesthetic fields. The prehistoric field gathers the flows of matter and images toward a central or hollowed-out region and begins to distribute them across the body, the hearth, the cave, the house, and so on. However, it is also possible for one of these images to take on a kinetic dominance and thus increasingly subordinate centripetal functionalism to a hierarchical aesthetic formalism. This is what began to happen in the ancient world, and is the subject of our next chapter.

SECTION B

The Formal Image

The second major kinesthetic field to rise to dominance in the West is the formal field. This second type of kinesthetic field increasingly rises to dominance over the course of the ancient period from around 5,000 BCE to 500 CE, alongside the rise of monumental architecture, mythic sculpture, epic poetry, dramatic theater, and monophonic music. All art has form or shape, of course, but during the ancient period, the form of the work of art became the dominant ordering principle, model, and idea guiding all the arts.

In contrast to the prehistoric world of predominantly experimental and functional aesthetic orders in the continual mutation and redistribution of matter, formal aesthetics aimed to secure and regiment the patterns that emerged from these functions. For example, in contrast to the apophanian images of prehistoric cave painting that connected heterogenous images—from the Greek words ἅπτω (*háptō*, "to join") and φαίνω (*phaínō*, "to show," "to appear")—the ancient field of affects and images was defined by a new formal order that increasingly privileged a single kinetic pattern over all the others. Instead of allowing the field to continually mutate functionally, formalism introduces and maintains a relatively hierarchical order of images. In its most extreme articulation, only the "mental" affect of *pure thought* itself or "abstract form" alone is able to understand the insensible formal order of the sensible image. In this way, abstract form is said be a *contemplatation* of the image with a disinterested attitude independent of the sensuous work of art itself. Once formalism emerges in this classical aesthetic field, it returns again and again, in the medieval, modern, and even contemporary periods.

Chapter 6 offers a preliminary kinetic theory of form, considered conceptually. In chapters 7 and 8, however, it is considered more historically. In short, aesthetic formalism occurs when either a particular concrete pattern or an abstract mental pattern becomes a model and takes on an organizational dominance over all the other affects in a functional aesthetic field. The theory of kinesthetic form is defined by three interrelated features: centrifugal motion, formation, and modeling.

CHAPTER 6
Centrifugal Form

All aesthetic experience owes its origins to materiality, and no form of it, not even the immaterial kind, can erase its material origins.

—James Porter[1]

This chapter argues that aesthetic form emerges from matter in motion. Form is not different in kind than matter and function but only different in kinetic distribution or regime. The form of something is always a circulation of motion, but in a specific pattern. Form is not a geometric essence like a circle, or an empirical object like a round plate, but, rather, a material kinetic process of rounding or roundness. [2] The form or shape of images and our ideas of images are both made possible by the continuous circulation of matter into a specific form or pattern, again and again. All things are kinomorphic, but when a fixed pattern is used as a model that forces all other flows into a single set of kinetic relations, there is a kind of formalism or dominance of form over kinetic formation. This is what occurred in the ancient world.

CENTRIFUGAL MOTION

The formal aesthetic field is defined by the continuous organization of matter at the periphery of the field by the center. This is possible because form acts first and foremost on an existing centripetal field of gathered images and entrained functions. In other words, an aesthetic field is assembled first as a series of events in a constellation, and then is preserved as folds in a field of circulation. Only then does it become possible to centrifugally reorganize or redistribute the affects in that field into more specific functional orders.

As more and more flows of matter are gathered into a centripetal region of accumulation, the distribution itself can become increasingly uneven and asymmetrical. For example, when one starts to collect heterogeneous matters in a pile, some of those matters end up at the bottom of the pile, some at the top of the pile, others at the periphery, and still others closer to the center. All the flows are equally heterogeneous, but they become differentiated and ordered topologically by the centripetal motion of collection. Just as multiple hearths become gathered into larger houses, so larger houses are accumulated into cities that can give birth to a whole new centralized architectural form ordered around a central temple-palace-citadel occupied by the warrior-priest-king.

Centrifugal motion and the form of the center therefore emerge as products of centripetal collection, and not spontaneously from some foreign, outside, transcendent power or immaterial creator. Thus form is not opposed to function, matter, or content; rather it is a type of kinetic distribution that emerges from matter and relies on its accumulation. Once a central accumulation occurs, it becomes possible for the form of the center to take on a new primacy and agency with respect to the periphery. For a center to redistribute matter from on high along the periphery below, there must first be a centripetal accumulation substantial enough to support it. In short, every great center requires an even greater periphery.

CONCRETE FORM

There are two kinds of forms: concrete and abstract. Concrete form is perceptual and sensuous. It is the shape, outline, or figure of the exterior surface of an image. Concrete form is nothing other than the unique collection of lines, colors, sounds, smells, and tastes that compose the material and sensuous exterior shape or pattern of the image. Since concrete images can contain other images, there are both inner and outer exterior forms that can order and shape each other. The historical emergence of aesthetic formalism is made possible by four kinetic moves, three of which are concrete and one of which is abstract. We will consider the concrete forms first, and then the abstract one in the next section.

Furthermore, we outline these four moves as conceptual in this chapter and as historical in chapter 7. These four moves are the material kinetic conditions for the emergence of a formal center that can react back on and subordinate the periphery.

Division

The first kinetic move that makes possible the emergence of aesthetic formalism is the division of the work of art into inner and outer dimensions. In contrast to the continuous circulation of centripetal material flows spiraling toward a central region without a single point of accumulation and without an absolute exterior limit, form introduces a kinetic division between an absolute exterior and interior, between a periphery of as-yet-uncollected flows and a central region of already collected flows. In the functional aesthetic field, there is no radical interior or exterior of sensation but rather a continuous spiral of life, death, and rebirth; of accumulation, destruction, and re-accumulation.

However, once a sustained central accumulation of images is achieved, the difference between the center and the periphery becomes increasingly asymmetrical. The collected flows of images inside the body, hearth, cave, house, bowl, and drum become affectively distinguished from those that are not collected. The bowl holds the milk and grain accumulated through centripetal labor, but the cows and grass, which are relatively peripheral and inferior products, remain uncollected. Interior or contained images are not only differentiated from noncontained or uncontained ones but also increasingly valorized, preserved, and treated as the ends or aims of the uncontained flows. In this way, the flows that are contained appear to have the center as their natural trajectory and presupposed purpose. In other words, the hollow interior of accumulation begins to function as an active site of attraction or apparent causality, and no longer as a passive site or mere repository.

What we call the exterior form of an image is the shape or appearance that comes into being when the process of accumulation reaches the limit of its container. For example, when houses are collected into the largest possible organization, they look like a city. As the city grows, it extends the shape of its exterior walls to interiorize more houses. Although the surface shape of the periphery emerges only on the condition of centripetal accumulation, once the accumulation is maximized at the center, the form of the product appears retroactively to have been what was becoming accumulated all along.[3] The ancient city becomes a model of the world itself, an ultimate and natural telos. In short, centripetal accumulation makes possible an asymmetry between accumulated and unaccumulated matters, interior and exterior images, and thus a division between inner and outer form.

Hierarchy

The second kinesthetic move occurs when the difference between interior and exterior forms is further differentiated by their degrees of centripetal attraction. Every stable region of centripetal accumulation has images and affects that are more or less interior with respect to the most centralized affects. In prehistoric societies, the fields of images in this centralized accumulation were regularly destroyed through social practice (burial, potlatch), natural disaster (flood), the need to migrate (owing to agricultural exhaustion of the soil), and other reasons. However, in ancient societies, increased sedentism, militarism, metallurgy, writing, and so on began to allow for unchecked accumulations and this increased hierarchical asymmetries between accumulated images.

Before Plato's great chain of being there was already an ancient kinetic chain of motion defined by the hierarchical order of containment, capture, and accumulation. The countryside is subordinated to the interiority of the city, which is in turn subordinated to the interiority of the warrior-priest-king, who is in turn subordinated to the gods. Each further interiority requires an even greater and wider accumulation and subordination of the periphery. The periphery therefore exists to greater or lesser degrees *for the sake of that which it contains*. However, the chain of containment is also a chain of relative immobility that refers all the way back to the ultimate interiority at the center of all being: the divine. If the surface appearance of the image is the shape of the container, then the interior is the apparent cause, agent, or image that produces or dictates this appearance. Form thus becomes ranked or ordered by the greatness and value of the images that have been accumulated (inner affects) over those that have not (outer affects).

Reorganization

The third kinetic move follows from the second. Once an absolute limit of interiority or accumulation of images is reached (a center), all the vast periphery is capable of being reordered around this central point. If there is an absolute form of accumulation that contains nothing but itself, then everything else is for the sake of it, while it is for the sake of nothing else.

Not only does a hierarchy emerge between the shape or appearance of what is accumulated inside the field of sensation and the external shape of the limit of this accumulation on the outside, but the inside begins to take power over the shape of the outside. In other words, one specific affect or group of affects takes on a central priority that reacts back on the periphery and redistributes it on a new basis. The prehistorical functional feedback loop of centripetal accumulation remains in place, but it is now subordinated and reorganized on the basis of a single set of central images that both require and reorder those on the periphery, like the ripples made by a stone dropped into a pond.

Instead of treating the interior and exterior as a single functional cycle of heterogeneous flows of matter, the ancient aesthetic field distinguishes an inner form or pattern from an exterior one, and it tries to make the latter match the former. The exterior appearance of the city should match the splendor of its inner life: its warrior-priest-king, god, or the power of its people. The exterior surface of appearance should match the inner essence and power that has been accumulated. The best accumulations should be placed in the most beautiful bowl, for example. The importance, value, and greatness of that which is accumulated should be expressed in the external appearance of that which holds it.

ABSTRACT FORM

The second kind of form is abstract form. In the first three kinesthetic moves, form still appears as the *exterior shape* of the image, even if it is something inside other images. Form still remains something relatively external. For example, grain is accumulated in a pot (division), stored grain is valued more than unstored grain (hierarchy), and the exterior appearance of the pot or work of art should reflect the value of its contents (reorganization). The vessel is still filled with something, even if the value or beauty of its form is still superior to the mere appearance of the vessel's exterior.

However, in the fourth kinesthetic move, the concrete inner form of the vessel is replaced with an *abstract form*. As containers increasingly outlast the material images they contain, the container itself—specifically, the interior of the container—takes on a new importance. Milk is consumed, but the bowl lives on. The seasons change, but the house and city stay in the same place. Wild animals migrate, but domestic ones remain in their pens. In other words, the hollowed-out center of the vessel itself takes on an aesthetic primacy beyond the actual concrete things stored in it.

In this incredible kinetic reversal, inner form no longer appears concretely but, rather, as the trace of something that was once inside. The external form of the aesthetic field appears as the shape left by something now missing—an absent inner form whose appearance can only be determined negatively by the impression it left on the inner and exterior forms. Although missing, this abstract inner form becomes the centrifugal and retroactive "cause" of the periphery appearance. Absence becomes more aesthetically primary than presence.

However, abstract inner form is not actually either ideal or immaterial. Absence to the other senses does not mean absent to the mind. As we have seen in the first three kinesthetic moves, the sensation of form emerges entirely from matter in motion. The experience of the "missing form" is an experience made possible by the material condition of the hollowed-out centripetal object. As such, abstract form has an affective status, albeit a strange one. Abstract inner form is something known or thought independently of its concrete expression or content in a work of art. Abstract form is not a particular curve, color, or sound but, rather, an affect of the mind: an idea of ratio, harmonic order, proportion, meter, or interval that can be comprehended independently of any particular exterior surface of appearance.

However, form is not strictly insensible, since it is an affectation or image of the mind extracted or derived from material images. The whole sensuous body, including thought, is part of the functional kinesthetic field and is thus an affected thought. However, when the body is removed from a field and is no longer sensing an order of images, it can still be affected by that field as thought. No doubt, early humans had a similar sensation regarding their memories of the dead. Thought is just an affective residue of sensation in the mind. The mental image is an image that lives on as the sensory residue of that which is missing.

Although mental images were prehistorically subordinated as mere parts to the more robust whole field of functional images, the ancient period performs the most stunning kinetic inversion of this residue in all of history. Instead of treating this affective echo or memory of images in the mind as a product of the more primary centripetal and functional process of sensory accumulation, thought ends up becoming the more interior and fundamental image around which all the others appear to be ordered. During the ancient period, the whole of sensation was not just reorganized around the aim of stimulating and producing mental images, but also mental images became the creative origin of art, sensation, appearance, and exteriority itself. Kinetically speaking, the most interior, residual, and "absent" image produced in the aesthetic field becomes the inverted origin and organizing power of all other images in the field. Thought or idea thus becomes available to the other senses through the work of art.

Creation

The fourth centrifugal and kinetic move, then, occurs when the periphery image is not only reorganized but also recreated by a single mental image of abstracted form: harmony, proportion, ratio. The sensory presence of the absent model shapes the exterior forms of appearance. In this fourth movement, only mental images are extracted from the field of images and used as perfect ratios and intervals or as a models from which to produce faithful copies.

The kinetic inversion is radical. Even the senses of the human body become reorganized around this new centralized and centrifugal inner sense of thought. Some senses, like touch, smell, and taste, occur on the surface of the human body, while others occur on the interior, like sight, sound, and thought. As we argued in the previous chapter, the human sensorium is made possible by the centripetal and bipedal collection of material flows into the convex sense organs of the body. The convex retina, the convex ear cone, the hollow mouth and nose are, like so many little caves and bowls, to gather up the flows of matter. Insofar as sight, sound, and thought are topologically more interior and less exposed to the exterior of body, so they also appear as the result or aim of the exterior surface of the body. The shape of the ear, for example, appears convex and bowl-like or shell-like for the sake of the internal sonic images. Once enough images (sounds, sights, and thoughts) have been stored in the brain, they are capable of then re-creating sensation in their own image.

This is the kinetic and affective foundation of idealism. Historically, matter is centripetally collected into a sensuous body. The body then gathers and internalizes matter into the body through sensitive surfaces, the most interior and residual sense of which then comes to think that its own mental image preceded the other sensations that support it. In its extreme and absurd version, human thought is projected onto a god of pure thought that retroactively creates human thought itself.[4] In short, the most residual product becomes the center of a whole new kinetic reorganization of images.

MODEL AND MOLD

Aesthetic formalism finds its first philosophical expression in ancient cosmologies in which an invisible god becomes the craftsman or modeler of a copied world. Prehistoric aesthetics first created the centripetal field and filled it with images, but ancient aesthetics merely presupposed the prior existence of this centripetal field and then reorganized it around an absent but creative central model.

In Greek mythology, for example, the bowl—sustainer of all life—was said to be modeled on Aphrodite's breast. Having encountered the prior existence of the prehistoric bowl, the Greeks postulated the missing model that must have formed the bowl in the first place. In Greek mythology, the hollow created by prehistoric humans became the mold of a divine breast, and not the other way around. The kinetic inversion here is explicit: the abstract is not modeled on the concrete; rather, the concrete is modeled on the abstract. Aphrodite's perfectly rounded breast is the inner form of the bowl whose exterior surface is only the mold or sensuous form of this eternal and unchanging perfection. Aphrodite was in turn born in a rounded scallop shell, which in turn emerged from the rounded bubbles or foam (*aphros*) of the ocean. The divine geometric form thus comes to be seen in the ancient period not as a composite or abstraction from images but, rather, the other way around. Sensuous images are copies of a more primary divine model.

However, in several archaic Greek mythologies we can still see the traces of the sensuous origins of the divine. For example, Zeus, the king and father of all the Olympian gods, emerged from the empty hollow of a cave. In fear that her baby would be eaten by her husband, Cronos, Rhea consulted with Gaia and together they conspired to deceive Cronos by giving him a swaddled rock instead of a baby. The baby was then hidden in a cave on the island of Crete, where Gia's earth sprites danced and played loud music to cover up the sound of baby Zeus's crying. Zeus was invaginated in a cave on Crete, as Lucretius writes (*vagitum in Creta*), where he was raised by a goat and protected by the dancing (*choreo*) of the Curetes.[5] The *chora* is not only the creative "space" of the cave from which new images emerge but also the fundamentally hidden space produced by enclosure. The *chora* covers, obscures, and folds the flows of life together in its enclosure. However, the *chora* is not simply passive fecundity; it is the creative and pedetic movement of matter itself, exemplified in the dancing (*chorea*) of the Curetes. They therefore conceal not by negativity but rather by an excess of noise and motion. From the centripetal hollow of the cave comes a divine central power.

We can also see traces of the sensuous origins of the divine in the collective fragments of the Orphic songs/hymns. Several fragments describe creation as coming from an egg wrapped in a spiraling snake coil;[6] others claim that it was Chaos who created the egg;[7] still others claim that it was Night (Nyx) and Ether who produced the silver egg from which the world was born.[8] Despite the differences between these fragments, Orphic cosmogony is largely described as one in which chaotic or formless motion is centripetally gathered together into a creative world-egg.[9] Before creation, everything is preformed by the form of the egg.

In Plato's *Timaeus*, however, these older myths are fully inverted in favor of the divine creation of the sensuous. For example, the central character of Plato's dialogue, Timaeus recounts a cosmology in which a divine craftsman (*demiurgos*), or "wax-modeler,"[10] creates the world as an imperfect copy based on a perfect divine model: "Now so long as the craftsman keeps looking to what's in a self-same condition, using some such thing as a model, and fashions its look and power, then of necessity everything brought to a finish in this way is beautiful; but if he should look to what has come to be, using a begotten model, the thing isn't beautiful."[11] In the ancient art of lost-wax casting, a wax model is first created around which a plaster mold is made. When the hot metal is poured into the mold, the wax is melted or "lost." The mold is removed, with only the metal copy remaining (figure 6.1). Plato's ontology is based on a kinesthetic formalism. His idea extends to all of being what is actually a specific historical and aesthetic practice of wax-cast modeling. If we see a sensuous exterior surface, it is precisely because there must have been an original model that has been

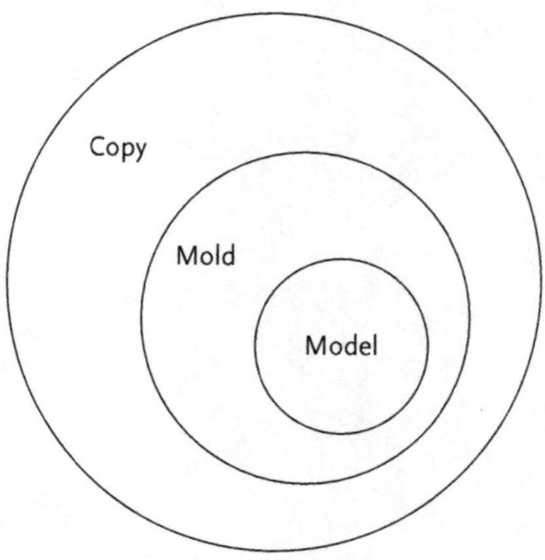

Figure 6.1 Lost-wax casting process: model, mold, copy

lost to the senses and can only be determined negatively by the affections of the mind.

The original model was perfect, but through the materiality of the casting process, the copy becomes imperfect. For Plato, all of being is hollowed out and filled with a more primary and creative center that radiates outward and determines its periphery in centrifugal fashion.

CONCLUSION

This chapter has given us only a conceptual introduction for understanding the four defining moves of kinesthetic formalism. The important takeaway is that form may be both concrete and abstract, but it is certainly not static, transcendent, unreal, or illusion. Formalism simply emerges as a different way of organizing and distributing images on a field of sensation by division, hierarchy, reorganization, and centrifugal creation. Functional entrainment, however, is still required as the material support for the field to be organized. In the next chapter, we move on to the more detailed historical and aesthetic expression of this process in ancient art.

CHAPTER 7
The Ancient Image, I

The ancient image is defined by the emergence and dominance of centrifugal form in both its concrete and its abstract dimensions. However, it should be said that the arts of the ancient West also developed along divergent historical and geographical trajectories. There is no single line of development for all art over such a broad period. Furthermore, even with respect to any single art, like writing, there is no clear linear evolution. New techniques arise, but the old ones do not necessarily disappear. Multiple writing techniques are used simultaneously by different groups of people. Geographically, for example, the Greeks did not begin using written language until much later than either Mesopotamia or Egypt.

The rise to dominance of ancient formalism took thousands of years of nonlinear development. Ancient art did not abolish prehistoric aesthetics and centripetally ordered images but rather took them up and used them in a new way. For example, ancient sculpture was still done by carving stone; paintings were still made on the walls of enclosures; houses were still built with walls and ceilings with hearths in the center; drumming, the body, the cave, and wind instruments all persisted alongside the newer arts of the ancient period. In some cases, there were more abrupt changes, like the invention of writing, but in most cases, there was more of an incremental development of aesthetic orders that can only be grouped together with the benefit of hindsight.

There is no single art of the ancient period that unilaterally determines all the others. As we have seen in the previous chapter, the very idea of modeling is one that emerges historically in and through a number of concurrent and divergent ancient practices. Although all ancient arts relied on modeling, no one art was modeled on another. Rather, just like the prehistoric arts, the ancient arts entered a kinesthetic resonance pattern or regime of motion. In the ancient period this was broadly, although not exclusively, a centrifugal regime of motion.

This chapter and the next argue that a kinesthetic shift occurred in the arts from a centripetal to a more centrifugal pattern of motion over the course of the ancient period. This claim is supported by looking closely at the kinetic patterns produced by six major aesthetic fields that define the ancient image: written verse, tragedy, metallurgy (covered in this chapter), and the city, the chordophone, and pharmakon (covered in Chapter 8). The argument of this chapter is that each of these major fields is defined predominately by a distinctly centrifugal pattern of motion and a formal aesthetics.

WRITTEN VERSE

The first major centrifugal aesthetic field of the ancient period is that of written verse. Written language has its own material and centrifugal structure, which we have dealt with elsewhere,[1] but this written language also makes possible a whole new field of written poetic verse. Poetic verse is an ordered distribution of images that first began as song in the Paleolithic. During the ancient period, however, prehistoric oral poetry, song, and performance began to take on an increasingly centralized, homogenized, and centrifugally ordered structure in writing.

For example, the written verse of *The Epic of Gilgamesh, The Iliad, The Tanakh*, and *The Aeneid*, like the emergence of writing itself, follows four kinesthetic operations that increasingly order poetic speech in a new aesthetic field defined by the centrifugal motion of increasingly abstract forms.

Form and Formula

The first kinesthetic operation of written verse is that of division between repetitive verse and nonrepetitive prose. Before writing, all storytelling, song, and poetic technique emerged though oral tradition. Orality precedes writing by thousands of years and is defined primarily by centripetal patterns of motion, as I showed in "Being and Motion."[2] Oral verse and song are distributions of sonic images that are gathered by individual poets from the wide periphery of the world, by different interpretations of events, and are brought to a collective site of shared performance. The kinetic materiality of oral poetry and song is simple: sound waves from the periphery (stories of travel, stories heard elsewhere, or stories simply created elsewhere) are gathered in a hollowed-out vessel like a cave, a hearth, a house, or even the hollowed-out vessel made by the participant bodies themselves as they gather in a circle to sing to each other or as a chorus. Verses are shared experiences that are subject to retelling, embellishment, reinterpretation, and mutation. The material structure of poetic performance thus gives rise to three major poetic devices, which increasingly make possible a kinetic division between verse and prose, between poet and audience, and between the center and periphery of the poetic image.

Repetition

The first poetic device is that of repetition. Over the course of thousands of years, certain patterns in oral storytelling and song were transformed into increasingly formulaic rhythmic orders of sound. The exact or proximate repetition of poetic phrases was used as a way to aid in the memorization, transmission, and performance of oral poetry. Repetition was a means of increasing the centripetal capacity of the poet to remember lines.[3] Using predetermined, repeated turns of phrase and rhythmic segments not only helped the singer to remember the song but also provided a basis for extemporaneous modification of the song. For example, Homeric epithets from *The Iliad* and *The Odyssey*, like "kléos áphthiton" ("everlasting glory"), "rosy-fingered dawn," "swift-footed Achilles," and "the wine-dark sea" serve this function.

Kinesthetically, this poetic repetition produces a centripetal accumulation of some sounds and not others, some poetic images and not others. By repeating a series of sonic images, one is literally returning to a sonic starting point, creating a sonic loop or refrain. The loop or sonic fold of sound lays out a poetic field of anticipation within which slight modifications emerge as continual modifications and expansions. It draws a sonic circle. Poetic "parallelism" creates two halves that fit together to create a whole poetic phrase within which subsequent linear developments occur. For example, the Akkadian *Epic of Atrahasis* uses repetition and parallelism precisely to this effect.

misil massarti musum ibassi
bitum lawi ilu ulidi
It was night, halfway through the watch,

The temple was surrounded, but the god did
not know.
misil massarti musum ibassi
Ekur lawi Enlil ulidi
It was night, halfway through the watch,
Ekur was surrounded, but Enlil did not know.[4]

The centripetal accumulation of sonic patterns in the brain and body of an increasing number of settled urban bodies is not only functional; it is also formal or "oral-formulaic," as Milman Parry says. What began as a functional aesthetic device for centripetal accumulation (memorization and transmission) ended up producing an increasing division between the oral verse, defined by these techniques, and oral prose, which did not use them. The gap between poetry and common language began to widen. Kinetically, this division marks a material and sonic distinction between the center of more frequently repeated sounds or words in the song and a periphery of less repeated sounds that add to or modify or are added to the more primary ones. Newer songs repeat older ones but also add something new in an ever-expanding sonic accumulation of the past alongside the present.

Participation

The second poetic device is that of participation. According to Walter Ong, oral poetry is defined by the participation of the audience and eventually the agonistic relationship of poets competing for the center stage.[5] The earliest spaces of performance are defined by the centripetal gathering and encircling of bodies that create a central opening within which the choral voices resonate back on one another. From the beginnings of collective song in the resonance chamber of the cave, hut, and hearth, the empty centripetal space became increasingly occupied by individual singers and poets who distinguished themselves from the periphery by a formal virtuosity of poetic verse. Once the space of the center became occupied by the poet skilled in the use of extemporaneous formulaic verse, the aesthetic field became increasingly divided between chorus and poet. In ancient epics, for example, the narrative structure even begins to change as the poet increasingly identifies himself with the hero through first-person perspective.[6] This division also introduced an antagonism between poets competing for the center of performance.

Therefore, in the devices of repetition and participation increasingly found in ancient oral poetry, we see a growing kinesthetic distinction between the center and periphery, poet and audience, verse and prose. With increasing urbanization, the heterogeneous periphery of songs and tales became centripetally accumulated around a central performer, along with the competition between performers for the center of performance.

Model and Copy

The second kinesthetic operation is that of hierarchy. Once the heterogeneous stories of the oral periphery were increasingly consolidated by the centralized forms of the performing poet, it was possible for poets to formally synthesize their performance even further into a single written document. What emerged first as heterogeneous stories told about many different people by many different people became synthesized into a single story told by a single poet. What was once a fully synesthetic experience of smells, tastes, sights, sounds, and heat of the hearth and gathered bodies now became a silent inscription on a mute surface. The story became mute in writing. Only with these conditions was oral story inscribed onto tablet or parchment.

However, once this occurred, a new asymmetrical motion was introduced between the single and centrally held written text (reduced to visual sensation) and the multiple and peripheral copies and performances of that text in their full sensation. Even the subtle variations of the poet's own performance could now be contrasted with the original accuracy and authenticity of the written document. This is the case for the Western and Eastern ancient epics, from *Gilgamesh* to the *Tanakh*.[7] The difference between the original written *Epic of Gilgamesh*, for example, and the copies made of it and then performed based on these copies is an increasingly asymmetrical one. The original is temporally more primary, spatially more centralized (written and preserved in the city), kinetically less changed by copying, and materially more durable and immortal than the human bodies that sustain the poem through performance. Through centralization, the written model becomes superior to the written and performed copies. The written model and its poetic form live on beyond its peripheral and centripetal performances, and thus—through an interesting kinetic inversion—become more powerful, creative, and primary than its own conditions.

Homogenization of Performance

The third kinesthetic operation is that of reorganization. Another interesting kinetic inversion takes place when the heterogeneous oral periphery becomes formalized into the written center and then copied back out to the periphery; the periphery thus becomes homogenized and formalized. Given the emergence of a material-kinetic hierarchy of written verse over oral performance, the periphery can then be subordinated to and homogenized by the identical model of the written text. Long after the author of the original written verse is gone, the model remains to faithfully guide future copies.

The plural and oral origins of the written poem are not only consolidated in the central text but are, in fact, retold as copies and reperformances of the central synthesis itself. "The poet who composes with only the spoken word a poem of any length must be able to fit his words into the *mold* of his verse after a fixed pattern," as Parry writes.[8] The irony is that this central story is a story that was never told; rather, it is a synthesis of actually told stories that had itself become a story. In other words, the performance is not a copy but just an image among other images, yet from this plurality of simulacra one of these images comes to take on a kinetic primacy as an "original" whose forms become fixed in stone and paper. In the case of epic poetry, the material persistence of the medium gives birth to the idea that there was actually a single person who really lived the story and now lives on in the song and the immortality of the indestructible writings.

Immortality

The fourth kinesthetic operation is that of creation. Once the periphery is reorganized by the material kinetic conditions of copying the "original" verse, the stage is set for the ultimate kinetic inversion. Over the course of thousands of years, the periphery appears to always have been reorganized around the central written document. The written product or text now appears to have actually preceded the oral and plural telling of the story itself. The form of the written poem appears to have always been there before any copy or performance of poem. Songs change, but the written document remains the same for an immemorial past and indefinite future.

The kinetic inversion here is amazing. The nonexistent epic hero actually appears to have preexisted the stories told about him. In fact, the stories are told precisely to ensure his immortality, his everlasting glory, or *kleos*. The very idea of eternity is thus derived from the notion of the immortality of that which never existed. In this final stage, immortality now precedes that which granted it its immortality: written verse. The aesthetic forms of writing and repetition appear to have preexisted their oral function. Speech becomes subordinated to the written word, just as the form of the verse itself subordinates the heroic content of inscription. The poetic figure of the epic hero, man-god, or divine being—although created through an oral accumulation—now appears to have preceded and even created the people, forms, and stories that sustain the immortality of the divine itself. Written form becomes identified with a largely mental and thus impoverished image of abstract repetition, ratio, and meter.

While poetic form begins historically as a concrete pattern of sonic repetition and performative accumulation, it slowly begins to take on an affective independence in the mind as a memory of a divine form or pattern without concrete shape. In other words, once they are produced, the idea of ratio, meter, and rhythmic pattern become mental formulae that can be thought of and imposed on a story independent of its exact content. Content can now be molded by form.

Historically, the invention of written aesthetic form is derived first from oral accumulation into the formulae of, for example, Homeric hymn; then, once written down, it standardizes the performance of the periphery until dactylic hexameter becomes a form imposed and copied elsewhere, without the same content. Only after all this can Plato retroactively say that the truth of art is simply the thought or idea of the meter itself, independent of the stories told in the meter. Abstract form thus emerges from concrete form as its impoverished but dominating mental image. For Plato, the stories can only be perversions or false poetic copies of the pure aesthetic contentless form of the meter. In other words, the impoverished mental image of meter alone becomes the central model from which all the now seemingly inferior copies are made in various ways. Virgil, Lucretius, Catullus, Ovid, and Cicero are simply concrete versions of the immortal and ideal form of the hexametric poem itself.

TRAGEDY

The second major centrifugal aesthetic field of the ancient period is that of tragedy. The birth of tragedy follows a similar series of kinesthetic operations as that of written verse, moving from the centripetal accumulation of relatively unformed collective voices to the emergence of an abstract theatrical form imposed back on them.

In its broadest definition, the basic kinesthetic elements of ancient tragedy occurred well before Greek tragedy.[9] The birth of tragedy is a long and incremental process that only passes through Classical Greek theater.[10] This is apparent in the ancient Greek word τραγῳδία, *tragōidía* itself, which comes from τράγος (*trágos*, "male goat") and ᾠδή (*ōidḗ*, "song"), a reference to the goat-satyrs of archaic Dionysian fertility festivals. The structure of these Dionysian festivals has its origins much further back in the annual rebirth rituals of older vegetation deities, the *eniautos daimon*, or the Year-King.[11]

Before moving on to the four common kinesthetic operations of tragedy, let's look quickly at the basic structure of three pre-Greek dramatic rituals—the Sumerian, the Babylonian, and the Egyptian—to see their continuity with Greek tragedy. Only then will we be in a position to locate the common kinetic operations at work in each.

The Sumerian Decent of Innana

The first dramatic ritual is the Sumerian New Year Festival that celebrated the life, death, and regeneration of the bull-moon-fertility goddess Innana, and is an early precursor of Dionysian fertility dramas. Both are defined by the centripetal gathering of the people around a central figure of sacrifice, resurrection, and divine marriage. According to "The Descent of Inanna to the Netherworld," the oldest recorded ritual dramatization of the lunar myth,[12] Inanna went to visit her sister in the deep. As she descended, she was stripped of her jewelry and regalia at each of the seven gates of the underworld. When she reached the deep, Ereshkigal fastened "the eye of death" on her and Inanna hung like a carcass on a hook. After three days, Inanna was released, but she was forced to sacrifice someone else to take her place for half the year. She chose her husband Dumuzi. Each year he dies and is resurrected, but with this Inanna "placed Dumuzi in the hands of the eternal."[13] Sumerians dramatized this event by sacrificing or castrating the priest of Inanna, resurrecting him, and then remarrying him to the priestess.

The name Cybele was given to an Anatolian expression of Innana, adopted by the Greeks around the sixth century and later by the Romans. Like Innana, Cybele had a son-lover (Attis).

> Whatever their origin, the relationship between Cybele and Attis confirms once again the image of the sacred marriage between the goddess and the god or king who once personified the year god and was sacrificed and dismembered in person or in mime at the spring fertility ritual. Whether the son-lover is Attis, Dionysos or Zagreus, the imagery of dismemberment and death followed by resurrection is the same. The sickle used for the castration of Attis and the flint knife used by the priests of Cybele point back to the sickle of the Old European male companion

> of the Mother Goddess of the Neolithic era. The castrated high priest of Cybele was regarded as Attis himself, and in Rome was called Archigallus. The shadowy lineaments of the old vegetation and initiation rites come into focus: it is more than likely that castration, like circumcision, was at one time substituted for the ritual killing of the king or high priest. Originally, Cybele may have had a single high priest and king, her "son-lover," who was at first killed but whose genitals in a later era were offered in sacrifice instead of his life.[14]

Castration is actually a symbol of fertility. Instead of sacrificing the god-king, his fertile organ was sacrificed in order to make the world fertile again. In Rome,

> on 24 March, the Day of Blood, the day of lamentation for the death of Attis, the Taurobolium, or sacrifice of the bull, took place and his genitals were offered to the goddess. This was the day when the priests flagellated and lacerated themselves, sprinkling the altar and the effigy of Attis with their blood, and when devotees castrated themselves. These rites represented the dismemberment of the god, the life-force of the earth, similarly enacted in the Dionysian and Orphic rituals, and most probably also in the Canaanite rituals detested by the prophets.[15]

Generation and creation are thus associated with the flow of vital fluids: blood, the *sangre semita,* the creative flow of seeds that gives life to the earth.

Attis is depicted as a shepherd like Dumuzi, the son-lover of Innana, just like Dionysus with his satyrs. "Sun-rays or ears of corn or fruit emerge from [Attis's] cap, proclaiming him both a solar god and a god of regeneration; this imagery is shared with the rites of Eleusis. In his rituals he was called 'the cornstalk' or 'the ear of wheat,' and his symbols were the pine-cone and the pomegranate. Like Dumuzi and Tammuz, he was lord of cattle, sheep and plants."[16] Innana, Dumuzi, Cybele, Attis, and Dionysus are also linked to the Oracle of Delphi by their engraved images at the site, the sacrifice of the bull, the pinecone *thyrsus*, and the festivals of fertility and intoxication.

The Babylonian Akitu

The same festival structure of collectively burying the dead fertility divinity who is then resurrected and married to the priest or king was carried on by the Babylonians in the New Year Akitu "barley cutting/sowing" Festival. On the twenty-first day of Adar and the first day of Nisannu, people from all over the periphery gather in the central city of Babylon and its central temple. A decorated or masked priest of Marduk (Ésagila) reads prayers of forgiveness, and the people collectively respond to these with prayers of their own. This is repeated for two more days, after which the *Epic of Creation Enuma Elish*, in which Marduk slays the goddess Tiamat, would be recited by the priest.

On the fifth day, the king submits himself to a masked priest dressed as Marduk. The Ésagila then strips the king of his regalia, just as Innana was. On the sixth day, all the statued gods of the temple are centrifugally unleashed on the city, producing chaos and capturing Marduk. The king then leads his people in a procession of song and dance to free Marduk. On the ninth day, a victory procession of song and dance led by the king goes to the "House of Akitu" where Marduk will slay Tiamat. On the night of the tenth day, Marduk marries Ishtar, the Babylonian Innana. The king now plays the role of Marduk by marrying the priestess of the Esagila, thus ensuring fertility and another complete cycle or circle of seasons.

In the ritual of Innana and Akitu we thus clearly see first a centripetal motion of gathering around the center, and then the emergence of a poet-priest from this center.

The Egyptian Festival of Osiris

The same festival structure was repeated in the death and resurrection of Osiris, the god of fertility and regeneration. The festival begins at Abydos with the collective burying of Osiris by planting seeds in "Osiris Beds" formed in the shape of Osiris, filled with soil, and sown with seed.[17] The five days of the public drama depict the murder and dismemberment of Osiris and his resurrection by Isis, who breathes life into him. An actor dressed as the god Wepwawet ("opener of the way") leads the festival procession.[18] This ritual theater is not only functional—planting, plowing, and so on—but also introduces a new formalization to aesthetic sensation performed by a masked god.[19]

Chapter 17 of the *Egyptian Book of the Dead* describes the centripetal gathering of wheat, smashed into a paste, and pressed into sixteen molds (the dismembered parts of Osiris), which are then placed in a silver chest near the statue of the god. Water is slowly added until the parts can be kneaded into a single mold of Osiris and buried. The mold thus gives birth to a concrete form that, through decomposition and death, disappears and is centrifugally resurrected outward into all the life of the earth in spring. The kinetic process is thus inverted: The god is no longer a concrete form produced by a concrete mold but rather a model that must have pre-existed the mold itself, since it was a mold of the god! The wheat mold of Osiris is infinitely creative and fertile, but also irreducible to a single concrete mold, or copy.[20]

The division between tragedy and comedy is derived precisely from this more primary dramatic movement of the death of the fertility god and his or her resurrection and marriage.[21] Tragedy moves down toward death, and comedy moves back up toward light and rebirth. They are two sides of the same circular dramatic motion within which a central and centrifugal figure can emerge.

The Chorus

The first kinesthetic operation of tragedy is thus the division introduced between the central priest-king-actor and the peripheral people-chorus. Prehistoric song, dance, and ritual theater were defined by an increasingly centripetal accumulation of performance until around 3,500 BCE, when this activity was stockpiled into urban, written, and theatrical forms. This occurred when a single aesthetic figure was isolated from the periphery and came to occupy the center in the form of a priest, king, actor, or poet. In lyric poetry, the periphery eventually becomes the audience, but in choral poetry and theatre, the periphery becomes a performative resonance chamber responding to a central point, mimicking the kinesthetic relation between the poet and the audience. In Sumer, Babylon, and Egypt, the masked priest, king, and god became distinct from the people—leading the procession, performing the sacred marriage, and reciting divine texts. The single, central priest said a prayer and the multiple, peripheral people responded in turn. This is the chorus.

In Greece, Arion of Corinth and the dithyramb increasingly transformed the wild Dionysian festivals of immemorial past into the circular chorus. The chorus in turn was increasingly divided between a central *coryphaeus*, leader, or head who directed the chorus, and a peripheral group that was directed. The Greek chorus, as the Greek word *chora* suggests, was an active kinetic space that made or opened a space by moving. The kinetics of the chorus here are clear: The chorus gathers around centripetally and opens up a space within which it is possible for a central figure to emerge, distinct from the periphery. Choric movement is further divided into strophic and antistrophic "turns" as the chorus moves and turns from east to west, then west to east, then finally to an epode where a single *coryphaeus* steps forward from the center and sings solo. This makes possible a further individualization of the epode-song, by Stesichorus and Archilochus in the seventh century BCE, into a new verse form of iambic trimeter, followed by a verse of iambic dimeter.

Since the Greek word *chora* also means "countryside," the centripetal gathering of the countryside to the center of the *polis*, or city, takes on a kinopolitical meaning as well: as a division or distinction between the urban and rural, center and periphery, the speaker and the singer, Dionysus and Apollo, women and men. The two are brought together and united in a single, continuous centripetal enfolding that gives birth to a central figure: the actor.

The Actor

The second kinesthetic operation of tragedy is the introduction of an increasing hierarchy between this new center and its periphery. The central figure emerges from the chorus not as just another choric figure but also as a personage of superior form: priest, king, god, or actor. Just as the bard emerges into a position of hierarchical centrality and competitive advantage over his or her audience through formal technique—repetition, formulae, and so on—so the priest, king, god, or masked actor takes on a leading, directive, or superior kinetic position with respect to the periphery. This is nowhere more obvious than in the Greek invention of the *mechane*, a crane that raised the actor, like the god, into the sky (*deus ex machina*) above the stage and peripheral chorus.

According to Aristotle, the transition from improvisational chorus and ritual procession to the addition of individual performers, has to do with the superior nature of the individual's form. "The reason is that in old days free citizens themselves formed the choruses; it was difficult, therefore, for a large number to sing together like virtuosi, so they sang enharmonic songs."[22] It was therefore the introduction of concrete poetic forms like repetition and meter that created a hierarchy between poetry and prose, poet and audience, virtuosi and novice.

Slowly, to this single skilled actor was added another and another, until the chorus was increasingly diminished and served only as a response to the more beautiful form of the central actors. It is worth quoting Aristotle's history of this transition at length:

> It certainly began in improvisations—as did also comedy; the one originating with the authors of the dithyramb, the other with those of the phallic songs, which still survive as institutions in many of our cities. . . . The number of actors was first increased to two by Aeschylus, who curtailed the business of the Chorus, and made the dialogue take the leading part in the play. A third actor and scenery were due to Sophocles. Tragedy acquired also its magnitude. Discarding short stories and a ludicrous diction, through its passing out of its satyric stage, it assumed, though only at a late point in its progress, a tone of dignity; and its metre changed then from trochaic to iambic. The reason for their original use of the trochaic tetrameter was that their poetry was satyric and more connected with dancing than it now is. As soon, however, as a spoken part came in, the very nature of the thing found the appropriate metre. The iambic, we know, is the most speakable of metres, as is shown by the fact that we very often fall into it in conversation, whereas we rarely talk hexameters, and only when we depart from the speaking tone of voice. Another change was a plurality of episodes. As for the remaining matters, the embellishments and the account of their introduction, these must be taken as said, as it would probably be a long piece of work to go through the details.[23]

With the introduction of more actors, the form of tragedy changed from more musical to less musical. The ecstatic and pedetic kinetics of the satyric dance performance were not only increasingly centralized on stage but also a new central dramatic form was invented that was superior to the one of the periphery. The new form of the center thus subordinated the periphery. To this we should add that the writing down of theater also produced a new form that reacted back on the periphery by allowing for repeat performances, thereby homogenizing the periphery as described earlier in the case of written verse. The central individual was elevated above the collective periphery: the triumph of Apollo over Dionysus, light over dark. As Nietzsche writes, "Apollo himself the glorious divine image of the *principium individuationis*, through whose gestures and eyes all the joy and wisdom of 'illusion,' together with its beauty, speak to us."[24]

The Theater

The third kinesthetic operation of tragedy is the reorganization of the choral periphery by the new form of the center. This occurred already as a result of the hierarchy of central form (iambic versus trochaic, written vs. oral, scripted vs. improvisational, actors over chorus) over a relatively less formal periphery. In addition to these formal hierarchies, we should add the complete reorganization of theatrical motion into the theater apparatus itself.

The verticality of the Babylonian ziggurat, where the priest spoke from on high, and the Egyptian palace, from where the god-head emanated, are finally consolidated in the Greek amphitheater—the ultimate centrifugal theatrical device. This occurs through an absolutely ingenuous kinetic inversion of the holy mountain. Where the centrifugal voice of the priest-king moving downward and outward was often blown away by the wind, the theatrical pit amplifies it upward and outward. It retains the same centrifugal kinesthetics of the center, but it inverts its verticality.

By maximizing a centripetal accumulation of bodies into a recessed and sloped pit, the Greek *theatron*, literally "seeing place," captures sound and vision in a total enclosure. Once enclosed and concentrated, a central stage or orchestra radiates its motion out and around radially. The ancient Greek ὀρχήστρα (*orkhḗstra*, meaning "to dance") comes from the proto-Indo-European root **ergh-*, meaning "to set in motion," "stir up," "raise," plus the suffix **-tra*, denoting "place." The orchestra is thus literally the place where motion is raised up and distributed outward and centrifugally through dance. The orchestra was circular, and its radial rows of seats expanded outward in concentric circles to the periphery (figure 7.1). People from all over the periphery were literally funneled by the city-state, as a civic duty, to a

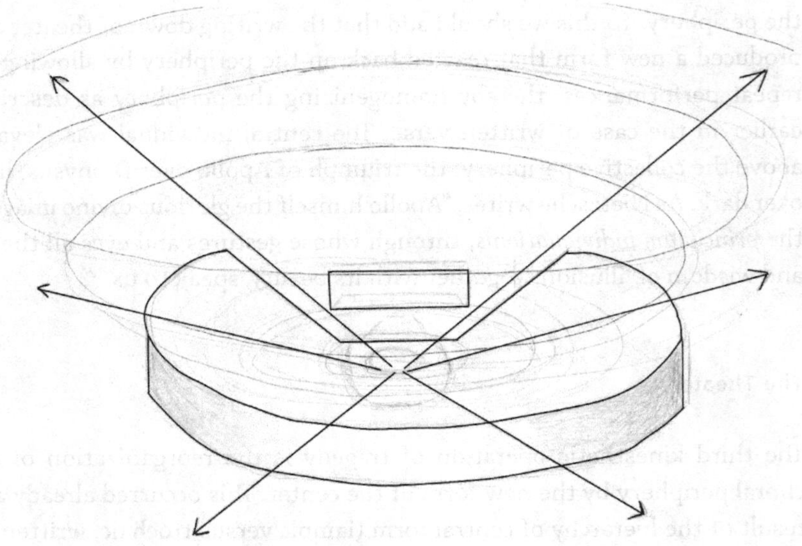

Figure 7.1 Centrifugal Kinetics of the Ancient Greek Amphitheater

central location in order to be kinesthetically reordered and reformed by a centrifugal pattern of motion.

The Mask

The fourth kinesthetic operation of tragedy is the recreation of the periphery itself in the formal image of the center. This is accomplished by the use of the theatrical mask. The use of theatrical masks goes back to Sumer, Babylon, and Egyptian rituals. Priests and kings in these civilizations all used masks to perform various rituals and ceremonies, from the Babylonian Ésagila's performance of Marduk to the Egyptian priest's performance of Wepwawet in the drama of Osiris. We should also include here Egyptian, Mycenaean, and other ritual and funerary masks, all of which were directly related to either divine expression or life after death.

The kinesthetic question is thus: How does a sensory image of that which seems to exceed sensation (the pure immortal form of a god or the soul) emerge from purely immanent sensory flows of matter? This question is not answered by the ritualist theory of drama but, rather, presupposes it. From a sensuous and concrete formal cycle of seasons and vegetative rebirth, how is something like a distinct and unchanging deity extracted and formalized in the first place?

In all these cases, it is the invention of the mask that both provides a concrete form to the central ritual-tragic figure and introduces a mental image of an abstract form of the deity or soul. The later (abstract) has its grounding in the former (concrete). It is possible to get ahold of this idea following the four kinesthetic operations of ancient aesthetics in the material kinetic production of the mask itself.

The theatrical mask is made by creating an empty vessel or mold from a centripetally gathered flow of clay, leather, and pigment onto the curved surface of the mask, dividing a central interior from a peripheral exterior. The mask is a bowl for the head. Second, something with concrete form (actor or priest) is placed inside the vessel or mold and thereby gains an importance or hierarchical status beyond those not wearing the mask. Third, this new concrete form is used to reorganize and reshape the periphery according to this central and centrifugal form or pattern. For example, the chorus or audience responds and is affectively ordered by the actor's or priest's masked actions and words. Finally, the ultimate kinetic inversion occurs when the contents of the mask are removed and sensation is confronted with the disjunction between the concrete form of the now-inanimate mask and the missing but animate deity who lives beyond the mask or statue.

The concrete form of the mold thus gives birth to the idea that the mold must have been of an immemorial original whose abstract form appears only as a mental image: memory, belief, idea. In a dramatic kinetic inversion, the centripetal flows that were once folded to create the mental image of an abstract form now seem to have the original cause or face on which the mask was molded. This is certainly a matter of a historically increasing degree of abstraction culminating in the most abstract of all sky gods—Yahweh or Plato's craftsman—who can no longer be portrayed in any mask or statue and who is the sole creator of the world, not just vegetation.

Only on the condition of these four kinesthetic operations is it possible for Aristotle to write that the purpose of tragedy is catharsis.[25] For Plato, all the arts are limited by their mere mimesis of abstract forms and can thus only fail to attain any knowledge of those forms. For Aristotle, however, tragedy has a kind of negative power to use concrete sensuous forms in order to purge or purify us of our irrational emotions like pity and fear. By purging the periphery of its irrational emotions, the audience is at least brought closer to the realization that there are pure abstract mental images that exceed the other senses and that are superior to them (behind the mask of sensation).

METALLURGY

The third major centrifugal aesthetic field of the ancient period is that of metallurgy. In the broadest sense, the ancient art of metallurgy is the process of ordering a liquid flow of formless matter into a durable aesthetic form. In this sense, this technique goes all the way back to the natural sculptural techniques of the cave and the early human accumulation of liquid earth into clay bowls and vessels. Even as early as 11,000 BCE, humans were already smelting copper beads.[26] However, in addition to this centripetal accumulation and cold shaping of matter, what begins to emerge during the ancient period is a new pattern of centrifugal motion that reacts back upon the accumulated matters and starts to form them.

Metallurgy provides the historical and material conditions for the rising primacy of concrete form over liquid unformed contents, as well as over more abstract theories of form based on these conditions. Just as in written verse, form is neither immaterial nor transcendent but, rather, is kinetic and historical. This kinetic emergence of form in metallurgy is defined by four operations: the invention of the cast, the mold, the copy, and the idea of form. The great sculptures of the ancient world, such as the metal weapons and tools of Mesopotamia, the death masks of the Egyptians, the athletic bronze statues of Greece, and the bronze emperors of Rome, are all defined by these four kinesthetic operations that increasingly order the sculptural arts into a new aesthetic field defined by the centrifugal motion of form.

Smelting

The first kinesthetic operation of metallurgy is that of a division between the solid and heterogeneous mineral periphery and the liquefiable and homogeneous metal center accumulated in the bowl, vessel, or hollow container. That which is liquid and containable becomes increasingly distinct from that which is solid and uncontained. Smelting allows a new form of mineral accumulation that was never before possible. It permits a centripetal extraction and accumulation of a liquid mineral into a concentrated central location. This difference becomes increasingly dramatic after the introduction of tin and lead smelting, around 6,500 BCE.

This is not just a matter of a difference in the materials of accumulation; there is also an important kinetic difference. With the accumulation of all previous materials, the contained flows remained fluid—water, milk, grain, and so on. Neolithic pottery marks the limit of centripetal accumulation insofar as it only contains liquids in its hollow center that no longer retain the shape of the pot when they are removed. Only the pot itself retains the shape of the clay that made it, but it is unable to do the same for other matters.

However, with the introduction of metallurgic extraction and smelting, a unique kind of flow is accumulated—a flow that can retain the shape or concrete form of its central accumulation. Since tin and lead have lower melting points, they were first melted in hearths, the oldest human site of centripetal accumulation. Kinetically, the invention of smelting allowed for the form of the center to persist beyond its containment in that center. With smelting, the heterogeneous periphery of composite minerals is increasingly distinguished from a liquid core or center that persists in form without the aid of the same center to hold it. Metallurgy is thus an art only possible through centralization and differentiation between metal centers and nonmetal peripheries: extraction.

The Mold

The second kinesthetic operation of metallurgy is that of the introduction of a hierarchy or asymmetry between the central form of accumulated metal and the peripherally formed copies molded by the hollow form. Historically, the first molds or casts were introduced around the fifth millennium BCE for tin, lead, and, more importantly, copper for beads, tools, and weapons.[27] Eventually bronze alloys emerged around 45,00 BCE.[28] While smelting distinguishes a liquid or liquefiable center from a nonliquid periphery, the mold introduces a hierarchy between the creative forming center and the created or formed periphery. The mold hollows out a center in a solid and makes possible for the first time the presence and appearance of a hidden or absent form—the negative center, or interior of the pot within which the metal is collected. However, the center of accumulation no longer functions negatively as just another empty space, as it had with previous fluid accumulations; rather, it serves as a positive and creative matrix, womb, or materialization (from the Latin *mātrix* ["dam," "womb"], from *māter* ["mother"]) capable of shaping durable metallic flows.

This is a radical kinetic discovery. With metal, the invisible center of things can become visible. The center becomes the provider of military power through weapons, of innovation through tools, and of beauty through jewelry. The metal products produced through molding are powerful. However, since metal can always be melted down again into the cast, it is the hollow form that remains truly powerful and creative, even after its more materially durable metal contents have been destroyed and returned to fluids. This is the hierarchy and asymmetry introduced by the mold over and above that which is molded (the flow of metal). The smelted flow is formed and reformed by the mold.

The Model

The third kinesthetic operation of metallurgy is that of the reorganization of the material periphery according to a single and central model. While the first historical molds were created directly by carving or removing a center or hollow centripetally from the outside in, later molds were created from an impression of a model centrifugally from the inside out.

First a central model is made, and then a clay periphery is centripetally accumulated around it, thus impressed from the inside. In this way, not only does the center become distinct from the periphery but also a triple hierarchy is introduced between the model, the mold, and the copy. The model is the original, the creator, the perfect form from which an imperfect mold is created and from which subsequent imperfect copies are created. The model center thus reorganizes a periphery of imperfect copies through the mold. The periphery begins as heterogeneous mineral composites and even heterogeneous objects, but through smelting minerals it is homogenized as liquid metals and through the model, mold objects are homogenized as replicas or copies of an original. In short, the kinesthetic field becomes increasingly ordered by various positive and concrete forms or models.

The technique of lost-wax casting was the first to introduce this modeling process into sculpture, jewelry, art, weapons, and tools.

> The statuettes were cast solid by the direct lost-wax process: wax was cut, rolled, pinched, and carved into shape; wax body parts were warmed up so they could be stuck together; and the resulting wax model was invested with (covered by) a clay mold. After the mold was baked to melt out the wax, the bronze was poured into the now-hollowed mold. A base or an attachment for insertion into a stone base was usually made in wax and cast along with the figurine, sometimes serving first as the funnel into which the molten bronze was poured.[29]

The oldest known examples of the lost-wax technique are the objects discovered in the Cave of the Treasure (Nahal Mishmar) hoard in southern Israel, dated back to around 3700 BCE. The cave hoard includes hundreds of copper and bronze objects, weapons, several scepters, crowns, and other tools.[30] One of the earliest surviving lost-wax castings is a small lion pendant from Uruk IV and a copper frog from around 3,200 BCE.[31] The Egyptians used the lost-wax technique beginning around the mid-third millennium for early dynastic bracelets, gold jewelry, and eventually larger statues.[32] After 1,500 BCE, Egyptian craftsmen began molding glass as well.[33] Around the ninth and eighth centuries BCE, the Greeks began casting simple bronze figurines and geometric bronze cauldrons for sanctuaries.[34]

However, in order to produce larger bronze statues the Greeks also began hammering sheets of bronze around the exterior of wooden models, in a technique known as *sphyrelaton*, or "hammer-driven."[35] According to literary sources, around 700 BCE the Greeks learned from the Egyptians how to cast large-scale bronze statuary.[36] By the late Archaic period (ca. 500–480 BCE), *sphyrelaton* went out of use when lost-wax casting became the major technique for producing bronze statuary.[37] With this new technique, the Greeks produced thousands upon thousands of innovative bronze works. Around half of all Greek statues were bronze and the other half were marble. None of the marbles and fewer than two hundred bronze statues remain today.

The introduction of bronze cast sculptures thus not only introduced a new aesthetic primacy of form and model but also liberated the forms of sculpture from the previous restrictions imposed by marble and terra cotta. Since bronze has a greater tensile strength, freestanding sculptures were able to extend outward from their centers of gravity in *contrapposto* and ballet-type poses, like that of the *Bronze Warrior from Riace* and the *Artemision Bronze*. Bronze gave a new dynamism to sculptural form, but it also allowed for the introduction of finer detail in the material, thus making possible a new level of emotional expression in the metal, such as in the sadness of *Dying Gaul* or the defeated look of the *Seated Boxer*.[38] The *contrapposto* of Polykleitos's *Doryphoros* and the nude flesh of Praxiteles's *Aphrodite of Knidos* are often contrasted with early Archaic, rigid, geometric, "less humanistic" sculptures. Paradoxically, this so-called humanistic turn in Greek sculpture was only made possible by the nonhuman and highly technical properties of metallurgic bronze and beeswax.

At the same time that sculpture became more tensile and dynamic in form, it also became more homogenized in reproduction. It was once believed that existing Greek statues were original single-cast works, but recent discoveries confirm that numerous copies were made in workshops from the molds and were distributed widely. "The whole nature of the medium is reproduction," as bronze-sculpture scholar Carol Mattusch writes.[39] With the metallurgic arts, form becomes aesthetically homogenizing, reproductive, and mimetic. The central model comes to dominate, reorganize, and reform the periphery as in the case of lost-wax casting (see Figure 7.2).

Divinity

The fourth and final kinesthetic operation of metallurgy occurs when the abstract form or idea of the model creates the concrete model, mold, and copies. The mental affection of form is only a fragment of the whole concrete formal process of bronze casting and sculpture, and yet, at a certain point in history, it comes to be the dominant order of sculptural formation. However, this formal abstraction does not come from outside the work of art but, in fact, has its roots in the material kinetic conditions of sculptural metallurgy itself.

With the introduction of the lost-wax casting technique, the wax model is melted away by the casting process, leaving only its inner centrifugal impression on the clay mold. Thus the mold and its copies become copies or imitations of a *missing model*. They become copies without an original. However, the original is not completely absent; it persists in the mental

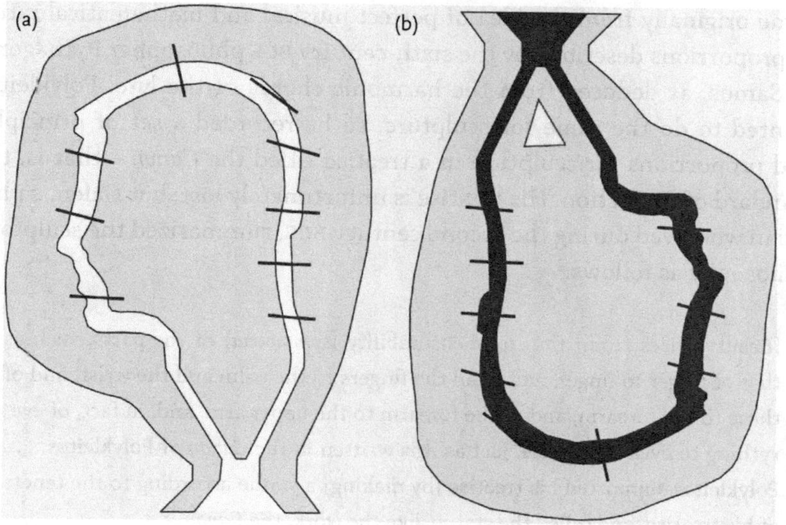

Figure 7.2 Two stages of the lost-wax method of bronze casting (after Sean A. Hemingway)
Source: From Helen Gardner and Fred S. Kleiner, *Gardner's Art Through the Ages: The Western Perspective*, 13th ed. (Boston.: Wadsworth Cengage Learning, 2010), p. 108.

affection of the spectator. Beyond the imitation or appearance of the concrete form, the spectator is affected by the idea of an original that is now missing from the world. This missing original thus takes on an increasingly divine or other-worldly quality, since the mental affect of its form does not appear in any exterior sensible concrete shape. In this ultimate kinetic inversion, the concrete form of the statue becomes the product of the abstract idea or memory of a divine missing model.

For the Greeks, the artist became a channel for the divine muses. The temple statuary rendered visible the invisibility of the divine forms of Inanna and Marduk, Isis and Osiris, Aphrodite and Zeus. The transcendent gods expressed themselves through the divine aesthetic inspiration (from the Latin word *spirer*, "to breathe into") of the artist who transmuted the abstract formal idea into concrete formal shape. The gods were believed to reside inside the statues and become identified with the missing hollowed-out center of the statue where the wax original used to be, and could then be directly worshiped and sacrificed to as such. From this formal center, the gods acted centrifugally outward toward the concrete periphery of the statue.

An exemplary case of this kinesthetic inversion occurs in the work of fifth-century BCE sculptor Polykleitos of Argos, who modeled his bronze cast statue *Doryphoros* (*Spear Bearer*) on the abstract form of a perfect human proportion between all its parts. The idea of perfect proportion came originally from the idea of perfect musical and mathematical ratios or proportions described by the sixth-century BCE philosopher Pythagoras of Samos, as deduced from the harmonic chords of the lyre. Polykleitos wanted to do the same for sculpture, so he recorded a set of principles and proportions for sculpture in a treatise titled the *Canon*—that is, the standard of perfection. His treatise is unfortunately lost, but Galen, a physician who lived during the second century BCE, summarized the sculptor's philosophy as follows:

> [Beauty arises from] the commensurability [*symmetria*] of the parts, such as that of finger to finger, and of all the fingers to the palm and the wrist, and of these to the forearm, and of the forearm to the upper arm, and, in fact, of everything to everything else, just as it is written in the *Canon* of Polykleitos. . . . Polykleitos supported his treatise [by making] a statue according to the tenets of his treatise, and called the statue, like the work, the *Canon*.[40]

This increasingly formal abstraction and kinetic inversion reached its zenith in the Hebrew god Yahweh, who is so immaterial that he can no longer reside in any concrete form at all. Yahweh, the single divine sky-god, emerged at the extreme limit of concrete form as a pure abstract form, speaking and appearing only in the most subtle forms of matter—wind, fire, smoke, breath.[41] No statue or idol could hold him or express his pure form because the statue remained too shapely, concrete, sensuous, fixed, and material. "The Most High does not live in houses made by human hands," as the Hebrew Bible says.[42]

Compared to the sensuous gods of the Greeks and Egyptians, Yahweh is a sensuous impoverishment, a mere gust of wind or smoke everywhere but nowhere in particular. His increase in power as the only creator-god is inversely proportional to his materiality as invisible air. The entire world becomes his statue, created and inspired by him, through him as him. The volcanos become his foundry, the earth and all sensuous reality become his copy and body. Just as in Plato's fourth-century dialogue *Timaeus* God becomes a divine craftsman and the world itself becomes a copy of an ideal form contemplated in the mind of the divine craftsman, so does the Hebrew God. In short, the lost-wax model of metallurgy becomes the material and kinesthetic condition of theology, idealism, and abstract formal aesthetics.

However, this kinesthetic pattern was not unique to written verse, tragedy, and metallurgy. These same four kinesthetic operations can be found throughout ancient architecture, music, cooking, medicine, and perfumery as well, as we will see in the next chapter.

CHAPTER 8

The Ancient Image, II

In demonstrating that a kinesthetic shift occurred in the arts from a centripetal to a more centrifugal pattern of motion over the course of the ancient period, this chapter continues the thesis begun in Chapter 7. In the previous chapter, we looked closely at this centrifugal pattern in the formal arts of written verse, tragedy, and metallurgy. In this chapter, we demonstrate the dominance of the same kinetic pattern in the arts of architecture, music, and medicine. The argument of this chapter is that each of these major fields is defined predominately by a distinctly centrifugal pattern of motion and a formal aesthetics. The argument, therefore, is that the origin of form is not eternal or immaterial but rather in the kinetic structure of matter itself.

THE CITY

Architecture is one of the oldest arts, but it became a predominantly urban art during the ancient period. During prehistory, early humans gathered together from the periphery to the center of the hearth, the house, and the village, but around 5,000 BCE a single village increasingly became the walled city and political center for all the surrounding villages. Architecturally, the city is not just another village; it is a capital village that becomes a formal model for the organization of villages, sustained by villages. Once it is produced centripetally, the city then centrifugally reacts back on the surrounding villages and reorganizes them hierarchically.

Ancient urban planning follows a clearly centrifugal model. Almost all cities in Mesopotamia, Egypt, Greece, and the Roman Empire were planned around the centralized storage and formation of stored matter. Even though many ancient cities were internally divided by grid patterns, the center always preceded the grid.[1] First, the center is determined as the axis and center of the universe; then, the lines radiating outward are measured and subdivided in the form of a grid. The cardinal directions on which the grid is based all begin at the *axis mundi* of the city, which was often modeled on the cosmos.[2] Just as the heavens moved, so did the city—in cosmopolitical periodicity. For example, the temple and palace were located in the center of Babylon. Outside the center was the housing for public officials and then for merchants. Beyond the wall were the neighborhoods, then the agricultural areas, and so on into the periphery: a vast centrifugal distribution of matters from those more formed at the center to those less formed at the periphery.

In short, the city becomes a model for an increasingly abstract form applied to the unformed and inferior matter of the agricultural periphery. Just as in written verse, tragedy, and metallurgy, this occurs in four kinesthetic operations that emerge, coexist, and rise to dominance over the course of thousands of years in the ancient world—the wall, the column, the arch, and formwork.

The Wall

The first kinesthetic operation of the city is the division between an inside center and an outside periphery. Outside the wall, there are relatively unformed flows of matter or raw materials that the city needs for survival, while inside the wall are relatively more organized and formed matters of urban life. However, the wall emerged not from the periphery of the village but from the central hollow of the Neolithic granary, and it expanded centrifugally toward the periphery. During the Neolithic, the village increasingly accumulated its centripetal surplus in an enclosed hollow granary whose supplies determined the life and death of the population. However, with the ancient architectural invention of the brick wall, more grain could be stored in a larger vertical tower, a larger population could be sustained, and a substantial social asymmetry could be maintained by anyone who could control the flow of grain.

Ancient sovereignty and the political power over life and death are, therefore, not abstract or ideological; they are grounded in the real material kinesthetic conditions of centripetal storage and the controlled release of grain flows. Once the warrior-priest-king took possession of the granary, it was increasingly transformed into a divine, political, and military enclosure or hollow around whose central capacity and contents everything turned. The granary became divine because it determined life and death. It became political because it was the site of taxation and remuneration. It became military because it determined the size and strength of the army.[3] It is no coincidence that this first walled granary was increasingly expanded to house the city's god and gods in the form of the temple, and to house one's military and political leader in the form of the palace and citadel. From these material structures the wall was extended centrifugally outward to the periphery of the city itself along concentric circles, in successive inner and outer forms.

The contents of the hollow interior became increasingly secured and controlled such that the central interior of these structures were a mysterious place with unknown contents. While the contents of the granary remained concrete as grain went in and grain came out, the temple, palace, and citadel that housed them grew increasingly less sensible and more hidden.

The concentric expansion of the wall made possible the division between inside and outside in all its urban, religious, military, and political registers, but the wall itself was made possible only on the condition of the brick. Kinesthetically, a brick is formed from a single, central mold or empty cast to which fluid matter is added and hardened. If one is using stone, then a model stone or proportion is determined and the others are copied as replicas. All bricks are copies that proliferate centrifugally outward from this single mold or model. While the model or mold remains relatively static and fixed, the finished bricks become mobile repetitions distributed to the periphery of the city. This is not a metaphor. Real material (clay) is molded according to a single central apparatus and then moved to the periphery of the city as a wall.

The exact historical emergence of walls is certainly a matter of degree without sharp historical breaks. For example, as early as 7,000 BCE the proto-city of Jericho was already making sun-dried mud bricks and vertically piling relatively homogenous stones into city walls and houses. However, these bricks were still very roughly shaped, did not require significantly skilled labor, and were not symbolically marked with the unifying seal of the king. Furthermore, Jericho and a few other cities were the exception to the rule. The vast majority of Neolithic villages did not start using bricks until around 3,000 BCE as they became cities.[4] However, if we were to locate an approximate break or shift in the architectural field, we could look to the invention and dominance of the kiln-fired brick, which closely parallels the rise of the cities and politics. The wall is their limological condition.[5] As archeologist M. L. Smith writes, "For urbanism baked bricks seem to have been a precondition."[6] Baked bricks allowed for a dramatic extension of the vertical and horizontal motion of border walls that was not possible through piling alone. The stacked bricks of the Egyptian pyramids, the Sumerian Ziggurats, and Roman temples and towers dwarf even the largest Neolithic megaliths. It is easy to say that without the stacking of bricks, there would have been no ancient empires.

The architectural inventions of the brick and the wall introduce an increasing kinesthetic division between center and periphery, interiority and exteriority, mold and matter, formed and unformed, depth and surface. The brick wall makes possible not only the massive verticalization of the granary in the form of the temple, palace, and citadel but also an increasingly large hollow or interiority inside these structures, contrasting their divine interiority with the mortal exterior. While the prehistoric fence structure still allows a porous communication between interior and exterior, the wall introduces a more dramatic division between center and periphery.[7]

The Column

The second kinesthetic operation of the city occurs in the increasing hierarchy between interior and exterior architectural spaces. This is made possible by the use of columns and colonnades that allow greater compression support and thus increasingly larger and higher interior spaces. Kinesthetically, the empty vessel of the sacred village shrine thus expands centrifugally outward, first with the wall and then with the column.

Columns in the form of trees, wooden posts, and megaliths were prehistoric materials and techniques, but in the ancient world, the column introduced a new architectural form of the expanded interior. Even in the wooden columns of ancient Minoan and Mycenaean architecture, tree trunks were transformed into load-bearing compression structures to vault the ceilings of the temples and palaces like the multileveled Knossos. The Minoan column thus has its literal roots in the centrifugal and vertical movement of the tree itself, growing upward and outward in all directions from a single taproot and trunk. The use of load-bearing columns allowed the Minoans to create large open-plan spaces. These permitted the circulation of light flows to the center to be formed and shaped by interior architectural molds and frescoes while also allowing for the release of smoke and burnt offerings from the hearth to ascend into the sky through an *oculus* ("eye") or opening in the ceiling.

The well-preserved Mycenaean *megaron* (great hall) at the Palace of Nestor at Pylos is exemplary of this centrifugal kinesthetics. The vast palace is composed of two stories, storerooms, workshops, baths, light wells, reception rooms, and a sewage system—all radiating out centrifugally from a single central throne room. In the center of the *megaron* sits an enormous circular hearth bowl that centripetally gathers the exterior to the center to be sacrificed. The hearth is surrounded by four tall columns that vault the central ceiling above the rest of the room and allow the smoke to ascend through a central *oculus*, through which the king sees upward and outward as the gods see downward and inward. From this divine columned center, motion is socially directed outward through the entire temple, city, and world. The eye and mind see; vision and thought then direct the body outward. The Minoan subject both shaped and was reshaped by the kinetic pattern of this architectural form.

We can see the same kinesthetic operations of ancient poetry and sculpture at work in ancient architecture. Peripheral matters are first increasingly gathered into a concrete center, where they are formed by a concrete interior mold but also vaporized into a vertical and smoky abstract form and then thrust back abstractly on a relatively unformed periphery. The architectural hierarchy of the central *megaron* over the peripheral rooms of the palace is achieved by its size, centrality, and verticality—all made possible by columns.

Ancient columns and colonnades are almost always sites for the centrifugal dissemination of hierarchical political power and divine commandment. Examples include the first stone columns made in Egypt by Imhotep for the mortuary precinct of the palace at Djoser (2,630–2,611 BCE);[8] and the Great Hypostyle Hall of Karnak (ca. 1,290–1,224 BCE) with its 134 enormous columns arranged in sixteen rows, sixty-six feet high (eighty feet high in the center) and twenty-two feet in diameter, and supporting a massive stone ceiling. This is the case not only because columns make possible an increasingly large interior space, which dwarfs some of the largest prehistoric caves, but also because they provide increasingly hidden and thus powerful or superior movement within this space.

In ancient Sumer, this hidden space was described as a "waiting room," where the priests and priestesses would await the arrival of the gods. In ancient Persia, the Hall of Hundred Columns at Persepolis was appropriately called the *apadana*, from the Sanskrit words *apadana*, meaning "to arrive" and *apa-dha*, "a hide-out or conceal," and from the Greek word *apotheke*, or "storehouse." The etymological link makes clear the centrifugal kinesthetics of the ancient granary as a storehouse expanded into a divine arrival point and a concealed source of divine/social power.

The Greek temple, or *noas*, also describes "a fold or concealment" (from the proto-Indo-European root *nes-*) in the central room of the temple that housed the statue of the god. The Greek acropolis is the central building of the city that both housed the god of the city and provided a fortified military defense center in case of invasion. Appropriately, the word *acropolis* comes from the word ἄκρος, *ákros*, meaning "highest," "at the extremity." This also makes clear the vertical and hierarchical structure of the building and its radiating peristyles (exterior colonnades).

However, there is not only a social and spatial hierarchy of the colonnaded city but also a kinesthetic hierarchy of the column's image. As the stone columns are made larger and larger during the ancient period, they are increasingly constructed of separate, homogeneously formed segments, disks, or bricks stacked around a central metal pole. Each disk is modeled on an original only slightly smaller as they are hierarchically graduated toward the sky, until the smallest-shaft disk gives way to the formal surplus of the decorated capital, which bears no resemblance to its inferior, segmented, and often-movable shaft disks.[9] The segmented stone column thus becomes a brick wall by other means: centralized, centrifugal, and hierarchical. The mobile disks are stacked but only around a central immobile and invisible core that holds them together.

The stone column is thus also structurally hierarchical insofar as it is ordered with a wide basis and plinth increasingly graduated up the shaft until it reaches the voluted and flowering acanthus capital in its Ionic and Corinthian orders. The capital peak distinguishes itself with an excess modeled on the natural flowering of the trees on which they were historically based, or with the divine or heroic statues that stand on them. The peaks of colonnades are in turn crowned by the decorative frieze, which is in turn crowned by the pediment of the gods. The column renders into a sensible image the hierarchy of vertical forms (capital, god, enclosed space) over horizontal forms (stylobate, human, and exterior space).

The Arch

The third kinesthetic operation of the city occurs in the form of the arch, which eventually reorganizes both the movement of the wall and the column into a new form. Although the arch emerged around 3,000 BCE in ancient Sumer to support underground drainage tunnels, and around 2,000 BCE in the arched doorways in the city of Ur, and again in the temple of Kuri-Galzu around 1,400 BCE, it was used only infrequently by the

Egyptians and Greeks, who still largely used the post-and-lintel structure.[10] For the Romans, however, the arch became absolutely central.

The arch form uses the brick-and-mortar technique of walling to span the new vertical space vaulted by the column. Without using bricks, the verticality and span between columns is limited by the weak tensile strength of the horizontal lintel stone. However, the form of the brick and column are reformed and reordered by the arch, whose strength comes not from a merely vertical or horizontal motion but also from a radial or distributive motion down and around an empty enclosure. The arch brings the brick ceiling into a continuity with the wall and brings the column into continuity with the wall through a mutual curvature or folding of one into the other. The column provides verticality and the bricks provide curvature and a redistribution of motion back down the columns into the walls. The reorganization affected by the arch is thus defined by three operations.

First, the arch creates a curved and enclosed space divided between the inside and the outside of the arch: the extrados and intrados. Motion is divided between that which moves and circulates inside the enclosure and that which does not.

Second, the arch creates a hierarchy of motion that moves not simply from the top down but also from the top to a central keystone that then redistributes motion and stress downward radially through the supporting voussoir stones, down the imposts, and outward through the abutments as thrust. The keystone takes the kinesthetic place of the hierarchal capital and is frequently distinguished as such from the voussoir stones by an enlarged size or decorated form. One of the most beautiful examples of this are the figures of Roma (goddess of Rome) and Fortuna (goddess of chance or fate) on the Roman arch of Titus (100 CE).[11]

Third, once the first two operations of the arch are in place, the arch can be extended in both directions, thus producing a barrel vault with an elevated ceiling much higher than any column-and-lintel structure could bear. Two barrel vaults can then intersect in a groin vault, which redistributes motion again to the four columned corners of the intersection.

Finally, in its most radical concrete formal expression, the arch can be rotated along its central axis to produce the centrifugal dome that radiates kinetic stress from the center equally in all directions to the periphery. In this final operation we can clearly see the centrifugal movement that was in the arch all along, but only in two dimensions. The Roman dome in the Pantheon (118–125 BCE) achieves the ultimate concrete form: the form that reorganizes and redistributes motion to all the other subaltern forms—walls, columns, arches, floors—at once (figure 8.1). Just like the Mycenaean *megaron*, the vaulted ceiling of the Pantheon opens up at its

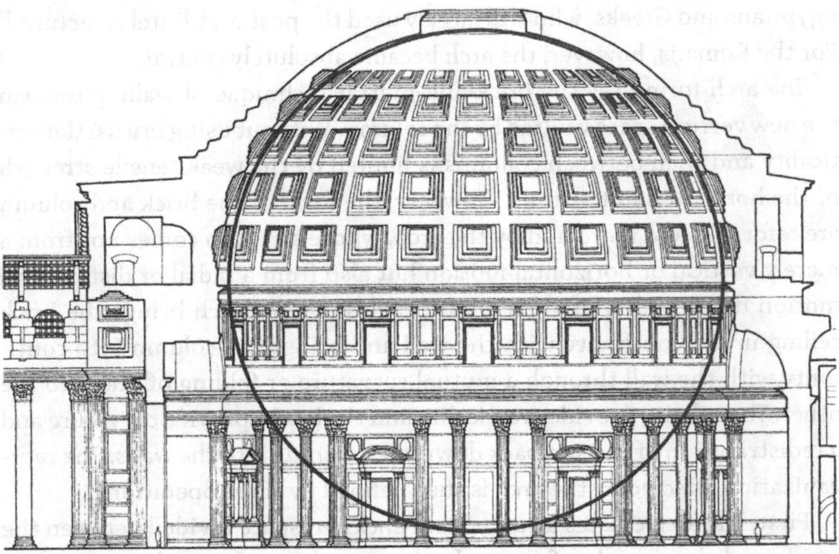

Figure 8.1 Cross-section of the Pantheon
Source: From Peter Sloterdijk and Wieland Hoban, *Globes: Macrospherology* (South Pasadena, CA: Semiotext(e); and Cambridge, MA: MIT Press, 2014), p. 415.

center to the sky through an enormous oculus. It achieves, like the Roman Empire itself, a maximum centripetal accumulation of the periphery and a maximum centrifugal radiation at the same time. The oculus lets in the flow of rain from the sky, along with all the gods. It then forms the rain into channels and redistributes it through a drainage system, just as it forms the invisible gods into statuary and radiates them along the periphery, each in their own enclave bordered by columns under the perfect geometric ratios of the square coffers radiating around the heavenly ceiling. The longitudinal and lateral ratios of the Pantheon form a perfect sphere, as their height and width are identical. The periphery is completely equidistant from the center of the room. The Pantheon is an orb—the first authentic spherical construction on earth. In this orb, ephemeral and divine light are given their perfect formal expression in the radial and centrifugal distribution of compression motion in the domed sphere.

The introduction of the arch thus makes possible the most dramatic centrifugal expansion of interior space in ancient history. Beginning from the elevated vertical center made possible by the column, the arch allows for a radical increase in the clear span between the walls, essentially hollowing out the room in all directions. This kinesthetic reorganization of matter means that all sensuous concrete images and forms are increasingly pushed to the periphery of the temple-palace. The center becomes filled with a purely invisible divine smoke (incense and burnt offerings) and ethereal smells (flowers and perfumes).

Formwork

The fourth and final kinesthetic operation of the city occurs in the creation and subordination of concrete form by abstract form. This move has its material-kinetic foundation in the use of formwork in arch construction. Formwork is the temporary wooden mold or scaffolding required to support arch construction and poured concrete. Once the stone or concrete is in place or set, the formwork is removed and the structure supports itself.

Just like the use of a wax original in bronze casting, the use of an enclosed token system in writing, and the use of a formal written synthesis around the epic hero or gods, a concrete material is placed in the center as a model and later removed after the periphery has been molded. This creates the unique kinesthetic effect that the concrete form is the result of a now-invisible memory or mental affection. It's just as the bowl is a mold of Aphrodite's breast, so written verse is a sonic and mnemonic mold of the muse's memorial voice, and the temple is a mold of the Gods' divine presence. In each case, the missing mold was cast from a missing and memorial model, now abstracted into a supposedly eternal or divine form.

The arch accomplishes this for the building by clearing out the cluttered columns and opening up an undivided and invisible center. The arch thus creates an increasingly abstract form, as if the radial interior of the enclosed vessel-temple-palace were only the mold of a more primary abstract and divine form in the center that had impressed itself on the periphery. The divine descends from the oculus into the open vertical center of the room and radiates out laws and power, and impresses his or her concrete form on the surrounding walls in forms such as statuary, written verse, or painting.

Formwork is the work of a concrete form that has been moved and removed. The Romans took full advantage of the mobility of formwork in their alternating or daisy-chain use of formwork arches, each distributing its motion to the next in a series of mutual supports like those found in their great aqueducts and bridges at Pont-du-Gard in Nîmes, France (ca. 16 CE); or the gateway at Porta Maggiore in Rome, Italy (ca. 50 CE); or the Flavian Amphitheater in Rome (ca. 70–80 CE). Formwork thus not only has concrete form but also has a more mobile concrete form than the molded arches it forms.

In formwork is thus the mobility that gives rise to the concrete form of the mold, and when the formwork is *re-moved*, it appears as an abstract and insensible memorial form. The memorial character of the arch form is well attested in Roman architecture like the Arch of Trajan, Benevento (ca. 114–118 CE), the Arch of Titus in Rome (ca. 81 CE), and many others erected all over the empire as centrifugal expressions of its political power and mobility, all the way to its peripheries and ports (see figure 8.2).

The arch is a memorial structure because it stands after the original model formwork has been removed. The arch itself is a memory or mental affection of a now missing form. The formwork, like time and speech, passes, moves, and decomposes, but the arch, like writing, persists and retains a concrete form derived from the initial creative act of the now abstracted and remembered model. The invisible abstract is remembered as the creator of the visible concrete.

This is why so many great arches, vaults, and domes are decorated with the idealized forms of deities and heroes. However, all these supposedly abstract forms have their origins in an originally removed and mobile formwork. Matter produces form, which then reshapes matter.

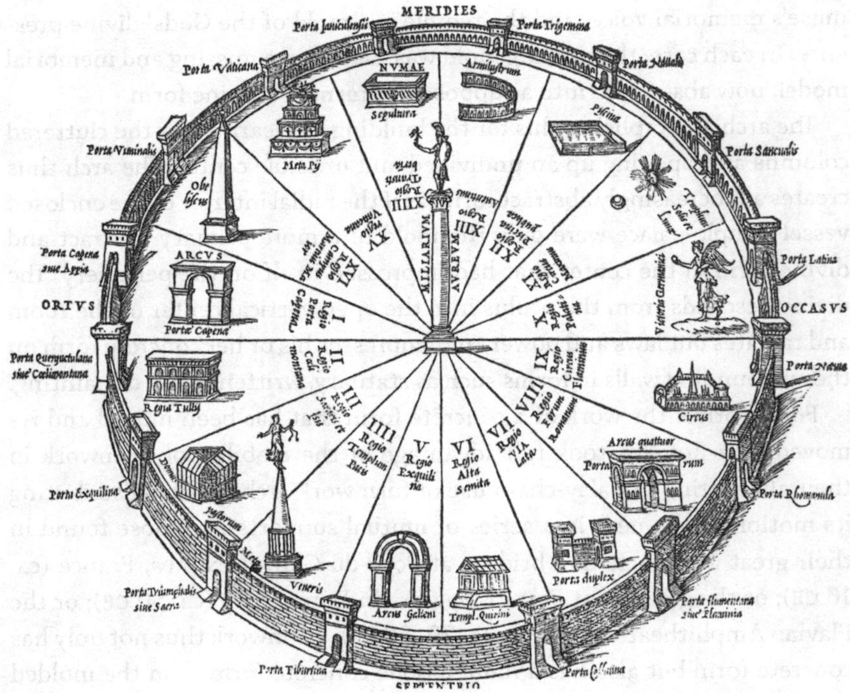

Figure 8.2 Rome as the center of the world; engraving (1527)
Source: From Peter Sloterdijk and Wieland Hoban, *Globes: Macrospherology* (South Pasadena, CA: Semiotext(e); and Cambridge, MA: MIT Press, 2014), p. 423.

THE CHORDOPHONE

Ancient musical kinesthetics are defined by the same introduction and rise to dominance of a centrifugal and formal regime of motion. This occurs most strikingly with the invention of the chordophone, or stringed instrument, like the harp and lyre, known to the Sumerians as early as 3,500 BCE,[12] to the Egyptians as early as 3,000 BCE,[13] to the Minoans around 2,000 BCE,[14] and eventually to the Greeks.

Body

Just like the other major ancient arts, the introduction of stringed instrumentation was defined by the four kinesthetic operations of centrifugal motion. The first operation acts directly on the material structure of the drum and flute, which were defined by the centripetal accumulation of matter flows into a hollowed-out vessel or resonance chamber. To this hollowed-out chamber the chordophone adds a vertical extension on which a series of strings are suspended between a bridge and a yoke. The sound box, bridge, and yoke form a new circular surface divided between the periphery (bridge/yoke/box-bowl) and the central strings that float vertically above the center. The first operation thus introduces a division between the central vibrating strings and the peripheral resonating bowl.

Strings

The second operation adds a hierarchical relationship between that which moves (the strings) and that which is moved (the bowl). The strings are plucked or strummed, and the body vibrates. A hierarchy thus emerges in the lyre between the vertical, rectilinear strings and the horizontal or curved drum body. One plays the string to move the body instead of moving the body directly to produce sound. The strings move the body not by the direct centripetal input of air or vibration by the player but rather, through the indirect vibration of the strings themselves.

The chordophone is thus defined by a relatively indirect vibration. The player moves the strings, but the hovering vertical strings are the ones that move and transform the body. In this sense, the lyre is a much more ethereal instrument than the prehistoric drum or flute, whose action was direct. After it is strummed, the lyre literally plays itself in its sustained vibrations in a way that the flute or drum cannot. The hierarchy of strings over wind and tympani is especially clear in the Greek myth of Apollo and his lyre, who brings clarity, truth, and reason to the Delphic nonsense of Cybil's tympani.[15] This, too, resonates with the musical structure of Greek tragedy: stringed Apollo over the fluted Satyrs.

Amplification

The third kinesthetic operation of the chordophone is its transformation and reorganization of the periphery by the center. The box or bowl of the chordophone becomes merely a resonating chamber subordinated to the strings. The chamber thus takes its shape for the sake of amplifying the strings through a single, central opening or hole that focuses and concentrates the sound of the hollow body.

Additionally, the pattern or form of the sounds produced by the various strings actually induces a change in the body of the listener. Different patterns or scales have different effects. The human listening body becomes another resonance chamber, like the box or bowl. The strings float above the body of the resonating chamber, vibrating it and reordering it in its own image. The strings themselves can even begin to vibrate independent of touch if the same frequency is played elsewhere. The transcendent mythology of the self-playing lyre from Apollo has its origins in precisely this kinetic phenomenon.

Like the Greek theater, there is a movement of sound through a centrifugal amplification system. Unlike the drum, flute, and idiophone, however, which are direct mechanisms meant to accumulate sound in their enclosure, the chordophone expels all sounds through a single, central hole. In this sense, the concentrated sonic center takes on a new kinetic primacy in relation to the periphery (figure 8.3).

Scales

The fourth operation of the chordophone occurs when the concrete forms and patterns of the strings are increasingly ordered into diatonic scales, or systems of musical notation, and are subordinated to the increasingly abstract forms of theatrical performance. What began as improvised and heterogeneous patterns start to take on a more fixed, concrete form of graphic and performative art. As early as 3,400 BCE, musical notation and the diatonic scale began to react back and homogenize and order the concrete forms of sound, and not just for the chordophone.[16]

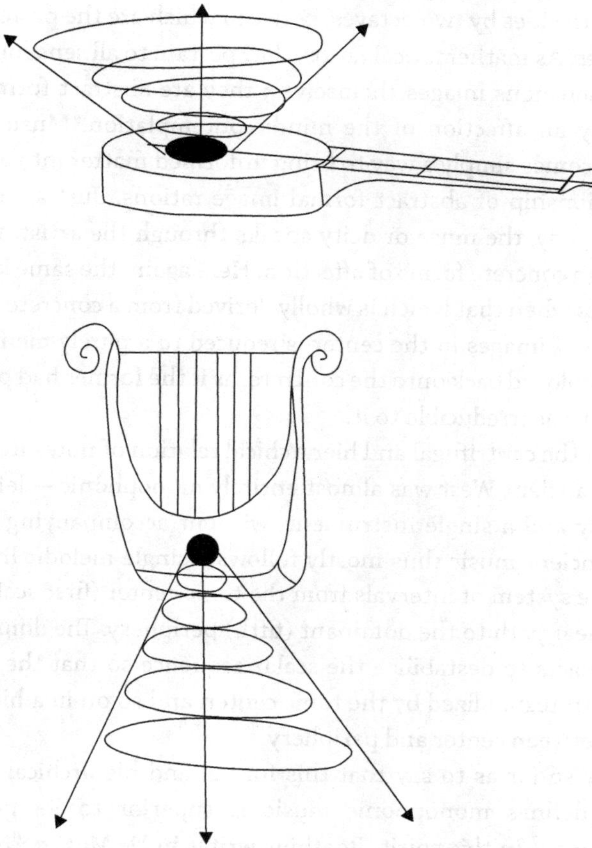

Figure 8.3 Centrifugal amplification in chordophone

Once the concrete, patterned vibrations of the strings enter the resonance chamber, their graphic recording and ordering into scales begins to invert the kinetic relationship. The scale and notation form now persist after the tones are removed from the resonance box. Through notation, predetermined diatonic relations, and the dramatic subordination of musical forms now seem to precede any concrete sonic form of the instrument. For example, as early as ancient Sumer, the lyre box contained instructions from the gods for playing predetermined forms for the souls of the dead. Predetermined ritual songs and forms of motion based on delimited scales, through ritual performance, are defied into abstract forms or ratios that precede the actual sonic images played in the concrete patterns.

This process reached its apex with the Pythagorean mathematization of fundamental musical intervals: 1:1 = unison, 2:1 = octave, 3:2 = fifth, 4:3 = fourth within the octave. The Pythagorean diatonic scale is thus limited on both sides by two octaves, between which are the perfect fourth and fifth tones. As mathematical ratios, they pertain to all sensuous images but are not sensuous images themselves; they are abstract forms understood only by an affection of the mind: contemplation. Music thus increasingly becomes simply a way to bring unformed matter into a properly formed relationship of abstract formal image rations. Just as in written verse and tragedy, the muse or deity speaks through the artist, who then works through concrete forms of affection. Here again, the same kinetic inversion occurs when that which is wholly derived from a concrete sensuous accumulation of images in the center is reduced to a purely mental image and then redeployed back onto the concrete, as if the former had preexisted the latter and was irreducible to it.

Because of the centrifugal and hierarchical relation of notes in the scale, music of the ancient West was almost entirely monophonic—defined by a single melody and a single instrument, without accompanying harmony or chords. Ancient music thus mostly follows a single melodic line within a determinate system of intervals from the tonic center (first scale degree) through a linear path to the dominant (fifth) periphery. The dominant periphery then acts to destabilize the scalar sequence so that the sequence may be in turn restabilized by the tonic center, and so on in a hierarchical circulation between center and periphery.

Plato goes so far as to say that this formal and hierarchical scalar relation that defines monophonic music is superior to its polyphonic destabilizations.[17] In this spirit, Boethius writes in *De Musica*, "Nothing is more proper to human nature than to abandon oneself to sweet modes and to be vexed by modes that are not."[18]As Plato writes in the *Timaeus*,

> Music as, through its sound, is useful for hearing, this much was given to us for the sake of attunement. And attunement, which has coursings akin to the circuits in our soul, has been given by the Muses to him who makes use of the Muses with his intellect—not for the purpose of irrational pleasure (which is what it's now thought to be useful for), but as an ally to the circuit of the soul within us once it's become untuned, for the purpose of bringing the soul into arrangement and concord with herself. Again, because the condition becomes unmeasured in us and deficient in grace for most, rhythm too was given to us by those same Muses as our companion in arms for the same reason.[19]

For Plato, it is because matter is kinetic and flows pedetically that our bodies become deformed or detune our soul out of its properly formed and hierarchical monophonic ordering. Music, following the proper ratio given in advance by the muses and untainted by sensuous enjoyment, is thus able to retune our soul in harmony with the divine, good, beautiful, and true. Music, for Plato, is thus stripped of its synesthetic foundations in sensuous images and is transformed into an abstract form: a pure mental image.

PHARMAKON

The final major ancient image we look at in this chapter is the pharmacological one. The ancient arts of cooking, perfume, and medicine—emphasizing the images of taste, smell, and touch—are also defined by a centrifugal distribution of images into dominant forms. This occurs primarily with the historical development of the recipe form: a determinate list of ingredients combined in a certain form, pattern, or ratio in order to transform/restore (poison or cure) a moving body. The recipe form goes back to at least 3,900 BCE, with the ancient Sumerian beer recipe handed down to humans by the god Enki in a hymn to Ninkasi, the goddess of brewing.[20] Throughout ancient Sumer, Akkadia, and Egypt, numerous recipes and facilities for beer- and bread-making have been found.[21]

A similar rise of medicinal recipes emerges from the code of Hammurabi and in ancient Egyptian medical treatises defining the body as a circulatory pattern of hot, wet, or dry flows to be formed by purges, enemas, washings, and perfumings.[22] Egyptian drugs ranged from leeks to the fat of the hippopotamus, and from pomegranates to fried mice and lapis lazuli.[23] The Greek *Hippocratic Corpus* is filled with recipes for foods and herbs, as well as instructions for actions, exercises, and medical operations.[24] Recipes for perfumes also existed in ancient Sumer (ca. 3,500 BCE),[25] Egypt (ca. 2,498–2,181 BCE),[26] Greece, and Rome.

What ancient cooking, perfume, and medicine all share in common is their increasing formalization into a recipe pattern or φάρμακον, *pharmakon,* which is a recipe form (ingredients and instructions), "a means of producing something" defined by its ability to both poison and heal.[27] The related Greek word *pharmakos,* meaning "sacrifice," is thus related to this production form insofar as sacrifice, like the recipe, is a supplement used to restore an original divine, medical, or sensory balance. The divine purified center is preserved only on the condition of the built-in sacrifice of the periphery. This increasing formalization of cooking, perfume, and medicine into the recipe form occurs through four material and centrifugal operations.

Essence

The first kinesthetic operation of this *pharmakon* is an increasing division between concentrated and unconcentrated matter flows. During the Neolithic, vast and diverse matters were gathered from the periphery into centralized vessels, but during the ancient period, these gathered matters underwent a transformation in their centers through cooking, fermentation, sublimation, and concentration that distinguished them from the original gathered materials.

The ancient world introduced the first formalizations of foods into ingredient combinations and cooking operations through these processing techniques. This had the first kinesthetic effect of producing a rarefaction and concentration of a pharmacological "form" in the vessel that was distinct from an evaporative periphery separated, "sacrificed," or melted off from this concentration. For example, curd is divided from whey (for making cheese), essential oils are divided from plant fibers (for use in perfumes), fats are melted from meats (for cooking), and solid matters are divided from their liquid evaporative aspects (concentration). Recipes are defined precisely by the division between essential and central flows collected at the central bottom of the vessel and the inessential peripheral flows burned off into vapors. That is, one part is sacrificed in order to purify the remaining flow. Cooking, medicine, and perfume all originate in the same kitchen hearth—the earliest alchemical laboratory.

Chemical Apparatus

The second kinesthetic operation is defined by an increasing hierarchy or importance and purity of the center over the periphery. Since the central essence takes time and includes a labor process of preparation evidenced in the vast chemical apparatus of Tepe Gawra in northeast Mesopotamia (ca. 3,500 BCE),[28] medicine, potions, perfumes, and prepared foods become more rare and rarified—and thus more expensive and valuable. Concentrated and centralized flows are often more nutritious, durable, preserved, and fragrant than their peripheral counterparts. Preserved foods, medicines, and perfumes were highly desired because they were portable concentrated versions of their unformed original matters.

For example, the perfumes of Cyprus were very expensive and were traded all over the Mediterranean for religious ceremonies and funerary rites. As the book of Exodus states, "You should make of these a holy anointed oil, a perfume mixture, the work of a perfumer; it shall be a holy anointing oil."[29] Noah sacrifices "of every clean animal and every clean bird . . . the Lord smelled the pleasing odor."[30] For the Greeks, sacrifice to the gods was also linked to the burning of fat and thus to a mix of cooking, incense, and divine medicine that occurs at the same time.

Regimen

The third kinesthetic operation is defined by the reorganization of the periphery by the *pharmakon*, or recipe form. This occurs through the use of regimen: a diet of food, a habit of exercise, a prescription of medications or perfumes. The idea of the regimen is that there are different concrete forms or patterns of images (ingestion and activity) that can produce corresponding patterns or forms of health in the body. What remains typically obscured in aesthetic theories that fetishize vision is clarified in the kinesthetics of the regimen: the fact that art is a mutually transformative process. For this reason, regimen, food, and perfume are rarely said to be representational but are simply interactive, as all art truly is.

The recipe form is therefore not just different from the body it is applied to; it also reacts back on the body and aims to reform and reorganize it in new ways (poison or cure). The central, concrete form of the *pharmakon* radiates outward from its concentrated center and diffuses itself into the periphery of the milieu, the body, the environment through evaporation, dissipation, and metabolic absorption. In radiating out, the *pharmakon* interacts with the body and environment, transforming them and being transformed by them. In short, the concrete form of recipe, through the regimen of its usage, reorganizes and reforms the body in a new way: poison and cure at once, effect and side effect at the same time.

Balance

The fourth and final kinesthetic operation is the retroactive creation of the periphery by the central recipe form. The *pharmakon* is defined by a particular concrete form, but this concrete form has as its counterpart a form of health with which it forms a perfect balance. In other words, the *pharmakon* is the cure to the illness that it presupposes. In a dramatic kinetic inversion, the formalized essence of health and regimen of the recipe form, which is produced through a centripetal concentration, reacts back on its periphery as if the body and world were defined by a fundamental imbalance or lack that needs to be filled by the *pharmakon* itself. The invention of this form of the *pharmakon* thus presupposes a world of imbalance, a world that the *pharmakon* seeks to restore to its original balance.

This type of *pharmakon* reaches a relative height of abstraction in the Greek *Hippocratic Corpus* (420–370 BCE), which characterizes the body and the world as made of a balance of elements and fluids that can get out of balance and move in the wrong way owing to the deficiency of matter itself. The *Corpus* thus renders explicit what was always implicit in the very idea of the recipe form. The recipe forms matter in such a way as to restore a presupposed healthy balance to the world and body.[31]

Sacrifice, or *pharmakos*, is therefore built into the system. Insofar as there was once a perfect model world and body that became imperfect and lacking, there is also a way of temporarily redistributing this lack in the form a recipe, a concentration, and an evaporative process that gives back the peripheral vapors to the heavens and reconcentrates the useful and missing essences back on earth. Burnt offerings and perfumes to the gods restore divine balance through vertical evaporation (divine enjoyment) and local condensation (terrestrial enjoyment). The gods smell the fat; the humans eat the meat. Cooked or fermented food ratios sacrifice their waste. For example, beer occurs when grain gets wet and ferments. The chunky waste product (mash) is sacrificed to Ninkasi so that the concentrated liquid may be centralized through the straw and enjoyed as intoxication. However, these are only temporary fixes for a fundamentally abstract problem of asymmetrical and centrifugal imbalance. The price of a cure is the cost of illness, and imbalance is a sacrifice to ensure balance.

In short, the mental image of a prior abstract form of health appears as a necessary prerequisite for the problem to which the concrete form of the recipe is the alimentary, medicinal, olfactory cure. The perfect ratio of the *pharmakon* restores the imperfect ratio of the world and body. This logic of balance reaches an extreme articulation in Plato's theory of olfaction, where human sensation itself is already an unstable and inferior copy of a perfect, model, and immobile world. Sensory images will not restore the imbalance; only contemplation of the insensible mental image or form of perfect health itself will accomplish this.[32]

CONCLUSION

All the new and dominant arts of the ancient West are defined by these four kinesthetic operations that emerge, recede, conflict, and coexist throughout the period: division, hierarchy, reorganization, and re-creation. Together, these four operations increasingly order the circulation of images into concrete forms or kinetic patterns. Once these concrete forms emerge through centripetal accumulation, however, they begin to react back on the affective functions that created them. Form begins to subordinate function, even though it kinetically relies on it for continued support. Finally, toward the end of the ancient period, form begins to seem as if it had actually preceded the concrete kinomorphic bodies from which it emerged. The mental image, memory, or idea of form begins to appear divine, epic, or ontologically pure and primary.

In short, abstract forms such as gods, heroes, or ideas seem to express themselves in concrete images while also existing beyond them. In fact—and this is the thesis of chapters 7 and 8—concrete and abstract form are nothing other than kinesthetic patterns of motion derived from the centrifugal activity of the model on the mold and the mold on the copies.

The older centripetal and functional aesthetic operations of prehistoric art such as vase/wall painting and rock sculpture continue to persist, mix, and merge with these new ancient arts.[33] However, once these ancient aesthetic forms begin to multiply, a whole new kinesthetic regime of motion begins to emerge with the rise of the medieval period to grapple with the plurality and relation of formal images among themselves. This is the subject of the next chapters on the medieval and early-modern image.

SECTION C

The Relational Image

The third major kinesthetic field to rise to dominance in the West is the relational field. This third field rose to increasing prevalence over the course of the medieval and early modern period, roughly from the fifth to the eighteenth centuries. Alongside the persistence of the major classical arts discussed in section B, a number of images also emerged during this period that began to follow a new aesthetic order: polyphonic music, monastic and Gothic architecture, perspective in painting, stained-glass windows and mosaics, epistle literature, and others.

This section argues that the kinetic relation between different aesthetic forms is defined by a "relational image." Relation, however, is not a distinct, autonomous flow that simply occurs between contrasting aesthetic forms. Kinesthetic relations permeate, define, and order the forms themselves. Abstract and concrete forms rise to dominance in antiquity, but the order and relation of these forms to one another only becomes the *primary focus* of art during the medieval and early-modern periods.

CHAPTER 9

Tensional Relation

Relational arts generate images emphasizing the difference and connection between different aesthetic fields. So far, we have seen how functional and formal fields of images emerged according to distinct kinetic patterns. However, what we have not yet seen is the kinetic pattern that begins to emerge *between* the aesthetic fields themselves. If the flow of matter, which constitutes images, is purely continuous, then we cannot simply posit the existence of spatial or temporal "gaps" between aesthetic fields that would differentiate them. There is no empty or nonsensuous background on which discrete functional or formal regimes are distinguished. Within and between the previous two aesthetic regimes there is, in fact, a third regime that links the two and yet keeps them apart: a regime of tensional relation.

The insight of medieval and Renaissance art is to have discovered and literally brought to light precisely the sensuous being of these connective and relational kinds of images. The medieval image thus renders sensible the material-kinetic relations that give aesthetic fields their distinctness and unity by means of various contrasts, tensions, and juxtapositions.

In this sense, what we are calling an aesthetic "relation" is not a secondary distribution of images that emerges only after or in response to the multiplication of aesthetic fields or patterns. Aesthetic fields themselves emerge as distinct fields, different from or similar to other fields, from yet a third relational aesthetic field. There are no such things as isolated aesthetic fields. All fields emerge from a flow of matter folded up into images that circulate into functional or formal fields—as we have seen in the previous chapters.

Similarly, there are no pure relations independent of the aesthetic fields they relate. Circulation is already *a circulation of images*. Both functional and formal fields of sensation thus presuppose this relational flow, but they also obscure it as a nonfunctional and nonformal difference. All aesthetic fields therefore presuppose this constitutive kinetic relation, but medieval and Renaissance art brings it to the explicit foreground of sensation as the constitutive kinetic condition of function and form in the arts.

This third aesthetic regime is quite different from the others that have come before. In prehistoric art, images were centripetally gathered into a functional aesthetic feedback loop, creating a relatively discrete work of art. In ancient art, images increasingly converged in concrete patterns or forms that began to react back and homogenize the centripetal images that produced them. However, the division between abstract and concrete forms poses a new problem for the work of art: What is the sensuous nature of the relation that allows the abstract or divine form to explicate or unfold itself in the concrete, and also allows the concrete to be implicated or enfolded in the abstract? In other words, what is the sensuous relation between two or more aesthetic fields?

In this chapter, we offer first a preliminary and more general definition of kinetic relation, which is then historically developed in the next two chapters. In short, kinesthetic relations or linkages keep distinct aesthetic forms or fields of images together and apart—distinct, contrasted, and yet moving together in ordered correlation. Relation is present in all works of art, but during the medieval and early modern period, relation becomes one of the most primary and constitutive features of the historical aesthetic field.

The goal of this chapter is to prepare a description of the conceptual and kinetic features that define this period of the image: tensional motion, illumination, and contrast.

TENSIONAL MOTION

The relational aesthetic field, dominant during this time, is defined by the continuous distribution of images back and forth between two or more formal aesthetic fields. Tensional, rigid, or linked motion is thus defined by a rigid or inelastic connection or relation between two or more formal circulations. Images of the Middle Ages are transport images that mediate between forms through rays of light, reflective surfaces, written correspondence across distances, and sonic diffractions.

What tensional motion adds to the previous two regimes is simply an inelastic link between regimes. Images, by nature, are always leaking from their fields of circulation and connecting to others. These leaking flows can create a tension or linked relation when the connecting flow is not only produced but also reproduced continuously, back and forth between two separate circulatory aesthetic fields. This holds the two circulations together and apart, granting them a degree of unity and autonomy without releasing them entirely from a mutual subordination to one another.[1]

For example, a formal and centrifugal circulation can be connected to a functional centripetal one through a third flow of images that links the two without reducing them to the same circulation. Tensional movement is a combinatory or mediating motion. Instead of all circulations simply rotating around a single center, multiple centers—each with its own centrifugal or centripetal motion—can move with and alongside each other in a de-centered but shared motion. Therefore, a rigid or tensional linked motion both keeps the center from bleeding by gradation into the periphery and keeps the center and periphery apart as distinct fields, all while participating in the same linked motion.

Tensional motion is, therefore, composed of at least three distinct aspects. The first is a centrifugal and centralized flow whose outward motion has gone out too far and has escaped its circulation. This flow remains distinct from centrifugal motion insofar as it remains a periphery to it, but also remains connected to it as a leaked or escaped flow.

The second aspect is another circulation that centripetally receives or accumulates this linking flow but also redirects it back to the previous circulation, producing a mutual connection or relation between two or more circulations, each of which requires some degree of centripetal and centrifugal motion. However, this mutual connection does not necessarily entail a symmetry of motion. The motions of each affect each other through the link, but some circulations are larger and more powerful than others.

To put this into the historico-aesthetic language of the Middle Ages, the Gothic cathedral ceiling radiates the centrifugal form of the heavenly mosaic outward and downward, but this radiation is only possible on the condition of a more primary flow of light let in from stained-glass clerestory windows rendering the ceiling fresco visible in the first place. The mosaic is formally impressive, but it appears only on the condition of a flow of light let in from the clerestory that links it to the form of the windows, the glass, and the outside. The invariable form of classical antiquity (the mosaic) thus becomes explicitly modulated and connected to others through the manipulation and art of light that illuminates the related forms: window form and mosaic form.

The third aspect of tensional motion is the connecting flow itself. The connecting link of light, for example, binds the two formal circulations together and keeps them apart. Without the affective relation of a linking flow of light or sound, a given kinesthetic field is incapable of sustaining a kinetic relation—symmetrical, asymmetrical, or elliptical—with other fields. In other words, it would remain cut off from its power of creative motion.

Additionally, the linkage goes both ways. A linking connection between abstract form and concrete form allows the concrete form to interpret, internalize, and redirect that motion elsewhere in both vertical descending relations and horizontal parallel relations.

Specific historical examples of this kinetic tension between forms follows in the next chapters.

ILLUMINATION

The relational aesthetic field is defined by the continuous distribution of light. Illumination is a flow of matter that relates different forms as contrasting dimensions of the same woven kinetic background. In the art of the Middle Ages, there is no radical division between aesthetic subject and object, divine and mortal, but, rather, a continuous distribution of light into degrees of lighter- and darker-colored regions. There are no lacks or absences, only degrees of shadow and illumination.

Without the movement of light, there is no distinction between visible forms. Light is what occurs in, through, and between all forms as a constitutive kinetic relation. If forms are not related, they are not differentiated; and if they are not differentiated, they are not distinct forms. Without a kinetics of light, even the most dramatic images disappear into the night, in which all cows are black. Light is what gives aesthetic relation.

Divine Illumination

Medieval Christian theories of light and illumination should, therefore, not be understood as metaphors or analogies for an immaterial nonkinetic or mental substance. Illumination must be absolutely kinetic and sensuous so it can relate the invisible divine to the visible terrestrial. Without movement there would be no communication, revelation, radiance, or clarity—no transmission of mental affection at all. Without the movement of light, image relations could never change, and sensation would be frozen.

The kinetic and sensuous character of light is attested to by many early church fathers. According to the fathers, even before the existence of scripture, God made himself visible in sensuous nature through light. Following Romans 1:20, "For the invisible things of him from the creation of the world are clearly seen, being understood by the things that are made," the third-century Christian theologian Origen (184–253 CE) writes. "I think that He who made all things in wisdom so created all the species of visible things upon the earth, that He placed in some of them some teaching and knowledge of things invisible and heavenly, whereby the human mind might mount to spiritual understanding and seek the grounds of things in heaven."[2] God and Nature are duplex, but for Origen there is also an asymmetry between them. God is the author both of scripture and of nature.

Tertullian (ca. 155–240 CE) expresses a similar aesthetic when he writes, "We conclude that God is known first through Nature, and then again, more particularly, by doctrine; by Nature in His works, and by doctrine in His revealed word."[3] We learn of God first, and naively, through our senses of his works but more particularly through his doctrine in the scripture. Basil the Great (ca. 330–379 CE) declared the material world to be "a training place for rational souls and a school for attaining the knowledge of God because through the visible and perceptible objects it provides guidance to the mind for the contemplation of the invisible."[4]

In the Latin West, it was Augustine above all who played the decisive role in formulating the doctrine of illumination: "The mind needs to be enlightened by light from outside itself, so that it can participate in truth, because it is not itself the nature of truth. You will light my lamp, Lord."[5] Humans cannot see the nature of truth or the Word on the page of the book without the lamp lit by God. God is not reducible to the light, but by releasing a flow of light, he illuminates a shared, linked realm where the truth can appear. "If we both see that what you say is true, and we both see that what I say is true, then where do we see that? Not I in you, nor you in me, but both of us in that unalterable truth that is above our minds."[6]

Truth therefore occurs not simply in one mind or another, subject or object, but in a zone or flood of light. Illumination is a fluid wave or ocean of light. It is the atmospheric condition of visibility within which the sensuous text and other forms can become visible. The light itself is not what is visible but, rather, the flows of light within which the visible, like the text, becomes visible. The natural world and our senses are constantly changing, but the very conditions of their visible changing do not change: light itself, visibility itself. "God has given us sensible signs and spoken words," Augustine writes, "to show us something of the divine."[7]

The Metaphysics of Light

In his *The Divine Names*,[8] Pseudo-Dionysius describes a metaphysics of light in the late fifth and early sixth centuries that had direct influence on the later writings and the Gothic architecture of the famous Abbot Suger (ca. 1081–1151).[9] He argued that "light is the visual image of God."[10] However, the kinesthetics of illumination reaches its highest and most original philosophical formulation in Robert Grosseteste's (ca. 1168–1253) book *De Luce* (*On Light*). In this work, as in *On Corporeal Change and Light* and in *On the Operations of the Sun*, Grosseteste raises the kinoptics of light to its highest possible aesthetic status as the primary corporeity of all being, sensation, and image. In contrast to the classical theory of illumination as "an incorporeal light that is reason and idea" put forward by Plato and Plotinus,[11] Grosseteste presents *a corporeal physics and optics* of kinetic light as purely relational.

In the beginning of creation, for Grosseteste, there is first light. Since "it is light which possesses of its very nature the function of multiplying itself and diffusing itself instantaneously in all directions,"[12] Grosseteste reasons, light "introduces dimensions into matter [giving it extension in space] . . . and acts through the power of this same light . . . as corporeity itself."[13] "Thus light, which is the first form created in first matter, multiplied itself by its very nature an infinite number of times on all sides and spread itself out uniformly in every direction."[14] Function and form thus emerge out of the purely relational flow of light. Light gives birth to them through relations or folds internal to light itself.

According to Grosseteste, after a single point of light spreads out centrifugally to its infinite limit, it returns to the center, folding back over itself and concentrating matter further and further into nine spheres or folds, of which the inner is more concentrated than the outer, and each is in linked tension with the others. Each is not a different substance but rather simply a different degree of condensation and rarefaction of light (*lux*) itself—"so the first body, through the multiplication of its light, is every body that comes after it. Earth is all the higher bodies because all the higher lights come together in it."[15] Every thing is light, but just folded, distributed, and related by degrees of light.

Incredibly, we see all three tensional kinetic operations in the theory of divine illumination. First, God is suffused with pure and perfectly circular (or spherical) inner light, and as pure visibility is invisible as such. From this pure visibility he releases an externalized flow or ray of light outward. Second, this externalized flow of light illuminates the medium of the created sensuous world. Third, this flow of light enters through the senses of the body, folding itself up into a pure interiority such that humans discover it as "their own." Divine illumination thus relies on a triple fold within the same continuous movement of tensional illumination. The flows of light permeate everything like a fluid but also differentiate within this pure continuity folds or kinetic regions held together and apart by the interrelational flow of light itself. This idea of divine kinesthetic illumination remains dominant up to the period of the Renaissance, after which point an increasingly natural illumination emerges and converges with that of the divine. The sun's light becomes identical to God's light.

Kinoptics

Illumination is, therefore, the kinetic process by which the three components of tensional motion reproduce the linkage that holds forms together, keeps them apart, and relativizes their shared motion. Illumination is distinct from the previous two kinesthetic processes that define the image as functional or formal. In the case of the image's primary determination as function, the flows of motion curve inward in an indefinite spiral process of centripetal internalization. In the case of form, the curved flows are closed off into concentric circles, thereby generating formal centers and centralizations. The image's primary determination as relation, however, emphasizes the linkage between distinct circulations and therefore retains a constitutive and luminous structure. All visible forms are bathed in light. They swim and float in it. Illumination simply means that the image cannot be reduced to a single form or even to multiple forms; the formal centers are connected through and move through linking flows of light.

Just as the flows that produce function are not discrete curves but, rather, form a continuous spiral, and the flows that produce form are not discrete circles but, rather, are continuously rotating spheres, so illumination is a continuous enfolding and unfolding that links multiple forms through a single continuous process. Illumination is continuous because relation is immanent to its *relata*. Once the connecting linkage between formal circulations is forged, the relation appears as the primary determination or condition for the forms it relates. It becomes the relatively inflexible flow that binds, separates, and regulates the kinetic reproduction of the forms. Although the connection between forms is kinesthetically derived from and presupposes the existence of the flows and forms being connected in the first place, it appears retroactively primary in a way similar to that of the centrifugal motion of form itself. Once the linkage is forged, the forms appear to be continuations of the linkage itself. Its flows become the constitutive power that moves through them and makes possible all their functions.

Without the relations between forms that allow them to act on one another and persist, there is no motion and no existence. Thus, kinesthetic relation not only becomes synonymous with relation in general but also relation becomes constitutive of the forms being related, as folds in a unifying luminous relation. However, the new unity produced by relation is a folded or contrasting unity of light and dark that also keeps the related forms distinctly and separately folded.

CONTRAST

The medieval kinesthetics of relation are, lastly, defined by contrast. In antiquity, the image moved largely in a centrifugal pattern from a central model to a peripheral copy and aimed for a kinetic balance between the two. The Greek theory of mimesis or imitation was defined not by an identical duplication of an original but by a specific relation of resemblance, analogy, or proportion between the model and the copy. The beautiful copy is the one that is different from the model in just the right ratio. The aim of good mimetic art is to distribute the image in the right balance such that the field of sensation becomes organized in concentric peripheries around a perfect center.

The arts of antiquity thus attempted to resolve the problem of contrast in terms of balance, proportion, or Pythagorean ratio. Accordingly, the relations of ancient art, however, remain binary divisions between the center and periphery—between the model and the copy. The Greeks bridged this chasm between forms by balance, unity, centrality, and linearity: *contrapposto,* musical interval, column spacing, and so on. They solved the problem of formal relation by simply introducing a gap or space between the unchanging forms whose overall distribution is one of balanced proportions, like the great *Canon* of Polykleitos or the golden ratio of the Parthenon.

In contrast, the medieval arts of relation privilege the tension and difference between forms. In addition to fixed ratios and balances, they highlight the movement of formal asymmetries. The contests and *agon* of the Greeks that is resolved through victory (Nike) and balance (Praxidike) is, in the Middle Ages, largely replaced by the sensuous beauty of the struggle itself: the flickering play of light through glass, the suffering of the passion of Christ in the darkened cella, the chewing (rumination) of the book, and so on. Kinesthetically, there are no divisions, gaps, or empty spaces—only bifurcations and folds, only distributions and patterns of relation in motion.

Aesthetic contrast distributes images in a pattern that keeps forms directly and continuously related to one another and at a distance from one another. Contrast emphasizes not the mimetic proportion between the center and periphery, model and copy, or golden ratio, but, rather, the movement of relation between them—that thin line between light and dark, inside and outside, that actively relates the two. Kinesthetic contrast is emphasized in the arts of the medieval surface like the reflecting mirror, the stained-glass window, the thin page of the book, the chiaroscuro canvas, and so on, as we hope to demonstrate in more detail in chapters 10 and 11 of section C.

CHAPTER 10
The Medieval Image, I

The medieval image is defined by the emergence and dominance of tensional relations between aesthetic forms. The forms and functions of ancient and prehistoric aesthetics did not fall away but rather persisted into the Middle Ages as *relata* of the more primary kinesthetic relations that constitute them. For example, sculpture, mosaics, stringed and wind instruments, arches, painting, written verse, and *pharmakon* all persisted into the Middle Ages, but they were also transformed by it. The classical age of centrifugal formalism and the reproduction of models began to give way to a new age of contrasts, tensions, and shifting relations between forms. There was no single revolution of the Middle Ages but rather a continuous variation of images shifting into and out of molds, contrasting and reorganizing them.

There is also no single dominant art of the Middle Ages, only a multiplicity of aesthetic practices that tend to resonate or follow the same tensional kinetic pattern described in Chapter 9.[1] The great accomplishment of medieval, Renaissance, and early modern kinesthetics (which we refer to very broadly as simply the "Middle Ages"—i.e., "between antiquity and modernity") was to introduce a new primacy of kinetic relation into aesthetic forms and functions. Instead of subordinating form to function (prehistoric art) or function to form (ancient art), the art of the Middle Ages subordinates both form and function to a more primary constitutive kinetic tension, contrast, and relation between forms and functions.

This chapter, like the previous historical chapters, is not an encyclopedic attempt to cover all the arts of the Middle and early modern ages, the social-historical context in which they occur, or even an interpretation of their meaning. For this, there are numerous other excellent scholarly sources.[2] Hermeneutics, symbology, semiotics, and other varieties of constructivist methodology are ubiquitous in art history, but this book aims to add a new and distinctly kinetic methodology to the mix.

Here I want to present a kind of realist aesthetics (versus constructivist) and kinetic materialism (versus formal idealism) that focuses on the material kinetic structure of the work of art itself, inclusive of milieu and viewer. What I call "kinesthetics" is a return to the works of art themselves as fields of images, affects, and sensations. The original contribution of this chapter more specifically is to offer a focused study of the material-kinetic conditions of the dominant aesthetic field of relation during the Middle Ages.

The argument of this chapter and the next is that during the Middle Ages, the aesthetic field was defined by a tensional and relational regime of motion. Each of these chapters marshals support for this thesis by looking closely at six major arts of the Middle Ages: glassworks, the church, distillation, perspective, the keyboard, and epistolography. Although empirically quite different and distributed over hundreds of years, each follows the same kinetic pattern or regime.

GLASS WORK

The first major tensional aesthetic field of the Middle Ages is that of glass work. During this period, glass became a crucial medium of illumination. Glass makes visible and sensuous the play and movement of light that distributes and relates kinesthetic forms. Glass is able to make visible the flow of light without blocking it completely or simply letting it pass through. Glass has its historical origins much further back than the Middle Ages. The first glass production (of beads) began in Mesopotamia and ancient Egypt around 3,500 BCE. By 1,500 BCE, glass production had expanded through the ancient world to Crete and the Mycenaeans, and by the first century BCE, glass-blowing had been developed, increasing the speed, affordability, and mass reproducibility of glass vessels.

In the ancient world, the glass arts remained almost entirely subordinated to centrifugal techniques such as "core-forming" the glass around a central hard-sand core model and shaping glass from the inside out through blowing. The glass arts also remained largely subordinated to the vessel form, although around the first century, the Romans did begin to use clear cast-glass windows for some important buildings. However, around the fourth and fifth centuries, glass began to take on a whole new kinesthetic distribution. The blown vessel was cut open, laid flat, and broken up into *tesserae* [cubes] for mosaics and cut shapes for stained-glass windows. The use of glass in mosaic, stained-glass windows, and mirrors then made possible a whole new aesthetics of luminous relation that rose as a predominant image in the Middle Ages.

Mosaics

Like glass, mosaic had its origins in Mesopotamia in the third millennium BCE. The earliest mosaics are those found in the temple building in Abra and consist of colored stones, shells, and ivory. Around the second millennium BCE, the first glazed tiles and bricks were used in Mesopotamia,[3] and around 1400 to 1200 BCE, Mycenaean pebble mosaics emerged at the great citadel of Tiryns.[4] From here the Greeks, and eventually the Romans, continued to use mosaics in both patterned and figurative forms in temples, palaces, and residences for the rest of the ancient period. However, it was not until the Middle Ages that mosaic became the material par excellence of aesthetic order. This new valorization was possible only because of two important and intertwined kinesthetic and material changes to the medium of mosaic itself.

Verticality

The first medieval change to the mosaic form occurred when it was finally liberated from the floor and raised up to the arch-vaulted ceilings of the church. Throughout the ancient world, the mosaic remained first and foremost a technique used for decorating floors and less often walls, although later the Romans did experiment with ceiling mosaics at *Domus Aurea* (64 CE). On the floor, the materials of the *tesserae* were limited to durable materials such as marble, whose black-and-white color range was used to create patterns and figures. Other colored stones were also introduced to produce a polychromatic mosaic, as in *The Beauty of Durrës* (fourth century BCE) in Albania.

The vaulting of the mosaic also introduces a new vertical relation of distance between the viewer and the mosaic. This verticality draws aesthetic focus to the distance, difference, and perspectival relation between the viewer and the viewed. This distance filled with air, smoke, and light must be "seen through" as a sensuous medium or image in its own right. The vertical mosaic now appeared to have the form it did because of the distance and lighting itself; light and distance became more explicitly and inextricably part of the image. The light image was meant to be seen through and to shape the forms of the mosaic itself.

The problem of making faces out of blocks of color at a distance, however, could not be solved through the tiny *tesserae* in the classical *opus vermiculatum* style often modeled on the higher art of painting. It had to be solved by increasing the viewer's vertical relation (distance) to the *tesserae.* Instead of making the *tesserae* smaller, the medieval artist invented a whole new kinesthetic relation to it through vertical relation. Instead of acting directly on the object, the medieval artist *acted on the relations of visibility* themselves. Blocks of color became distinct or unified figural forms and patterns only in and through the relation of distance and through the mediums of light, air, and smoke. A whole new atmospheric aesthetics emerged.

Luminosity

The second medieval change to the mosaic form occurred when the material of the mosaic was finally liberated from stone to glass. This is directly related to the verticalization of the aesthetic relation. Once the mosaic was vertically elevated to the ceiling, there was no longer a material need to keep using the same old durable foot-trodden opaque stones. Thus it was not until the vaulting of the mosaic in early Christian churches (ca. 400–500 CE) that the *tesserae* could be made of more fragile glass.

The mosaic was not elevated so that it could be made of glass, and it was not made of glass so that it could be elevated. Rather, the two kinesthetic operations of verticality and luminosity occurred at the same time, reinforcing the apparent necessity of one another. While verticality solved the problem of fine detail through the relation of distance, it encountered another problem of visibility through the relational flows of light that had to illuminate the now-distant mosaic. This problem was in turn solved by the introduction of new luminous relations of shimmering glass *tesserae*.

By covering the curved domes, aches, and walls of the church with reflective glass and highly reflective gold *tesserae*, light could be directed and amplified upward to the self-illuminating mosaic. As the sixth century Byzantine historian Procopios wrote of the golden Hagia Sophia church in Constantinople, "It was singularly full of light and sunshine; you would declare that the place was not lighted by the sun from without, but that the rays are produced within itself, such an abundance of light is poured into this church."[5] Again, the aesthetic problem of how to see the vaulted mosaic was solved not by a direct action on the content of the mosaic (characteristic of ancient formal aesthetics) but, rather, by transforming the relational matrix of light and illumination that rendered its forms distinct yet connected in the first place.

Reflection

A related change occurred in the placement of the *tesserae* as well. On the floor, the mosaic was subordinated to the horizontal plane required by the need for a flat walking surface. On the ceiling, however, the *tesserae* could be placed at nonhorizontal angles, allowing for new shapes, textures, and, most important, a new distribution of light and relation. By placing mosaics on curved surfaces, light could be directed elsewhere in a vast network of self-illuminating mosaic forms.

However, a basic understanding of optics reveals that even a curved surface of glass is more likely to catch light if its *tesserae* are placed at various nonconforming angles. Since the sun is not a static entity, light and its source are in constant motion. Visible form and color are in constant transformation and mutation because of the mobility of sunlight and even of firelight. In architecture, as in mosaic, one cannot expect a homogenous flat or even a curved surface to always have the same illumination, color, or formal relations. Luminous relations are in constant flux and flow. Therefore, the kinoptic aim was not to block or completely control this flow but, rather, to respond to it, to gather and redirect some of it where possible. By varying the angle of glass *tesserae* placement, at least some of a continuously changing daily and seasonal light source could be captured at almost any time during the day.

Tension

Light is in constant motion and thus offers a continuously changing matrix of relations. However, aesthetic forms such as the mosaic images of angels on the triumphal arch of Santa Maria Maggiore, Rome (ca. 432–440 CE) are fixed and mounted. The flow of light and the flow of glass enter into a tension—distinct from one another but also linked by the diurnal arch of the sun. As the brightness and color of light moves and changes across the adjacent clerestory windows, the angelic forms fade, dissolve, divide, and merge again. Two angels standing close to one another may appear as one or two, depending on the relations of light. Relations of light therefore distribute and color forms. These relations occur in the tension of rays of light, like luminous, geometric polygons traced out in sunbeams across the vault.

Light itself is not a form, but it is the relational matrix in which forms emerge and are distributed. For example, the tension between distinct colored tiles depends on the relations of light and distance between the tiles. The more light floods the vault or the closer one stands to the mosaic, the more visible and stark the difference between *tesserae.* The held tensions of plaster between different *tesserae* that define the form of Christ, for example, in *The Good Shepherd* in the Mausoleum of Galla Placidia, Ravenna, Italy (ca. 425; figure 10.1) are the same plaster tensions that define his sheep. Christ and his sheep are distinguished not only by the tensions of plaster that hold them apart but also by the color of the *tesserae* of Christ's luminous yellow robes and the sheep's white wool. Forms are related and distinguished by color and colors are related and distinguished by light. Therefore, light itself introduces both a luminous tension between light and dark and a tension between different colored *tesserae.* In *The Good Shepherd* mosaic in particular, the tensional relations of light are explicitly thematized by Christ's yellow halo and yellow robes, which use the same yellow *tesserae* as outline the concrete forms of the rocks nearby and two of the closest sheep. Light here gives *and* relates form.

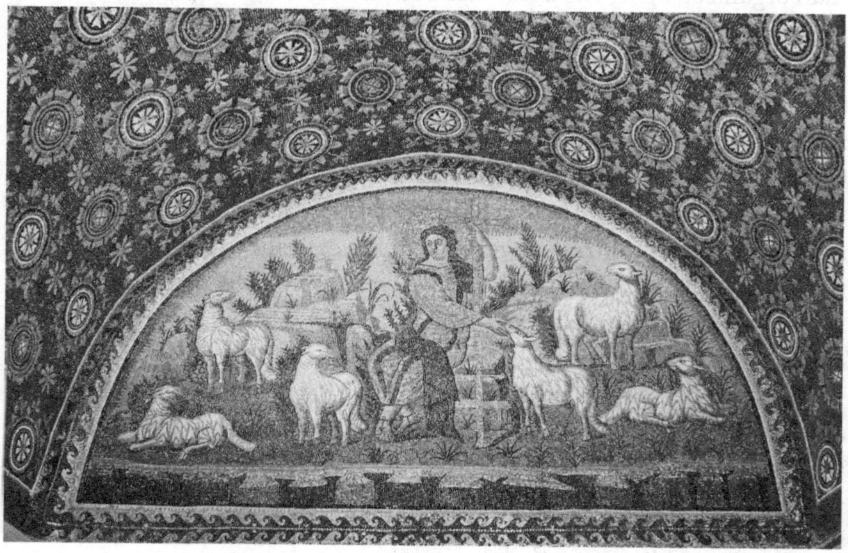

Figure 10.1 The *Good Shepherd* mosaic, in the Mausoleum of Galla Placidia, Ravenna, Italy
Source: Photo by Petar Milošević, Creative Commons Attribution-Share Alike 4.0 International license

Darker tiles are used to show distance and depth, and strongly contrasting tiles are used to create blending at a distance, like the shaded face of the Virgin in the Deesis panel in Hagia Sophia, which is made with blue tiles.[6] Blending is achieved first and foremost through contrast and tension. It brings forms into proximity but keeps them apart in tension by distributing itself unevenly—in this case, among the rocks and sheep.

In mosaic art, kinetic tensions abound that expose the sensuous relations between formal images. The tensions between individual *tesserae* and their colored forms, the tensions between floor and ceiling, and the tensions between dark and light tiles unevenly placed in the plaster.

Stained-Glass Windows

The use of stained-glass windows emphasizes a similar luminous kinesthetic relation of forms. Although the ancients had been staining glass for nearly a millennia, medieval Christians were the first to begin using it in church windows. As an art form, stained glass reached its height in the Middle Ages. As early as the fourth and fifth centuries, Christian churches began using thinly sliced alabaster stone set into wooden frames, producing a stained-glass-like effect.[7] Already here we can locate the search for a relational and luminous matter, a matter shot through with invisible but nonetheless material kinetic tensions. Form becomes nothing other than organized light. The difference between forms is simply the contrasts and tensions between light and dark: pure colors.

Stained-glass windows were built on the luminous discovery of the mosaic and then radicalized in Gothic architecture. Instead of simply reflecting light, stained-glass windows transmit, filter, and transform light. Glass no longer covers the wall but, instead, replaces it. It is, as Abbot Suger calls it, a *lux nova* (new light). This *lux nova* is a kinetic transmutation of form through the contrasting relation of the exterior with the interior. The thinner the limit, the more explicit the relation between two contrasting regions. At some point all matter can be shaved so thin that it becomes diaphanous, like alabaster.

As Hugh of Saint-Victor (1096–1142) writes, "Stained-glass windows are the Holy Scriptures . . . and since their brilliance lets the splendor of the True Light pass into the church, they enlighten those inside."[8] The very form of the interior becomes shaped, formed, and related to the exterior of God's light through the window of scripture. We must, however, resist the theological analogy and see the reality and literal materiality of this sensuous invention. As images, the windows literally, as William

Durandus, bishop of Mende, says, "expel the wind and the rain, that is, all things hurtful, but transmit the light of the True Sun, that is, God, into the hearts of the faithful."[9] The bodies inside are literally illuminated, literally protected with the sun. Stained-glass-window art thus introduces at least three main kinesthetic tensions.

Color Cells

The first tension is between the pieces of color glass held by the lead bars that keeps them together and apart. As Romanesque and early Gothic architecture (950 to 1240) increasingly enlarged the windows, the cellular tensions of the stained-glass segments multiplied and diversified. Just as in a mosaic, it is the contrast between color cells of light that generates the form of the image. Unity appears through fragmentation: a fragmented whole of relations. The stained-glass image appears only because the whole of the light is not given. It is only because light is fragmented in degrees and colors that its contrasts and relations appear.

Interior/Exterior

The second tension is between the lighted exterior of the church and the darkened center. It is only in such a high contrast between light and dark that the stained-glass images can appear at all. If the interior were as light as the exterior, the form would disappear into a single black or deeply colored portal. Only in the highest contrast do the deep colors of the stained glass appear the brightest and clearest. Extreme contrast thus brings out the implicit relations of light required for all visible form. In early stained-glass work, before the Gothic expansion and multiplication of windows, for example, the interiors of churches were deliberately dark so that the deeper and darker colors of the stained glass would still appear luminous. However, once more light was let in through the windows in the fourteenth and fifteenth centuries, the stained-glass colors had to become lighter and more muted to still appear luminous against a lighter background.

Flickering Light

The third tension occurs within the movement of natural light itself as it flickers with the movement of trees or birds outside, or as it lightens and darkens with cloud cover, and so on. The stained-glass window does not block the light but, rather, allows us the contrast of its flickering, shimmering, and alteration directly through the mutation of the color shades in the window. Coloration is the tension between light and shadow. If we were outside we would see the clouds, the trees, the birds, and so on, but inside we see only the pure contrasts of flickering light and shadow. We see the light go black and then return, without seeing why. In other words, we see the scope of the relation as forms appear and disappear in the light without looking directly into the blinding sun itself. We see fragmented blocks of color brought together through their contrasts and as they flicker and fade across the walls and mosaics.

Mirrors

The use of mirrors emphasizes a similar luminous kinesthetic relation of forms. The oldest polished obsidian stones go back to the Neolithic; the Egyptians and Greeks used polished copper mirrors in antiquity; even convex blown-glass bulbs backed with lead were used in ancient Rome. However, from the medieval period to the eighteenth century, the reflective use of light and illumination took on an increasingly important kinetic role in the arts.

Mirrors thus build on the luminous discovery made by glass mosaic and stained glass, and they radicalize it in Baroque architecture. From the early medieval period, we see not a radical break in the dominant kinesthetic regime of tensional illumination but rather an increasing intensity and multiplication of it: from the glimmer of mosaic and gilded halos of Byzantine and Romanesque, to the color light of the Gothic window, to the radical reflectivity and light bath of the Baroque. This final stage was made possible, in particular, by the introduction of lead-backed flat glass in sixteenth-century Venice, derived from the German method of flattening blown-glass cylinders. For the first time, clear leaded glass made possible a crystal-clear reflection of light and image. This brought about several new aesthetic tensions in the relational matrix of form itself: light.

Image and World

The first kinesthetic tension introduced by the use of mirrors in architecture and decoration is between image and world image. This is not a metaphorical or conceptual tension between reality and appearance but, rather, an actual, material, and kinetic tensional network of reflected light. Ambient light literally bounces off the world in a vast network of intersecting polygonal lines and angles, studied in depth by the burgeoning Renaissance science of optics.[10] Within this network of luminous flows of light concrete forms emerge as those regions in the light that absorb and reflect at specific angles and in specific patterns. Visible kinesthetic forms are thus patterns of relations of light. Glass and, above all, the mirror thus enter into this network as a special form whose surface reflects almost all the light that hits it, thus reproducing a complete image of the network.

This special reflective form—a form of forms—introduces an explicit kinoptical tension between the flows of light that compose the world and those that shimmer on the surface of the mirror. The world image and the mirror image are relatively distinct because if they were not distinct, they would be identical and thus have no reflection at all. However, the mirror image also reflects the world but only because the same flows of light that touch the world also touch the mirror. In other words, resemblance is not an ontological or purely psychological phenomenon; it is a material, luminous and relational one. Light gives form through relations of reflection. Visible forms exist only because of the specific intersection of flows of light. Thus image and world are immediately contrasted not as oppositions but as held relations of tensions of reflection back and forth between world and image. The tension is irresolvable because it is the same flows of light within which both are formed and contrasted in the first place. Visible forms are nothing other than the mobility of relations of light in reflective tension. By contrasting something with itself through reflection, the mirror makes explicit the relations that hold both together and apart. The world of images (outside the mirror) and the images of the world (on the mirror) are seen as precisely that: all images. The world reflects light just as the mirror does.

This is nowhere more apparent than Jules Hardouin-Mansart's and Charles Le Brun's Galerie des Glaces (Hall of Mirrors) at the palace of the "Sun King," Louis XIV, in Versailles, France (ca. 1680). There, hundreds of enormous mirrors reflect the incoming light of the opposed windows overlooking the Versailles Gardens and the gilded and jeweled furnishings inside. The Hall of Mirrors at Versailles is a magnificent demonstration of kinesthetic contrast and tension at every level. On one level, it shows visible aesthetic forms to be nothing other than the kinetic relations of shifting and shimmering light bouncing off the reflective golden capitals and crystal chandeliers. Light from outside and light from the numerous chandeliers and golden lampposts reflects off almost every surface of the hall. Every form reflects and is reflected by every other form—a multiplication of images, nothing but images. Versailles is a perfect explication of the aesthetic matrix of its time. Form is exposed as a pure relation of light.

On another level, the hall also holds apart and together the contrasting *grand appartement du roi* (king's suite) with the *grand appartement de la reine* (queen's suite) and the contrasting *salon de la guerre* (north) and the *salon de la paix* (south). The hall is literally the relational linkage that binds the rooms together and apart. The hall is the liminal space of luminous transport and irresolvable tension between contrasting dimensions. In short, it is the architectural structure of relation par excellence.

The basilica form, with its long and increasingly vertical nave, as well as the monastic cloister, become halls of light. Male and female, peace and war, light and dark, are not ontological forms or oppositions but, rather, reflective surfaces bouncing light back and forth through the long mirrored hall. At Versailles, nature is reflected through the hall of light in art; art is reflected through the hall of light in nature. The whole universe becomes both illuminating and illuminated in a vast network of light images.

Infinite and Finite

The second kinesthetic tension introduced by the use of mirrors in art and architecture is between the infinite and the finite. Again, this is not an empty conceptual opposition or metaphorical tension, it is a real material kinetic contrast of flows of light between parallel mirrors. In mirrored Baroque rooms like those of the Salon de la Princesse built by Germain Boffrand in Hôtel de Soubise, Paris, France (1737–1740) and the Hall of Mirrors built by François de Cuvilliés in Nymphenburg Palace Park, Munich, Germany (early eighteenth century), the placement of parallel and angled mirrors exposes the material conditions of illuminated form: infinite optical reflection. Light is continually bouncing off objects and giving them visible form through reflection, but only in the case of the parallel mirrors does this optical condition of luminous relation become directly sensible. Only in parallel mirrors are finite images exposed for what they are: infinitely mobile flows of light.

The mathematical optics of this phenomenon were elaborated in depth by the Italian physicist Evangelista Torricelli (1608–1647). Each time a flow of light bounces back and forth between two or more mirrors, it produces a corresponding mirror image. The longer the distance the light image travels in its successive reflections, the farther away the reflective images appear to be in the mirror. After one second, a flow of light will have reflected about 300 million images, appearing infinite. However,

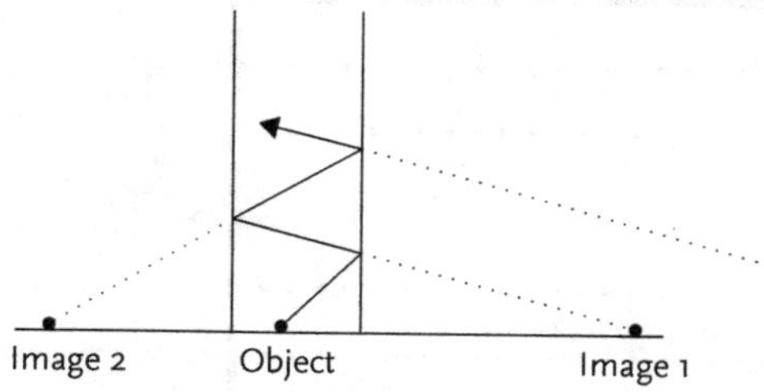

Figure 10.2 The kinesthetics of parallel mirrors

the process keeps multiplying images beyond our range of visibility indefinitely. Thus a finite surface is capable of enclosing an indefinite multiplicity of different images or perspectives (figure 10.2). All this is made into a sensible image owing to the relation of flows of light held in tension between two mirrors.

The kinesthetics of the mirror simply render visible and sensible the structure of optical relations and tensions that define all visible images and forms. Whether we see the relations explicitly or not, they remain the material kinetic conditions of all visible form.

THE CHURCH

The second major tensional aesthetic field of the Middle Ages is the church. The church redistributes images from their ancient centrifugal pattern of model, mold, and copy to increasingly more tensional and complex ones that hold together and apart multiple competing and interlocking aesthetic forms.

The Roman Empire officially adopted Christianity in 380 CE and began building Christian churches throughout the empire based on the architectural style of the Roman basilica or public meetinghouse where civil courts took place. The basilica is defined by its long, rectangular shape; central nave and aisles; and slightly raised platform and apse at each of the two ends. To this structure Christians added a transept that crossed the nave and added a dome to the apse, modeled on the Roman mausoleum and used early on at Old Saint Peter's Cathedral (321; figure 10.3) to mark the tomb of Saint Peter. The transept occurs again later in the north at Saint Gall (ca. 819) and increasingly becomes a model for northern basilicas. In Byzantine architecture, the Roman mausoleum and pantheon dome become the basis of more centrally planned churches such as the Hagia

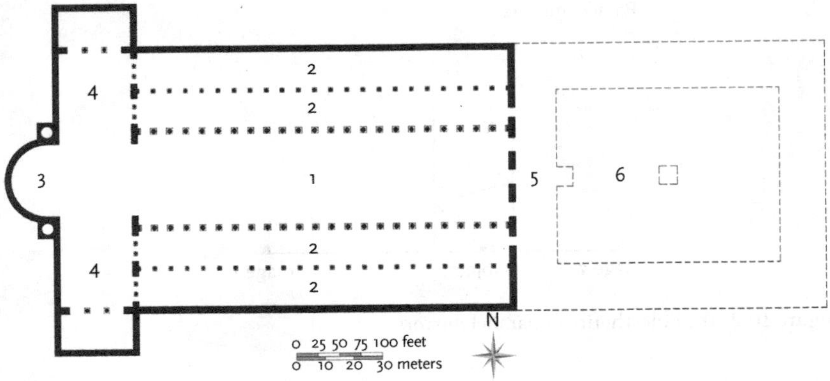

Figure 10.3 Restored cutaway view (top) and plan (bottom) of Old Saint Peter's Cathedral, Rome, Italy; begun ca. 319 by John Burge. (1) nave, (2) aisle, (3) apse, (4) transept, (5) narthex, (6) atrium
Source: From Helen Gardner and Fred S. Kleiner, *Gardner's Art Through the Ages: The Western Perspective*, 13th ed. (Boston: Wadsworth Cengage Learning, 2010), p. 218.

Sophia in Constantinople (537) and the Basilica of San Vitale in Ravenna, Italy (526).

Kinesthetically, Christian church architecture introduces two major tensional relations of illumination: horizontal tensions and vertical tensions.

Horizontal Tension

The transformation of the Roman basilica into a temple structure introduced radical new horizontal relations of light compared to the temples of antiquity. In contrast to the temples of ancient Sumer, Egypt, and Greece, which were all defined by a central and centrifugal darkened cella with a single vertical oculus, the Christian basilica is defined by a darkened interior, but also punctuated by horizontal windows toward the ceiling: the clerestory. In antiquity, the cella remained largely empty as the house for the statue of the god. Burnt offerings were sacrificed on the back porch (the Greek *opisthodomos*), and visual emphasis was placed on the exterior colonnade (peristyle) with its whole architectural order on display (pediment sculpture, volutes, and so on).

In the Christian basilica, however, the cella becomes inhabited and illuminated by horizontal shafts of light shimmering across the ceiling and illuminating patterns of swirling dust and smoke, and the sculpture of the god becomes the tomb of the saint (the mausoleum dome). In the Middle Ages, the binary opposition between the central darkened space (cella) and the luminous outside instead becomes a site of intersecting rays of light playing and forming images on the ceiling. To the central oculus of the ancient god is added a multiplicity of horizontal clerestory angels.

Vertical Tension

The second major tensional relations of illumination introduced by the Christian basilica were vertical tensions between the newly illuminated temple ceiling and the darkened nave floor. In the Christian basilica, the image of God can no longer be expressed as a direct image as it was in ancient statuary. Now God only appears as the pure matrix of flows of light flooding the ceiling and contrasting with the darkened floor. The God of the basilica appears as the paradoxical form of the "empty" vertical space of the vaulted apse and nave. The concrete forms of the floor appear only in the light of the relatively abstract forms of shimmering light on the ceiling.

Again, this is not a conceptual tension; it is a material kinetic contrast between reflective and nonreflective surfaces and the relative movement and distribution of light. Rays of light move more at the top and less on the floor. In fact, these rays only appear in such a striking way precisely because the floor is dark. A contrast or tension is held between the forms on the floor and those above. The play of light requires contrast and relative lack of light movement such that movement appears as relative to something not moving. This is how the basilica distributes the movement of images (figure 10.4). It is a theatre for the play of light. Every lit stage requires a darkened audience. In the basilica, however, the theatrical relation is verticalized—the stage becomes the ceiling and the audience becomes the floor.

In addition to the basilica, the church also expresses this tensional relation in its monasteries, pointed arches, flying buttresses, and new undulating column order. Let's look at the kinetic structure of each in turn.

The Monastery

The first major kinesthetic invention we will look at in church architecture is the monastery. Alongside the rise of the Christian basilica was the rise of the Christian monastery, which was built on the side of many basilicas. The monastery began with the invention of monasticism in Egypt in the fourth century. Hermits would lock themselves in single darkened cells clustered in the middle of the desert. While the Egyptian peasantry of the third century had a tradition of solitary desert hermitage. or *anachoresis*, as a form

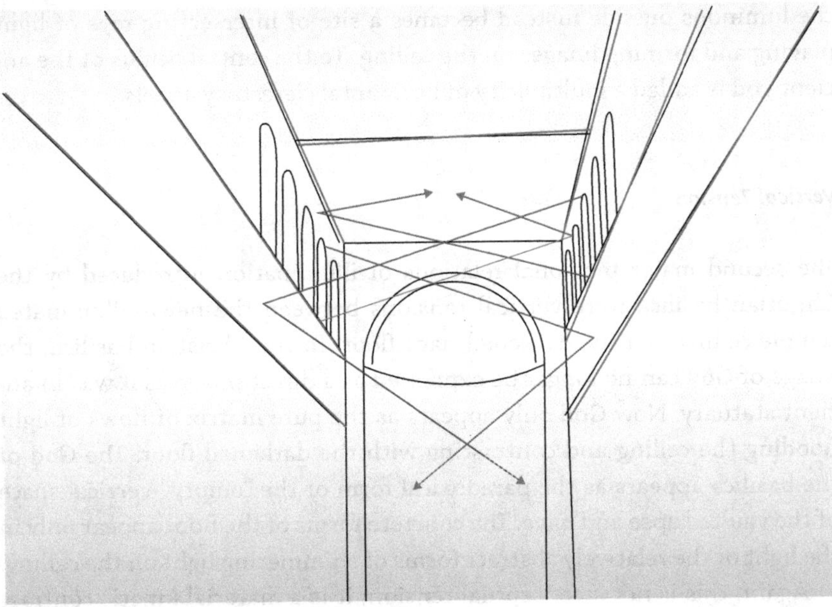

Figure 10.4 The kinetics of light in the basilica

of protest against tax collection and other social evils, in the late third century this practice was adapted by Saint Anthony into an ascetic Christian practice. In 370, Saint Basil linked asceticism, manual labor, and education under monastic rule for the first time.[11] Saint Athanasius brought these practices to Europe in 371, starting with several monasteries in Italy and then spreading to Spain, North Africa, and southern and central France.[12] By the seventh century, monasteries extended from Africa to Ireland, and introduced a radical new architecture of tensional enclosure between isolated cells.[13]

The goal of monasteries such as those used by Saint Pachomius and Saint Basil was not to confine mobility into cells but rather to invent and legislate a linked relation of mobilized confinement. Monks were not simply locked in their cells; they were also mobilized through cloistered hallways in all kinetic activity: prayer, food, drink, chastity, poverty, work, study, the renunciation of wealth, and so on. While eremitic monks, or "hermits," lived alone in a hut or cave, cenobitic monks, or "cenobites," lived together and thus required multiple cells.[14] Alternatively, the Carthusian Order, founded in 1084, combined eremetical and cenobitic life.

The Carthusian monks lived in cells arranged along three sides of a courtyard. Each cell had a room for work, a room for prayer, a bedroom, and a miniature garden. Meals were prepared by lay brothers and passed in to them through a hatch. The monks left their cells only at night to worship together in the basilica church.[15] The monastic system of linked enclosures distinguished the cells from one another. Every cell had its rank and classification (sleep, work, eat, pray, heal), and they all formed intervals in a continuous circulatory relay system guided by strict juridical limits.[16]

Cloister

In particular, the monastery introduces a new kinesthetic tension between linked and enclosed monastic cells through the relational structure of the cloister. The cloister is the tunnel of covered arcades that binds together and holds apart the cells from one another, as well as holding the cells apart from the central quad or garth in the courtyard. The images of light, sound, smell, and temperature are thus all divided between cells but also connected through the relational cloister hallway.

The cloister is therefore an inversion and multiplication of the Greek temple structure. This results in a transformation of the architectural space from centrifugal to tensional. On the one hand, the exterior peristyle around the ancient Greek temple that encircles and radiates outward from the darkened, enclosed cella is inverted. In the monastery, the peristyle instead occurs *inside the walls* of an open, *unroofed quad*. On the other hand, the darkened cella of the Greek temple is multiplied along the periphery as isolated individual cells or rooms. The Latin word *cell* comes precisely from the same word, *cella*, that describes the ancient inner temple.

However, this is not a simple or mere inversion, precisely because the cloister *adds the long arcade hall*. This hall holds the garth and cells together and apart through a series of arched, colonnaded, and windowed openings that let shafts of indirect light and air from the central court into the resonating chamber of the long hallway. The cloister becomes a hall of reflective shimmering light, especially when stained-glass windows are used. The cloister, like the Hall of Mirrors at Versailles, becomes a transport medium of luminous relations. The hall of light and color defines the contrasting visual forms of light in nature and the darkened monastery, as well as the distinct cells along the corridor, held together and apart in and through the light itself. Monastic mobility occurs in, through, and along an entrained flow of light and color that contrasts dramatically with the confinement of the darkened cells (figure 10.5).

Therefore, the open garth or monastic courtyard cannot be defined by a purely centrifugal radiation of direct sunlight but, rather, by a tensional hallway of indirect filtered and flickering light. In direct sun, there is only

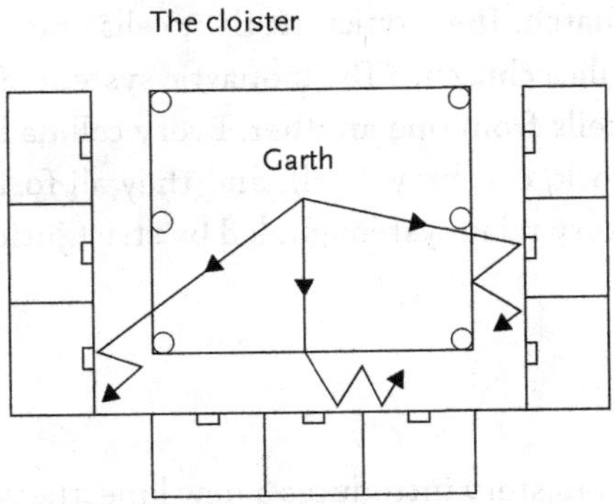

Figure 10.5 The kinetics of the cloister

blinding light, but in the diffused colored light of the cloister windows, light can be seen as coloration, as play, as the mobility and luminous flow that it is in its dust-illuminating shafts shaped by columns and modulated by window tracery. In other words, the cloister makes visible the luminous relations that define the visible forms of the tensional monastic *cellas*. Kinesthetic form is mobilized, multiplied, and even modulated through the cloister arcades and windows.

The Pointed Arch and the Flying Buttress

A second major kinesthetic invention in church architecture is the use of the pointed arch and the flying buttress. Although both the arch and the buttress originated in the ancient world, the medieval world multiplied and transformed them into a whole new regime of tensional motion that exposed the kinetic relations that constitute formal aesthetic differences.

Pointed Arch

Like Byzantine architecture, Romanesque architecture of the eleventh and twelfth centuries, with its thick walls, round arches, sturdy pillars, groin vaults, and large towers, relied on the strength of its walls to support its vaulted ceilings and domes. However, this posed a limitation on the height of the ceilings; the materials that could be used (flammable wood versus stone); and the size of the windows, which weakened the supporting walls.

In the late Romanesque, this problem was solved with the introduction of the pointed-arch rib vault. This allowed for the tension of the vertical weight to be focused more directly downward to the columned corners of the arches instead of radiating centrifugally outward into the walls, as with the Roman semicircular arch. First used in the transverse ribs of the vaults at Durham Cathedral in northern England, dating from 1128, this technique became a defining feature of Gothic architecture from the twelfth to the sixteenth centuries.

The use of the pointed arch and ribbed vaults allowed for at least two dramatic kinesthetic changes. First, with weight distributed directly on the columns and not as much on the walls, ceilings could be vaulted higher and could be made of stone. This made possible an even more dramatic vertical contrast between the darkened floor and illuminated ceiling and air above. Also, once the Romanesque wooden roofs were replaced by stone, the nave and choir ceilings were more directly visible and no longer obscured by wooden beams and rafters. The smooth surfaces of stone allowed for the addition of paintings, mosaics, relief sculptures, and other decorations that could be more clearly illuminated by the clerestory window lighting. The play of shimmering relational and formative light flows could now be seen across even higher and smooth decorated surfaces.

Second, the pointed arch made possible thinner walls that allowed for more and larger windows, creating an even more dramatic flood of light through the clerestory windows. Larger and taller stained-glass windows increased the colored light filtering into the church, as well. Instead of the simple tension between light and dark, the Gothic cathedral introduced a dynamic tension of multiple wavelengths of light that reflected innumerable shimmering surfaces: the *lux continua*. Psalm 36:9 reads: "In Thy light we shall see light." The Gothic cathedral became a slender skeletal structure of multiple luminous tensions between darker and lighter colorations moving through the central chamber. This idea reached full maturity at Amiens Cathedral in France (begun 1220), which rises an incredible 144 feet above the nave floor. The Amiens choir vaults resemble an enormous lantern, with brilliant light entering through the full length of the clerestory and triforium of the choir. The west facade of Reims Cathedral in France (ca. 1225–1290) became folded and porous, less a solid mass than an infinitely folded-up series of windows and caverns.

Abbot Suger similarly describes his remodeled Saint-Denis Cathedral as an "elegant and praise-worthy extension in [the form of] a circular string of chapels, by virtue of which the whole [church] would shine with the wonderful and uninterrupted light of most sacred windows, pervading the interior beauty."[17] Thinner, less weight-bearing walls also made possible a proliferation of window styles such as the rose window, which is a circular stained-glass window, and elaborate window tracery. These new windows further draw attention to the form of the colored light itself. From this perspective, all visible forms can thus be seen as produced through the sculpted distribution of light through a colored opening, a reflective and reflecting quasi-transparent surface. Visible form is multiplied and transformed as the movement of light is sculpted. Form becomes active, kinetic, and material, with its own agency just like the light through which it is distributed in and held distinct from other forms.

The Flying Buttress

The flying buttress was introduced in the late Romanesque at Durham Cathedral but was brought into full employment at Notre-Dame, Paris (nave and flying buttresses ca. 1180–1200) and utilized from the beginning at Chartres Cathedral, France (1194). To hold up the increasingly thinner and taller walls of the new Gothic cathedral, flying buttresses were introduced to counter the outward thrust of the nave vaults. Instead of making the walls thicker, a series of sloping buttress arches were extended from the interior to the exterior of the building, thus making vaults higher, windows larger, and walls even thinner and more porous.

The technologies of the column and pillar, deployed by all the ancient empires, are great testaments to circularity and verticality. However, they are dramatically limited by their lateral weakness. After a certain load/height, they begin to push outward. Thus the flying buttress is the great architectural invention of laterality and tension because it connects a series of circular columns with lateral linkages called flyers. The columns become shorter as they move outward, gradually displacing the tension outward and toward the ground. The flying buttress thus controls and directs the flight/flow of weight and motion downward to the ground.

Verticality thus becomes possible only on the condition of the network of horizontal linkages that support and define its form. Form (vertical vault) and function (weight-bearing kinetic support) are both subordinated to a lateral system of linkages of light (windows) and weight (column support) that flow or “fly” through the windows of the building. The flying buttress is a testimony to the fact that the centrifugal center body of the church can no longer support itself except by the tensional support from an exterior network of relations. The center becomes so large that it cannot illuminate itself except through the tensional rays of light let in through its increasingly large and multiple windows. This is reflected politically in the feudal suzerain system of vassalage.[18]

Sainte-Chapelle in Paris, France (1243–1248), for example, is like a pure luminous gem. Stained glass accounts for more than three-quarters of the Rayonnant Gothic structure. The chapel becomes nothing but infinite formal fluctuation defined by the shifting relations of piercing rays of flying light and buttresses.

The Undulating Order

A third major kinesthetic invention in church architecture is a new undulating architectural order, different from the Doric, Ionic, Corinthian, or combinatory orders. This order is associated with the seventeenth- and eighteenth-century Baroque period, and it originates from the Portuguese word *barroco*, meaning an irregularly shaped pearl.

Kinesthetically speaking, this is a perfect description of the irregular, undulating, shimmering, white architectural surfaces of the Baroque. While Gothic architecture produces tension through a vertical and horizontal contrast of light and dark, the Baroque produces luminous tension through the diagonal, oblique, or curved contrast between concave and convex shapes. Light is not simply filtered and formed through colored glass and projected onto smooth frescoed ceilings; it is increasingly bent, curved, and folded up across every surface of the church. This new undulating order introduces at least two new major tensional kinetic regimes: the shell and the wave.

The Shell

The shell is the first architectural structure, the home of the Paleozoic mollusk and the autonomous house. It is fundamentally defined by its curvature and undulation: convex when viewed from one side and concave when viewed from the other. In Baroque architecture, the shell becomes the surface par excellence from which all architecture and sculpture take form. The straight lines of Romanesque and Gothic architecture are replaced by the curved lines of the Baroque shell. All curves are convex, concave, or both.

Kinetically, this shape changed the whole distribution and play of light in the church. For example, by using stucco and plaster to sculpt the walls and ceiling of the church, the matrix of light rays can reflect and intersect in new dynamic patterns not previously possible. The Gothic may have allowed light in, but the Baroque puts it to work to illuminate the new multiplicity of stucco forms that line the walls and column capitals. Even columns, like those in Bernini's incredible *baldacchino* at Saint Peter's in Vatican City, Rome, Italy (1624–1633), become spiraled, undulating, and flowering from base to capital (figure 10.6).

However, the stucco forms, largely shells, flowers, vines, and other organic folded lacy matter are not simply illuminated. Every aesthetic field is filled with images that are both illuminated *and illuminating*. The whole distribution occurs through the movement of matter. Curves introduce concave shapes that gather light and reflect it back to a converging focal point before scattering it back out again, upside down. This produces the incredible optical effect that the air is glowing just inches from the curved stucco

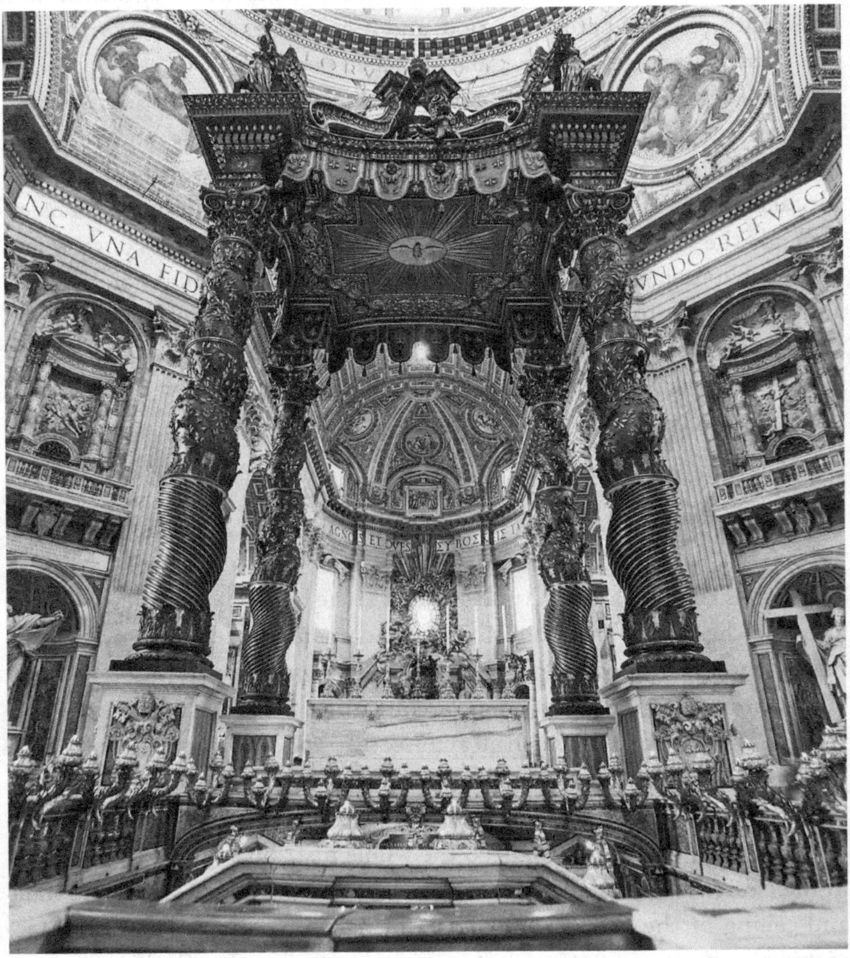

Figure 10.6 Detail of the ceiling, dome, and Bernini Baldacchino or Baldaquin at Saint Peter's Basilica, Rome, Italy
Source: Photo © Jorge Royan, Creative Commons 3.0

surface. Light is literally gathered, modulated, inverted, and redistributed in the concave shell.

Curves also introduce convex shapes that redistribute light in a much wider radial pattern that introduces light into far areas of church that typically lack contrast and shadow. Kinesthetically speaking, Baroque architecture simply continues the trend of illumination, contrast, and tension that began in the early medieval period, and pushes it to its limits by turning the entire surface of the church into a gilded, undulating, reflective surface. Giovanni Battista Gaulli's gilded nave vault of Il Gesù, Rome, Italy (1676–1679), the gilded ceiling of the Hall of Mirrors at Versailles, and many others are excellent examples of this.

In Borromini's San Carlo alle Quattro Fontane, in Rome, Italy (1638–1641), concave curvature is taken to its absolute limits. Here the choir dome is stretched into an oval, surrounded by four more domes on the side, and then every surface in and around them is filled with multiple receding coffered and scalloped concave structures, including shells, octagons, flowers, and scrolls. Borromini thus archives in San Carlo what the parallel mirror achieves elsewhere: an image of infinite multiplicity. Even more important, however, is the complex series of luminous converging focal points produced by so many concave surfaces. Recessed windows in the dome hide the origins of the light, and then the concave surfaces conjoin with it in a spectacular floating and formless luminosity. Here is the height and limit of architectural light—to bring into image the form of the formless, the pure relations of light that make possible the very distribution of aesthetic form itself. Light intersects with itself, producing a tension with itself, and thus exposes the very luminous matrix of all visible form as such. All this occurs through the kinesthetic circulation and tension of light rays (figure 10.7). In a true work of optical kinesthetics, Borromini reveals the material kinetic conditions of the visual image.

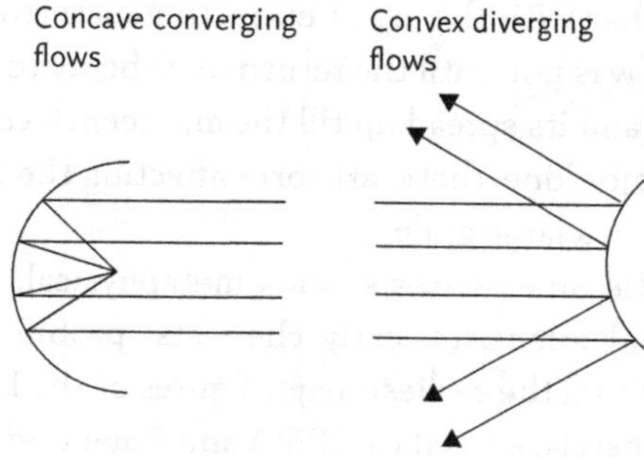

Figure 10.7 Flow of concave and convex light

The Wave

The wave multiples shells into a single continuous undulating surface. Concave and convex forms merge in alternation. The wave thus alternates divergent and convergent relations of light in tension and release. In the façade of San Carlo alle Quattro Fontane, Borromini goes beyond all his architectural predecessors and extends the church façade beyond and behind the lateral plan by turning it into a wave. The cornice ripples up and down and projects forward and back at the same time. On the lower level, the center bulges forward; on the upper level, it recedes into an enormous concave disk. What appears light from one angle appears dark from another, in alternating tension. In the dome of the Chapel of Saint Ivo (begun 1642), Borromini transformed the walls of the choir into concave recesses with stucco reliefs undulating and alternating all the way to the dome. The walls and ceiling become one continuous wave form.

In its most extreme articulation, the architectural waveform becomes so folded up and tiny that the recesses of the inner folds are hidden away from the light, like the micro-relief work on the ceiling of Bernini's Scala Regia (Royal Stairway) in Vatican City, Rome, Italy (1663–1666). The undulations of the wave become so small and so folded up that the contrast between light and dark becomes seemingly infinite and fractal, like a Koch snowflake. Pure continuum produces pure contrast, relation, and tension through folding.

DISTILLATION

The third major tensional aesthetic field of the Middle Ages is that of distillation. During this period, the distillation of chemical essences begins to take on an increasingly important role in the olfactory, gustatory, and medicinal arts. Ancient Greek alchemists writing before the fifth century CE were already familiar with the art of extracting distilled essences from liquids.[19] However, it was not until the return of alchemy to the Latin West in the tenth century and its spread up till the eighteenth century that distillation became a major kinesthetic art form affecting the images of food, drink, and medicine on a large scale.

While the discourse on essences sounds metaphysical, alchemists saw themselves as natural scientists, early chemists, probing the laws and structure of nature. From the earliest major figures of the Latin alchemical arts, like John of Rupescissa (died ca. 1366) and Raymond Lull (ca. 1234–1315), to the practicing alchemists of the modern scientific revolution, like

Robert Boyle (1627–1691) and Isaac Newton (1642–1726), alchemy could not be easily separated from the natural sciences.[20] What was produced during this time were distillations, extractions, and concentrations of certain chemical components from their organic composites, using various systems of evaporative cups and tubes (alembics).

The medieval kinesthetics of fluid essences is different from the ancient kinesthetics of the *pharamakon*. In the ancient *pharmakon*, a perfect centrifugal model of health was presupposed and various foods, medicines, and smells were used to return the dislocated peripheral body to its original centered state. However, in the distillation arts of the Middle Ages, the aim was not to move to and from a central model of sensation but, rather, to harness the purest material flows that constituted the form of a body (its chemical essence) and continuously reorganize this essence into another bodily form (*chrysopoeia*, or transmutation), or to extend the movement of a material essence indefinitely to allow for immortality.

In short, the centrifugal idea of a model body that has moved away from its center, as Plato says, is not entirely abandoned in distillation; rather, the model body appears to just be one image among many, all of which are connected through a continuous flow of "ether" or "essence." Form becomes deformed and variable: transformed. In the art of distillation, the essence of pure matter is unformed, but it becomes formed through its proper transmutation. Transmutation is therefore fundamentally a kinetic relation between two topological forms of the same, continuous modulation of matter. Forms are thus defined by their relations of tensions held together and apart by their relative patterns of motion within a fluid medium.[21]

The kinetics of the distillation apparatus itself is also defined by a tensional relation. Opposed to the ancient art of boiling liquids down to their concentrations, the proliferation and development of the alembic made possible a dual collection and transfer of the boiled-off vapors into separate containers (figure 10.8). The alembic thus allows for a double and linked

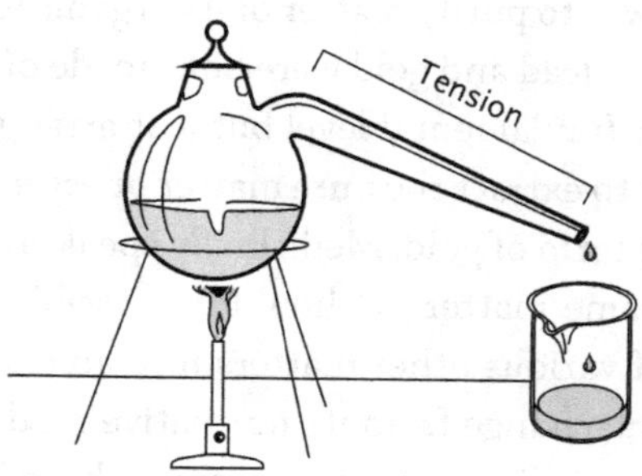

Figure 10.8 Kinetic tension of the alembic

concentration of vapors, on one side, and solids, on the other. Instead of a centrifugal radiation of vapors into the air that leaves a concentrated center, the alembic creates a rigid linking system between two heated or cooled containers. In short, more primary than the mixed composite forms of organic reality is the mobile, fluid, and transmutable essence that distinguishes them, connects them continuously to one another, and makes their form possible.

Additionally, the sensuous appearance of pure alchemical images was described by alchemists of the Middle Ages as pure light or glowing liquid without form.[22] This aesthetic description directly connects the arts of distillation and essence to the larger kinesthetics of light and illumination in the Middle Ages.[23] This luminous liquid of life, or *aqua vitae*, is the natural equivalent of the pure light of God in its healing powers. For example, both Robert Boyle and Isaac Newton argued that "luminiferous aether" was the formless matter of which visible light was made.

This is an alchemical notion that runs from the medieval period all the way through the nineteenth century and even up to the early twentieth century. It is no surprise that the English painter Joseph Wright of Derby would paint his famous work *The Alchymist* (1771; figure 10.9), equating the discovery of the "philosopher's stone," or pure distilled transmutable essence, with the discovery of luminous phosphorus. The full title of the painting is *The Alchymist, In Search of the Philosopher's Stone, Discovers Phosphorus, and prays for the successful Conclusion of his operation, as was the custom of the Ancient Chymical Astrologers*.

This kinesthetic essence image is discovered concretely in three arts of the Middle Ages: alchemy, the order of sanctity, and liquor.

Alchemy

The aim of alchemy was to purify matter of its organic forms through distillation. If, for example, lead and gold were both made of the same luminous matter at their most fundamental level but just arranged differently, then it should be possible to extract the pure matter or essence from lead and redistribute it into the form of gold. Medicinally speaking, if the human body is made from this same matter, it should be possible to distill or extract the pure essences of various other matters into an elixir that would allow the body to move and change from its normative model of health without decay—that is, immortality. This is an approach and methodology that differed quite dramatically from Plato's formalist *pharmakon*. Alchemical

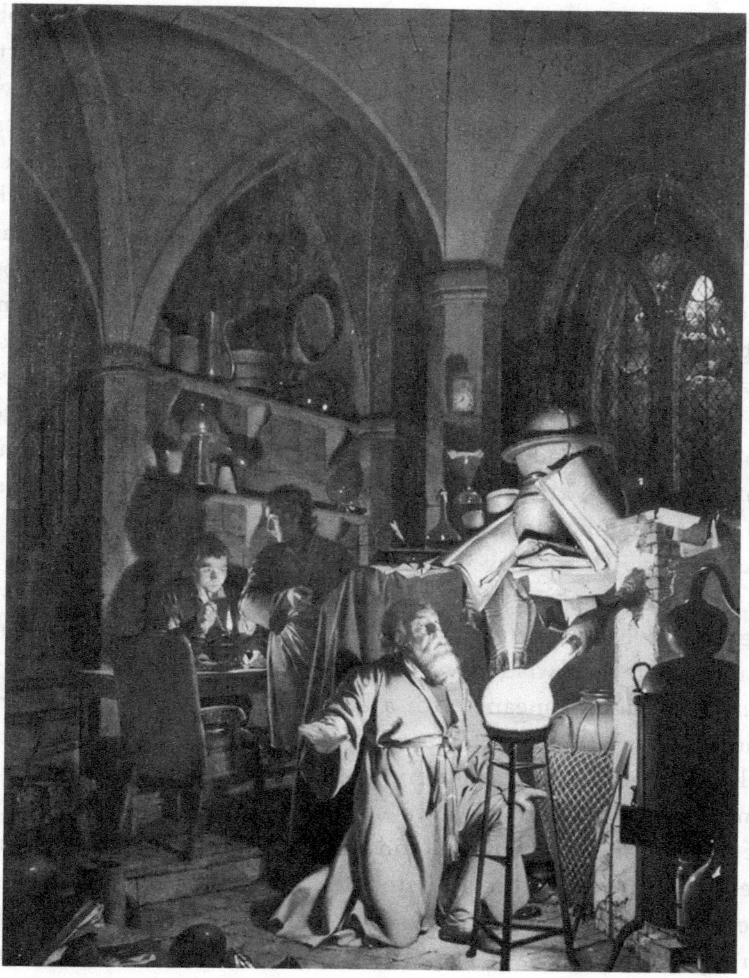

Figure 10.9 Joseph Wright, *The Alchymist, In Search of the Philosopher's Stone* (1771)
Source: Wikimedia, https://commons.wikimedia.org/wiki/File:Joseph_Wright_of_Derby_The_Alchemist.jpg#/media/File:Joseph_Wright_of_Derby_The_Alchemist.jpg

medicine was not just a return to a previous ideal form; it also moved the body forward and ahead toward immortality.

The Oder of Sanctity

The order of sanctity was an olfactory essence emitted by the bodies of saints and prayers, and identified with the soul itself in the Medieval Christian tradition. The smell of a body is the smell of its evaporative essence distilled, if you will, in the nose. Without positing any metaphysical entities here, it is entirely possible to describe the kinetic structure of olfactory religious experience. Rose water, perfume, and incense were all used in the ancient world as vapor images returning to their transcendent origin. While this practice without a doubt continued in the Middle Ages, a distinctly relational kinetic pattern was added in which sacred smell was no longer confined to the temple or even to saints but, additionally, permeated and suffused everything. Scent becomes something emanated by the divine itself, and smell becomes a form of relational knowledge with the divine. Just as alchemy distillation is associated with knowledge, so vapor images are better smelled than seen and become the primary image of knowledge. God communicates to us through his smell, and we communicate to him through our smell in the form of prayers.[24] Ethereal vapor becomes a connective circuit of relation and communication.

As the twelfth-century theologian Hildegard of Bingen wrote, "By our *nose* God displays the wisdom that lies like a fragrant sense of order in all works of art, just as we ought to know through our ability to smell what ever wisdom has to arrange."[25] In fact, the Latin word *sagax* ("sagacious") means both a keen sense of smell and a shrewd mind. Francis of Sales (1567–1622) similarly wrote that "mediation is like smelling first a carnation, then a rose, then rosemary, thyme, jasmine, orange flower, each one separately; contemplation is equivalent to smelling the scented liquid distilled from all those flowers put together."[26] Even into the seventeenth century, the "mystical odorist" Lorenzo Magalotti wrote of the importance of soaking the imagination in a continual bath of perfumes "from which the imagination, impregnated and satiated, will rise . . . imbuing the soul with vapors purified of every vestige of matter, which when they reach the mind act as a pure spiritual suffumigation, inundating it with so unique a harmony that all plurality is banished."[27]

The use of rosewater and rosaries, the latter originally made from roses, by the Christian church was a way of extracting a divine essence and connecting the believers in a mutual olfactory network. The use of incense was not just a return of the soul to heaven but also the smell and knowledge of the divine souls on earth. Unlike the gods of antiquity, the God of medieval Christendom did not appear directly on earth but, rather, communicated and appeared only indirectly through a linked tension of invisible smell images.

Liquor

The first documented creation of alcohol through distillation appeared in the medieval period (twelfth-century Italy).[28] Distillation produces not only new concentrated medicinal and olfactory images but also a new gustatory image of the distilled beverage. Distilled grains or plants have their own distinct taste image when separated from their organic form, which often bears no resemblance to the taste of the cooked grain itself. Once the essence of the food has been purified of its organic form, its "true nature" can finally be expressed in its liquor. By the fourteenth century, the consumption of distilled beverages had risen dramatically all over Europe for both medicinal and culinary purposes, as people began the quest for the pure and true taste of foods distilled from their organic forms. The appearance of such distilled beverages bore a close resemblance to clear, formless *aqua vitae* of the other products of alchemical and perfumed distillation, hence connecting them to health, immorality, wisdom, and illumination, as well. Organic forms come to be seen as merely the incarnations of a more primary and relational flow of vapor images.

The kinesthetic rise of glass, the church, and distillation arts produced a whole new kinesthetics of tensional relation. Something similar occurred in the other medieval arts of perspective, the keyboard, and epistolography, as we will see in the next chapter.

CHAPTER 11

The Medieval Image, II

This chapter continues the thesis of the previous chapter in demonstrating that a kinesthetic shift occurred in the arts from a centrifugal to a more tensional pattern of motion over the course of the Middle Ages. In this chapter, we continue to demonstrate the same thesis but in the arts of perspective, keyboard music, and epistolography. The argument of this chapter is that each of these major fields is defined predominately by a distinctly tensional pattern of motion and a relational aesthetics.

PERSPECTIVE

The first major relational aesthetic innovation we will look at in this chapter is that of perspective. The ancient Greeks and Romans used perspective and foreshortening in painting in a limited way, but these aspects did not become predominant or systematic until the introduction of geometric perspective in fifteenth-century Renaissance Italy. However, the argument of section C of part II is that what I call "kinetic perspective" in art is much broader than geometric or linear perspective. In addition to and more aesthetically primary than the use of geometric perspective is actually the use of kinetic or relational perspective, or the aesthetics of mobile light relations that shift according to the configuration of the subject-object-light matrix that allows the image or form to appear as linear in the first place.

In contrast to the relatively flat forms of ancient painting that floated in an empty space or background, medieval, Renaissance, and Baroque painting increasingly shifted priority to space itself. This not only includes the pictorial space of the canvas or the space between the viewer and the canvas; the space between everything related to the work of art now becomes the primary focus of painting. All three spaces form the kinesthetic trinity of perspective. Fixed geometric perspective alone, typically treated in theories of perspective, fails to capture the agency of light and the mobility of shifting perspective. In his *Trattato della pittura* (1509), Leonardo da Vinci accurately describes geometric perspective as limited and arbitrary *perspectiva accidentalis* soon after Filippo Brunelleschi's rediscovery. Art historian Edwin Panofsky remarks in his book *Perspective as Symbolic Form* that "the structure of an infinite, unchanging, and homogeneous space—in short, a purely mathematical space—is unlike the structure of psychophysiological space: 'Perception does not know the concept of infinity.'"[1]

The aim of linear perspective was to (1) abolish the curvature of the earth in the linear horizon; (2) to create a calculable, navigable, and predictable space like Mercator's flat-projection maps around the same time used by colonial powers; and (3) to posit a single one-eyed and immobile and colonial spectator for whom the earth was a grid of natural, scientific, and social domination.

However, linear perspective, like the fantasy of total aesthetic, and scientific and social mastery, is made possible only on the more primary condition of the relational agencies and motilities of the colonized matters themselves. Painting in the Middle Ages thus relies on the relational and kinetic materiality of light in order to overcome the problem of so-called representational space, as Panofsky describes it, by directly working with and shaping space itself as a relational medium of light and reflection.

Kinetic perspective in medieval and early modern painting is poorly understood as a representation of space but is, rather, a *movement and mutation of space itself* as a fluid medium of light and coloration. "Space," as the fifth-century Greek philosopher Proclus writes, "is nothing other than the finest light."[2] Perspective, as Panofsky rightly observes, "creates room for bodies to expand plastically and move gesturally, and yet at the same time it enables light to spread out in space and in a painterly way to dissolve the bodies."[3] Kinetic perspective holds forms apart and gives them a linked mobility, but at the same time dissolves them into a single fluid medium of light relations. Form emerges through light relations between objects and objects, subjects and subjects, and between subjects and objects, which themselves emerge as distinct forms only in and through the active mutation of a fluid luminous space. In other words, kinetic perspective is not reducible to a single geometric or linear perspective; it is the material kinetic condition of all points of view as such. Points occur only in and on more primary and continuous flows of light.

The art of what I call "kinetic perspective," distinct from linear perspective, emerges historically in at least three major painterly techniques: gilding, oil painting, and chiaroscuro. Let's look at each one in turn, not as a linear representation of abstract space but as a direct expression of the spatial illumination that makes linear perspective possible as in the first place.

Gilding

The first painterly technique is gilding. The earliest predominance of painting based on kinetic perspective occurred from around the fifth to the fifteenth centuries in Europe and is defined by the use of gilded and painting metal, mostly gold and silver. The technique of gilding goes back to ancient Egypt, Greece, and Rome, but during the Middle Ages it took on a dramatic predominance and ubiquity in painted works. Gilding is the application of a thin layer of gold or silver leaf, powder, or paint to surfaces like wood, stone, or metal. This gives them a thin coating of metal that is then burnished or polished to create a mirror-like reflective surface. Gold and silver are two of the most reflective metals.

This technique renders visible a whole new luminous and aesthetic set of relations previously obscured in most ancient tempera (egg-based) or encaustic (wax-based) painting. All pigment reflects light, but no pigment reflects as much light as polished mirror-like metal, whose coloration therefore changes much more dramatically and responsively to the tensional light matrix in which it is situated. This may not be obvious from seeing these works of art today, dulled by time, but historically gilded works of art would have been highly reflective surfaces. Specifically, this technique rose to predominance through the use three kinesthetic techniques: the halo, the icon, and the illuminated manuscript.

The Halo

The first kinesthetic technique of gilding that became increasingly prominent during this time is the gilded halo, or *nimbus* (from the Latin, "cloud") or *aureole* (from the Latin *aurea*, "golden"). A halo is the radiation of divine light from the head of a figure. The earliest uses of halos date back to the Egyptian use of the sun disk, or *shen*, above the heads of certain gods to indicate their immorality, and it can be traced through the golden crowns and radiating lines of light emanating from various Greek and Roman gods, heroes, and kings. However, the systematic use and development of gilded halos only began around the fifth century, with early Christian use applied not only to the image of Christ but also to other saints, kings, and holy figures.

The gilded halo introduces three new kinetic functions into painting. First, the halo renders visible the reciprocal action of light between the painting and the viewer. All paint reflects some light, but the polished halo reflects so much light that every minute change in the ambient lighting, including those introduced by the viewer as he or she moves around the room, is reflected in the shifting reflection and coloration of the halo.

The gilded halo allows the painting to become demonstrably active, mobile, responsive, and transformative. Every halo shimmers in the same fluid medium of light. The viewer sees the action of the world and his or her own body reflected in the painting at the same time as the action of the painting is reflected in the eye that absorbs its light. The whole network of reciprocal luminous action in which forms are related and distinguished is "kinetic perspective." No geometrical view can be as mobile and dynamic as the gilded mirror.

Second, this luminous relation materially expresses the *logos*, or word, of Christ not as a representation of Christ or a linguistic figure but rather as a direct communication of reflected light from the sun through the gilded halo to the world (and our eyes) and back again. Here, the *logos* of God has no representational content because it has become a purely kinesthetic process that occurs directly on the eye as much as on the world itself.

Third, the medieval multiplication of gilded halos on multiple figures in the same painting results in a polymorphic transformation of the painting itself. Instead of a single and centrifugal halo, multiple halos now enter into a tensional relation in which the light emanating from each one is slightly different from the others and yet part of the same luminous set of relations. This has a direct effect on the medieval form of the halo itself: It becomes polygonal. The ancient circular halo becomes triptych in the medieval "cruciform halo" that indicates the Holy Trinity of God, Christ, and Holy Spirit. It becomes squared for certain living earthly figures, such as Pope John VII.[4] It becomes triangular for the Trinity.[5] It becomes hexagonal, dotted, and even star-shaped, like a crown.[6] In short, the centrifugal circle gives way to the relations of light between multiple polygonal figures in tension.

The Icon

The second kinesthetic technique of gilding that becomes increasingly prominent during this time is the gilded icon. In addition to the use of gilded halos, the gilded icon is variously covered in gilded gold—its frame, background, foreground, or all three. On such shimmering backgrounds, the halo becomes just an outline around the figure's head identical to the reflective background itself. Similar effects as the gilded halo are extended to the whole painting, as in *The Ladder of Divine Ascent* (twelfth century; Saint Catherine's monastery) or painted sculpture in the case of the 6-foot-tall Gero Crucifix (ca. 970) covered in gilded gold.

It is no coincidence that many icons, defined by their active and responsive illuminations, are believed to be capable of miracles or otherworldly actions, like the Vladimir Virgin (late eleventh to early twelfth centuries). The luminous quality of the gilded icon reveals not only the particular luminous action of the work of art itself and the forms distinguished on it but, more important, reveals the fact that all painted forms participate in a tensional network of light relations. All painted aesthetic forms are defined by differences in coloration, which are produced by differing reflections of light—demonstrating the unique agency of matter in motion. As ambient light changes, so do the colors in the painting and the light that hits our eyes. The gilded icon has a material agency that is the real source of its spiritual agency.

Illuminated Manuscripts

The third kinesthetic technique of gilding that became increasingly prominent during this time is the illuminated manuscript. An illuminated manuscript is a handwritten book whose images and/or text have been inscribed with burnished gold and/or silver metal. For example, the oldest well-preserved painted manuscript containing biblical scenes is the early sixth-century *Vienna Genesis*. Like many other illuminated manuscripts of this period, the pages of the *Vienna Genesis* are made of fine calfskin dyed with rich purple, and the Greek text is inked in silver metal. The dark background creates a dramatic contrast of highly light-absorbing material (dark purple calfskin) and highly light-reflecting material (silver metal). Through this technique, the visual word of Christ is thus expressed in a literal luminescence. From the background of darkness and obscurity comes the living, moving, shimmering words of Christ.

In the sixth-century *Rossano Gospels*, Christ's halo and robes shimmer in gilded gold, illuminating the audience around him in front of Pilate. Even the covers of illuminated manuscripts were themselves illuminated in gold and shimmering gems, such as the famous front cover of the *Lindau Gospels* from Saint-Gall, Switzerland (ca. 870). Christ lies crucified in reflective gold. At its most intense, however, works such as the "Annunciation to the Shepherds" in the *Lectionary of Henry II* (1002–14) use a massive sheet of burnished gold as the background of the painting.

Illumination is therefore not about representation but rather about the expression of the letter and image themselves as luminous source. Through illumination the book becomes apparently active in the triune and triangular process of moving light from the celestial sun to the book, and then into the eye, and then from the movement of the eye and body to the book and back out into the world. The flows of light that emanate from God are the same flows reflected off the words and images of the book and internalized into the body and consumed by the eye and mouth.

The medieval tradition of *lectio divina* has no other meaning than this: to eat Christ's words with one's mouth and absorb his light with one's eyes. Kinetically, it is the same flow of light that keeps Christ, book, and reader apart and together in literal tension. Distinct aesthetic forms appear only within the triangulated continuity of the God-book-reader perspectival relation. Each is an emitter, consumer, and reflector of light. Each is an image. However, this kinesthetic dramatization is not limited to illuminated manuscripts; rather, it simply demonstrates the primacy of light relations inherent in but dimmer in all other books. All ink absorbs and reflects and thus actively participates in this kinesthetic process of perspectival image. Whatever linguistic or semiotic meaning the words may have is possible only on the more primary kinesthetic condition of its light relations.

Oil Paint

The second use of kinetic perspective emerged later, around the fifteenth century, and is defined by the use of oil paint. Painted perspective emerges not just in the geometrical representations of Brunelleschi but also more fundamentally and materially in the new use of light that oil paint makes possible. The paint itself, the frames, and the optical techniques of painting all give oil painting an explicitly active function, allowing it to change color and form based on the shifting perspectives of the viewer and ambient light. This occurs in three distinct techniques.

Varnish

Painting is an art of reflection, tension, and luminous relation. A painting reflects light from its surface and thereby renders visible distinct forms as patterns or waves of light. It takes ambient illumination, modulates it, and returns it back to the eye in a specific and dynamic pattern of triangulation that shifts according to the position and point of view of the spectator. This is its kinetic perspective. The same light is split into different colors in tension and in contrast with one another, but within the same flow of white light.

While this has always been the case with all painting since the prehistoric period, oil painting was the first to realize and make full use of this discovery. This occurred first of all in the use of oil itself as a medium for pigment, in contrast to egg yolk. While egg yolk dries quickly, is dull, and tends to flake, linseed and other oils dry slowly, with more luster and do not flake as easily. Oil paint is more reflective and more luminous, and thus renders the pigment and surface of the painting more luminous, as well. Furthermore, the addition of a coat of varnish to the exterior gives the painting a radiance and vibrancy unmatched by any tempera or encaustic. The whole surface of the painting becomes a single shimmering panel of light.

This was taken even further by Jan van Eyck (1390–1441), Peter Paul Rubens (1577–1640), and Rubens's teacher, Otho Venius (1556–1629), all of whom directly mixed their pigments with varnish.[7] By painting directly with varnish, the painter uses pigment that takes on a distinctly lustrous, thicker, and shimmering quality, even when dry. Even black pigment becomes reflective under these material conditions. In contrast to the mute colors and short brushstrokes of tempera paint, oil introduced multiple, textured layers of "intense tonality, the illusion of glowing light, and enamel-like surfaces."[8] The texture of these layers, like mosaic tiles, grabs onto stray flows of light and gives the painting a luminous topology of peaks and valleys.

Polyptych

The second technique in oil painting to make use of perspectival light relations is the increasingly important place of the polyptych retable, a frame or shelf enclosing painted panels above and behind an altar. Most of the early great oil paintings were done on multipanel altarpieces. This is not a coincidence. The material-kinetic structure of the polyptych altarpiece visually expresses and redoubles the triangular kinetic structure of the painting itself previously discussed.

To the triangulated relation "light source-painting-eye," the polyptych adds a "left-right-center" triangulation that harnesses the light flows of

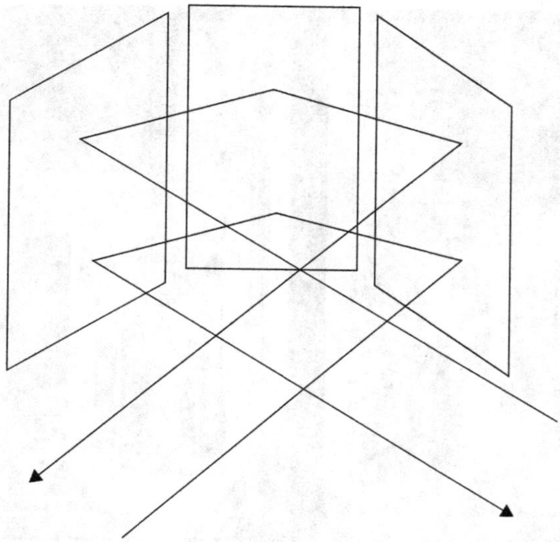

Figure 11.1 Polyptych illumination, painted on both sides

the first triangulation and further reflects and redirects them between its hinged panels. The tensional relation of the hinges that hold the wooden panels together and apart duplicates the tensional relations of light holding the distinct forms of light-painting-eye apart and the left-right-center of the polyptych. The material structure of the mobile, foldable, booklike polyptych panels thus merge the threefold traditions of gilded icons and illuminated manuscripts with that of oil painting.

Just like the manuscript, the polyptych is painted on both sides (figure 11.1). The two become duplicate images of each other. The divine word of God is reflected in the painted surface of the altar, just as it is reflected in the written surface of the Bible. Oil painting thus emerges directly out of the previous kinesthetic traditions of painting with light.

Many of these early oil paintings make explicit reference in the content of the painting to this same triangulated light relation. For example, in Melchior Broederlam's early oil painting *Retable de Champmol* (1399; figure 11.2), we can see the highly reflective mixture of gilded gold frames, haloes, brilliant oil coloration, and multipanel construction, all triangulated into a polygonal series of reflective surfaces within which distinct kinesthetic forms are held together and apart. In the upper left-hand corner of the left panel, we see the divine light of God shining down on the golden crown of the church and Mary reading the illuminated Bible, and then in the adjacent right panel of the *Visitation* we see the birth of the Christ child. The flow of light moves from heaven, through the shimmering book, to the

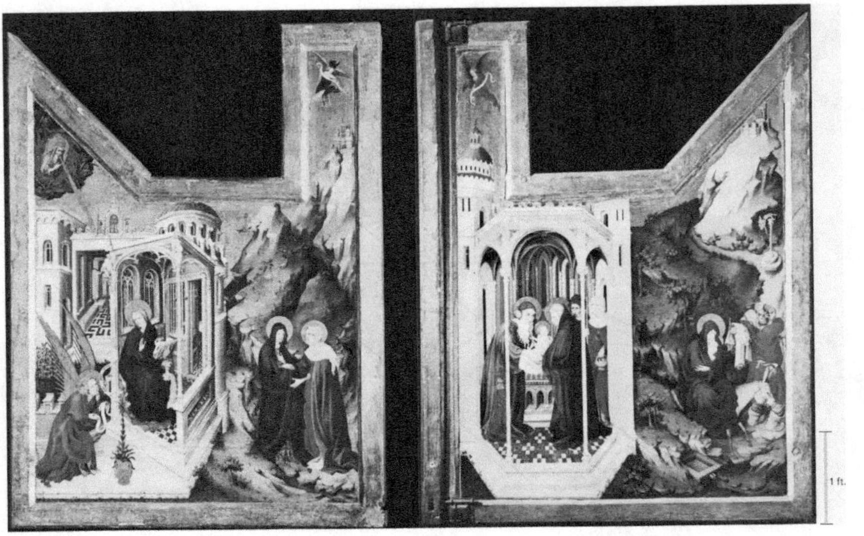

Figure 11.2 Melchior Broederlam, *Retable de Champmol* (1399), from the chapel of the Chartreuse de Champmol, Dijon, France. Oil on wood. Musée des Beaux-Arts, Dijon
Source: From Helen Gardner and Fred S. Kleiner, *Gardner's Art Through the Ages: The Western Perspective*, 13th edition (Boston.: Wadsworth Cengage Learning, 2010), p. 399.

haloed Christ body in the same way that light moves from the sun, to the altarpiece, and then to the eye of the viewer.

In Robert Campin's famous *Mérode Altarpiece* (ca. 1425–1428), a similar flow of light enters from the left panel, through the oculus of the central panel (supporting a small white figure carrying a cross) and onto Mary reading the Bible, and then only indirectly to the third panel, where Joseph is carving wood. Again, light moves from left to right via a polygonal triangular tension between God, surface (painting or book), and human body. Divine light paints the surface so that the surface can reflect this light and the eye or body can receive it indirectly, obliquely, and triangulated.

On the interior of Jan van Eyck's *Ghent Altarpiece* (1432), we find a central radiating pattern of light between gilded polyptych frames. The flow of light begins in the central panel emanating from the golden triple tiara of God the father, then moves outward to the next panels on the left (haloed Mary, queen of Heaven, reading a gilded Bible) and right (haloed John the Baptist reading a gilded Bible), then to the singing angels on the next left and right panels, and finally to Adam and Eve on the most distant and darkened smaller outer panels. Again, light moves through a tensional and hinged linkage between polyptych frames, from God to the surface of the book to the mortal naked bodies of humanity.

The third technique in oil painting that emphasizes light relations is the increasing use of mirrors and lenses. As early as the fourteenth century, there is textual evidence attesting to the increasing use of mirrors by artists to produce self-portraits and more accurate lighting and foreshortening. In 1381, Filippo Villani claimed, "Giotto painted, with the aid of mirrors, himself and his contemporary Dante on the altar retable of the chapel in the palace of the podestà."[9] The earliest images showing the mirror as an artist's tool date from 1402.[10]

Filarete (1400–1469) advised painters to use the mirror to instruct their painting from the beginning, saying "It is also good to draw with a mirror as I have said. If you have two of them reflecting in each other, it will be easier to draw whatever you want to do, that is, what you wish to portray . . . foreshortened forms can be drawn with greater ease if they are mediated by means of a mirror image rather than 'eyeballed.'"[11]

On the other hand, in his book *Della Pittura* (*On Painting*, 1435) Leon Alberti (1404–1472) advised painters to use the mirror just before completion to verify their work.[12] In his book, Alberti introduces painters to the idea of perspective in painting not as an opaque two-dimensional surface that represents a three-dimensional object but as a "window through which we look out onto a section of the world." Alberti calls the mirror the *iudex optimus*, the optimal judge, of paintings because it intensifies the picture's properties.[13] "So the things that are taken from Nature should be emended with the advice of the mirror."[14]

According to Antonio Manetti (1423–1497), the Italian painter Brunelleschi (1377–1446) discovered perspective in 1413 by using two types of reflecting surfaces: the burnished silver sections integrated in the painting itself, and the plane mirror used to view the picture (figure 11.3).[15] According to the art historian Giorgio Vasari (1511–1574), Brunelleschi painted Florentine buildings with a hole through them and held up a mirror in front of the painting. When the viewer looked through the painting at the mirror, he could see both the painting and the building and how the perspectives matched.

Additionally, instead of painting the sky in his painting of the Piazza Signoria, Brunelleschi used burnished silver polished like a mirror that would simply reflect the real sky in the background. In his San Giovanni panel, he went one step further and just cut out the sky part of the panel so that the actual sky could appear behind it. In this case, the triangular tension between light flow, canvas, and eye is brought out explicitly. Each holds the others together and apart, from a point of view. The eye sees

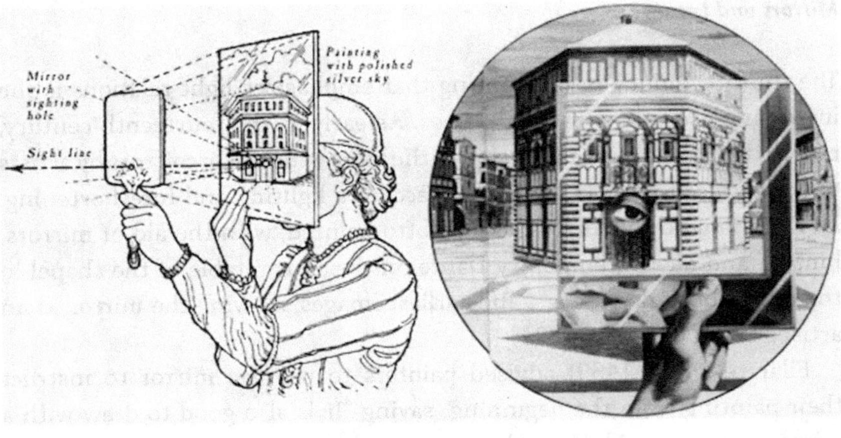

Figure 11.3 Brunelleschi's use of mirror to achieve perspective
Source: Unknown; image available at https://leonardosapprentice.files.wordpress.com/2013/10/brunelleschi.jpg

the painting only in the mirror and sees the building in the background alongside the eye, which sees itself as a mirror that also reflects light. Light, canvas, and eye all reflect and see light, and thus become mirrors. By using the mirror, Brunelleschi exposes himself not as a universal perspective from nowhere but as yet another kind of mirror. The illusion of God-like omniscience is overturned by the kinetic materiality of the eye itself as a mirror, and light itself as a relational and constitutive agency.

In his *Trattato della pittura* (*Treatise on Painting*, 1490s), Leonardo da Vinci (1452–1519) writes,

> You should take the mirror for your guide—that is to say a flat mirror—because on its surface the objects appear in many respects as in a painting. Thus you see, in a painting done on a flat surface, objects which appear in relief, and in the mirror—also a flat surface—they look the same. The picture has one plane surface and the same with the mirror. The picture is intangible, in so far as that which appears round and prominent cannot be grasped in the hands; and it is the same with the mirror. And since you can see that the mirror, by means of outlines, shadows and lights, makes objects appear in relief, you, who have in your colors far stronger lights and shades than those in the mirror, can certainly, if you compose your picture well, make that also look like a natural scene reflected in a large mirror.[16]

Like Filarete, Leonardo not only recommends but also actually uses the mirror as an exemplary flat model of nature from which to paint. The mirror is thus already a painting made of nature's reflected light. The painting made from this mirror is thus a painting of a painting, a mirror of a mirror. If the mirror becomes a painting, then the painting in turn becomes a mirror.

Accordingly, for Leonardo the painter himself becomes a mirror:

> The painter should be solitary and consider what he sees, discussing it with himself, selecting the most excellent parts of the appearance of what he sees, acting as the mirror which transmutes itself into as many colors as exist in the things placed before it. And if he does this he will be like a second nature.[17]
>
> And above all his mind should be equal in nature to the surface of the mirror, which assumes colours as various as those of the different objects.[18]

For Leonardo, everything becomes painting at the same time, as painting itself is revealed as mirroring or painting with light. The sun provides the flow of paint, nature then redirects and modulates itself into colors, which all reflect off one another like mirrors. Leonardo's use of mirrors makes explicit what was already implicit in the act of painting itself: that the human-made mirror, the canvas, the eye, and the mind are simply a further system of mirrors reflecting, modulating, and absorbing flows of light in their own way, outlining a vast tensional matrix of light relations within which visible forms emerge.[19] Each becomes an image in a vast matrix of reflecting images without center, origin, or representation.

From these early experiments, the use of mirrors by painters became ubiquitous. Hans Holbein the Younger (1497–1543); Pieter van Laer, known in Rome as II Bamboccio (1592–1642); Diego Velázquez (1599–1660); Rembrandt (1606–1669); Elisabeth Vigée Le Brun (1755–1842), in her 1786 *Jeanne Julie Louise Le Brun se regardant dans un miroir*; and many others all used mirrors in one way or another.

Velázquez's *Las Meninas* (1656; figure 11.4) in particular reveals the intricate tensional relations of light within which the canvas occurs as only one of many optical surfaces. The painting depicts not only a mirror image of the patrons on the back wall, which appears framed alongside other framed paintings on the wall but also includes the painter himself and the back of the canvas, which we are presumably looking at after its completion from the front. The flow of light moves from the window on the right down to the painter's canvas on the left, which then reflects it toward the mirror on the back wall, which then reflects it forward to the place of the unseen patrons (now the viewers), who in turn reflect their images to the mirror.

Every eye and face, every canvas, and every mirror in the painting becomes a reflective surface for the flows of light emanating from the

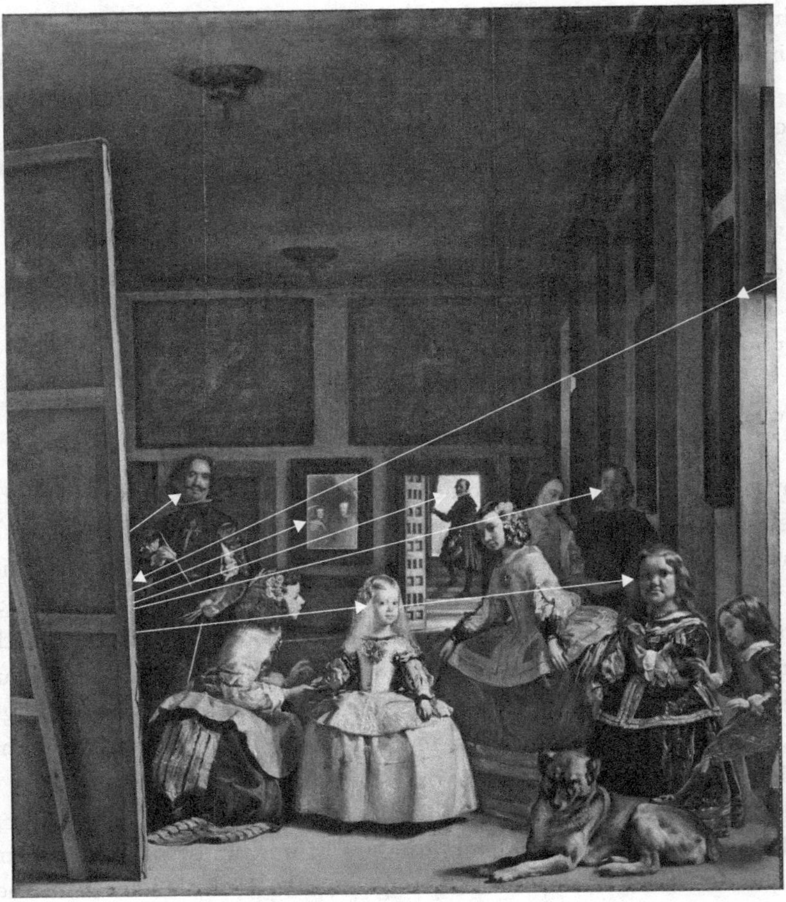

Figure 11.4 The reflective matrix of Velázquez's *Las Meninas* (1656)
Source: Wikimedia, https://commons.wikimedia.org/wiki/Category:Las_Meninas#/media/File:Las_Meninas,_by_Diego_Velázquez,_from_Prado_in_Google_Earth.jpg

window on the right and the hallway at the far back. Everything becomes a mirror, a reflection, an image. However, Velázquez's special genius was not only to have used such a scene of optical surfaces and flows to produce a series of colored images on a canvas—this had been done before—but also to have shown on the canvas itself the very material kinetic conditions of tensional light relations that made the canvas itself possible.

In the sixteenth century, the use of lenses and the *camera obscura*, which could project an inverted image from one side of the lens to a wall on the other side, was added to the use of mirrors. The design of the *camera obscura* was improved with the use of a mirror to correct the reversal and inversion of the image. If a mirror was placed at a 45-degree angle to the light before it was projected onto the wall, it would correct the inversion and reversion of the image.

The use of a mirror was suggested in a manuscript by the Venetian Ettore Ausonio (1520–1570), and later in a book by Giovanni Battista Benedetti, published in 1585. Although earlier writers had highlighted the value of the *camera obscura* to painters, knowledge of the *camera obscura* using a lens and a mirror became widespread with Della Porta's publication *Magia Naturalis* (1589).[20] Most famously, historians are confident, despite direct textual evidence, that Jan Vermeer (1632–75) used both mirrors and the *camera obscura* in order to reproduce light effects impossible to see with the unaided eye.[21]

The material-kinetic conditions of oil painting thus reveal the fact that oil painting itself already functions as a mirror and as a lens insofar as it modulates light into color and form. The eye and canvas become simply one more optical device alongside the others. Eye and canvas are revealed for what they in fact are: reflective surfaces, images in a vast matrix of tensional light relations within which visible forms appear together and apart. Kinetic perspective is more than a geometric point of view and a horizon; it is the vast network of relations of light within which points of view are possible in the first place.

Chiaroscuro

The third use of kinetic perspective emerged around the sixteenth and seventeenth centuries and is defined by the use of chiaroscuro. Chiaroscuro is not limited to oil painting; it also includes a wide range of luminous contrasts in a variety of media. Initially, in fact, the term "chiaroscuro" was first used to refer to the high light contrast introduced by using burnished metals on the dark-purple vellum background of illuminated manuscripts. "Chiaroscuro," from the Italian words *chiaro* ("light") and *oscuro* ("dark"), simply describes the tension or contrast between a highly reflective surface and a highly absorbent surface. Chiaroscuro is the name for the kinetic tension between contrasting wavelengths of light.

During the Renaissance, the term "chiaroscuro" was also applied to the technique of light modeling or shading in painting and drawing. During the sixteenth century, it was even applied to the use of two-color woodcuts and etchings. A final kind of chiaroscuro can be found in the compositional tenebrism of the Baroque, defined by an extreme contrast of dark and lighted regions of paint. From manuscript chiaroscuro to Spanish tenebrism, there is an increasingly dramatic use of luminous contrast, leading up to a breaking point defined by the eighteenth-century Rococo.

In all three major types of chiaroscuro (manuscript, woodcut, and painting), but to differing and increasing degrees, chiaroscuro performs several interrelated kinetic operations. First, the high contrasts of chiaroscuro make possible the appearance of depth between and within visible forms. The Greeks and Romans had already experimented with this, but the Middle Ages raised it to a whole new level of aesthetic preeminence and intensity never imagined in antiquity. The appearance of depth depends on the held tension between distinct colors or wavelengths of light. Colors are darker or lighter. The held differences between colors make possible the visibility of distinct forms.

The discovery of the Middle Ages, Renaissance, and Baroque periods was to realize that forms are made of colors, and that colors are made of contrasting relations of light. Form is thus a product of contrasting light relations. By using mirrors and lenses one could see for the first time the way in which subtle shadows emerge to distinguish forms. Visible forms do not float on an empty surface but, rather, are covered with hidden areas, shadows that obscure and render visible the formal image itself.

Thus a core paradox of formalism is uncovered in chiaroscuro: that in order for form to be visible, it must also be hidden. This is in part a material kinetic discovery made possible by the fact that in oil painting, it is much easier to paint dark colors first and then add lighter ones on top, in contrast to tempera. The base layer of dark and rich oil paint reveals what the ancients could not find in tempera: that darkness, shadow, and luminous contrasts are the conditions by which luminous forms appear luminous. Light only appears as light in relative darkness. While ancient kinesthetics was defined by the juxtaposition and multiplication of forms, Medieval through Baroque kinesthetics took as its starting point the dark or luminous gap between and around the forms, the matrix of light relations that gives the forms their order and appearance as visible forms.

Light is always in a matrix of light, a distribution of linked images and surfaces—eyes, canvases, bodies, windows, faces—that define perspective. Kinetic perspective is not a painted representation of a point of view but, rather, something that the paint does by virtue of its coloration and contrast, and by virtue of its reflective capacities in a larger network. Chiaroscuro is the ultimate discovery of the two most reflective surfaces: varnished black and white. White reflects the largest quantity of light, while black, although highly light-absorbent, when varnished takes on a mirror-like reflection because the light reflected comes almost entirely from the translucent gloss.

Depth and perspective are not "represented" on the canvas, but the canvas itself directly expresses the reflectivity of light, its own perspective, in the ultra-white skin of Caravaggio's *John the Baptist* (1604) and in the mirror black of Adam de Coster's *A Man Singing by Candlelight* (1625–35). If the genius of Velásquez was to show the great polygonal matrix of light relations in the content of the painting itself, then the genius of Tintoretto was to show the movement and kinomorphic activity of the light itself. In Tintoretto's *Last Supper* (1594; figure 11.5), the linear flows of light emanating from Christ's halo and the lamp also seem to be swerving and bending into angelic translucent forms of pure light flying around the ceiling. However, the process occurs only under the conditions of a general tenebrism within which light is capable of forming itself. The truth of discrete form is not the eternal model of antiquity but rather the series of contrasts between a multiplicity of polygonal light flows in shifting perspective.

THE KEYBOARD

The second major relational aesthetic innovation to look at in this chapter is that of the keyboard. The musical keyboard, reintroduced in the West in

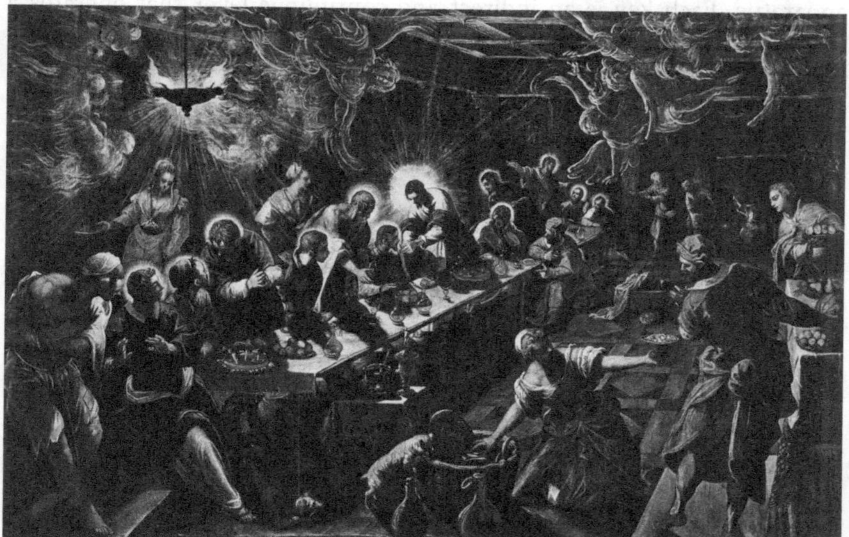

Figure 11.5 Tintoretto, *Last Supper* (1594)
Source: Wikimedia,https://commons.wikimedia.org/w/index.php?search=Tintoretto's+Last+Supper&title=Special:Search&profile=default&fulltext=1&uselang=en&searchToken=4aouzvu1d0zshx4urz6ys7ydp#/media/File:Jacopo_Tintoretto_-_The_Last_Supper_-_WGA22649.jpg

757 CE, after the dissolution of Rome, is a technical apparatus that allows for the production of sonic images through a series of indirect tensional mechanical relations. A keyboard can be used with an aerophone in the form of an organ, with a chordophone in the form of a clavichord, or with an idiophone or membranophone in the form of a prepared piano. In this sense, the keyboard is a unique musical apparatus whose kinetic action can be understood as distinct from the type of phonic structure it is connected to. Its kinetic action can also be seen as related to but distinct from that of the hammered dulcimer and other directly struck or plucked cordophones or directly blown aerophones.

The basic kinetic structure of the four major keyboard instruments of this period—organ, clavichord, harpsichord, and piano—is similar. In all four instruments, a row of keys is attached to a series of multipart levers whose linked tensional motion causes a sound to be made in the form of a plucked or hammered string or an opened or closed air valve. The keyboard mechanism operates through a direct and continuous depression of multiple, distinct, but interrelated levers whose motion corresponds directly to the continuous motion of the finger. The key is not a switch that sets off an automated process but, instead, a lever whose range of motion can be correlated to the manual motion of the hand at every point in the depression. Kinesthetically, the keyboard is therefore defined by the linked tensional motion that structures its indirect mechanical relation to the phonic apparatus. Each key is an indirect and continuous lever that operates a distinct tonally related sound. The keyboard is thus a kinetically relational instrument defined by two axes of relations: a vertical axis or linked mechanical train connecting each separate key to a separate sound, and a horizontal axis of tonally related sounds and mechanically related keys on a continuous board.

In contrast to the centrifugal kinetics of the ancient cordophone, whose strings hover above a center resonance chamber and radiate outward in all directions, the keyboard of the Middle Ages follows a tensional kinetics of linked levers, hinges, and pivot points whose actions are cellular and individually controlled. A kinesthetic network of linked tensional relations now becomes the material condition for the production of musical forms. Sonic images like melodies and harmony become products of a more kinetically primary process of tensional mediation and tonal individualization. Each string or pipe can now be individually sounded and muted independently of the others, with greater alacrity than ever before. Middle Ages keyboard kinesthetics contributed to the development of two major musical textures during this time: polyphony and homophony.

Polyphony

The first musical texture is polyphony, composed of two or more simultaneous lines of independent melody. Up until around the tenth century, almost all Western music was monophonic, defined by a single melodic line without accompanying harmony or chords. From the fifth to ninth centuries, the medieval music of monophonic plainchant thus remained kinetically similar to that of the ancient chorus—multiple voices all singing the same melodic line. Even ancient Greek water organ (*hydraulis*) and Roman portable organ (*organum*) music remained largely subordinated to the dominance of a single, central, hierarchical melodic line.

Organum

It was only with the eighth-century introduction to Europe of the pipe organ and its new kinetic structure that polyphonic music was able to emerge there. It is no coincidence that soon after the first pipe organ appeared in Western Europe (757 CE) and entered the church at Charlemagne's request in 812 CE that we also find the first treatises on polyphonic music: *Musica enchiriadis* and *Scolica enchiriadis* (ca. 900 CE). Further, it is no coincidence that the name first used to describe early polyphonic music was the Latin *organum*, from the Roman name for the organ. The two are materially and kinetically related. The organ creates polyphony, and polyphony creates *organum*.

With the addition of a pedal-controlled bellows, both hands were free to play the keyboard. Each hand was now capable of working independently on the keys. To a single melodic line a continuous bass line (*basso continuo*), chord, or another melodic line could be added. By contrast, the portative organ required the player to use one hand to pump the bellows, and thus limited one hand to playing largely single melodic lines. From its invention until the fourteenth century, the organ remained the only keyboard instrument, even if in some cases the keys were buttons or larger levers operated by the whole hand.

Polyphony was made possible by the new kinetics of the keyboard organ, but also emerged in vocal music around the same time and later spread to strings and other instruments. Like the kinetics of the keyboard, polyphony has a similar tensional or relational kinesthetic structure. By stacking and juxtaposing two or more monophonic lines, the wave patterns of sound images enter a diffraction pattern (figure 11.6), remaining distinct

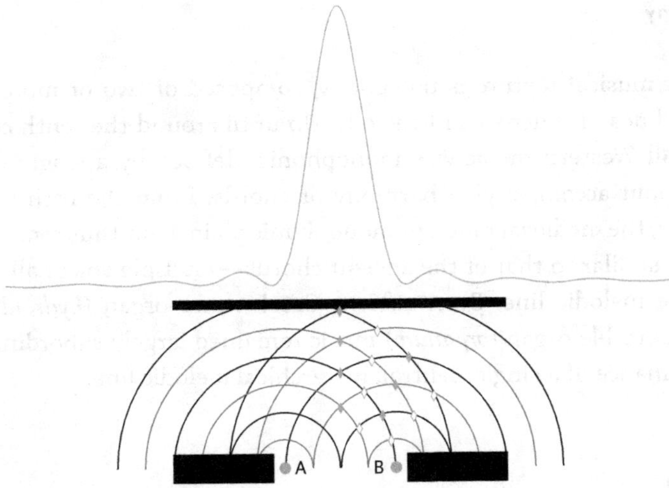

Figure 11.6 Sonic diffraction pattern of sound images
Source: Creative Commons 3.0. Siyavula Education, http://www.everythingmaths.co.za/science/grade-11/06-2d-and-3d-wavefronts/06-2d-and-3d-wavefronts-05.cnxmlplus
https://www.everythingmaths.co.za/science/grade-11/siyavula-physical-sciences-grade-11-caps

from one another but also creating a distinctly new tensional image by overlapping.

This new image or wave pattern is thus defined primarily by its relations of held tension between the two composite wave patterns. Each pipe, string, or voice emanates from its own distinctly modulated space, but instead of producing the same tone, each produces a different but harmonically related tone at the same time.

Motet

Conceptually, early church polyphony was thought to be a more perfect emulation of the angelic singing of God's glory. Just as the angels would never need to stop and take a breath, so the overlapping layers of polyphonic sound could create a continuous sonic fabric, even though each of its air flows or keys start and stop within the flow. Polyphony made possible new continuous musical forms only under the condition of a more primary process of diffraction and contrasting relations.

For example, just below the continuous and melismatic surface of Léonin and Pérotin's *Feast of Saint Stephen* (ca. twelfth century) lies a more primary sonic diffractive tension between the basso continuo and the melodic lines. To this sacred harmonic organum Guillaume de Machaut (ca. 1300–1377) and Guillaume Du Fay (ca. 1397–1474) added the secular upper voices of romantic poetry in their motets. Thus to the long, melismatic rhythms of the twelfth-century Notre-Dame motet, Machaut and Du Fay added a new sonic tension between long and short rhythmic lines of nonsynchronous polyphonic layers (isorhythms).

The polyphonic tension between sacred and secular poetry and melody in the same songs, as well as the introduction of nonsynchrony and the use of dissonant harmonies using thirds and sixths in motets such as "*O bone et dulcissime Jesu*" by Josquin Des Prez (ca. 1450–1521) resulted in a religious backlash against polyphony as the "devil's music" and the use of harmonic thirds as "evil," as opposed to the perfect fourths and fifths of Gregorian chant. Pope John XXII even banished polyphony from the Liturgy in 1322, and Pope John XXII warned against it in his *Docta Sanctorum Patrum* (1324).[22]

Kinetically, the rise of polyphony thus marks a transition from centrifugal monophony to a more tensional relation between simultaneous, overlapping, nonsynchronous, and even dissonant musical textures defined by their unique sonic diffraction patterns. This remained the dominant musical texture until the rise of homophony, beginning around the sixteenth and seventh centuries.

Homophony

The second musical texture made possible by medieval keyboard kinesthetics is homophony, defined by the contrasting relations between a primary melody and supporting harmony. Homophony adds to the tensions of the polyphonic diffraction matrix a distinct melodic line that highlights certain tones in the harmonic pattern. In this sense, homophony is a type of polyphony. Rather than dominating the polyphonic harmonies, the homophonic melody is like the crest of a sonic diffraction wave, rising and falling with the rhythmic waves that support it.

Harpsichord and Piano

Homophony rose to dominance with the widespread distribution of the clavichord and harpsichord between the sixteenth and eighteenth centuries. Although the clavichord and harpsichord were invented in the fourteenth century, they were not fully implemented in musical composition until the sixteenth century. However, their tensional and relational kinesthetics allowed for an increasing control over each string, groups of strings, mobility of the instrument, and liberation of the foot from the bellows for other sonic capacities, like muting, moving registers, and so on.

The invention of the piano with its hammered strings in the eighteenth century allowed for an even greater control over the individual volume of each note, as well as the instrument as a whole. This allowed for a new range of volumetric contrasts and sonic tensions that can be heard most strikingly in the *sforzandos, diminuendos, crescendos*, and *decrescendos* of Beethoven. Unlike the harpsichord and clavichord, the piano allowed for hammered chords to be struck, and for them to ring out while a homophonic melodic line was played on top of it.

In short, the stringed keyboard allowed for more independent musical control over more sounds, making room for a new homophonic melodic motion. Melodic forms and hierarchical tonal scales could proliferate, but only on the material kinetic condition of a supporting set of polyphonic relations that allowed it to emerge and not collapse into monophony. Only in this context could something like the sonata form emerge as a series of distinct but interrelated movements between homophonic melodies in tension and ultimately resolution. For there to be a resolution, there must first be a real tension between sonic images to be resolved.

Orchestra

In particular, the increasing use of multiple instruments in the ensemble made possible an incredible instrumental polyphony and homophony. This process began with the trio sonata (violin, viol, and harpsichord) in Gregorio Allegri's (1582–1652) sonatas, Dieterich Buxtehude's (ca. 1637–1707) Sonata in B Flat Major, and Johann Pachelbel's (1653–1706) *Suite in E Minor*, then with the increasing number of instruments used in larger ensembles like Giovanni Gabrieli's *Sacrae Symphoniae* (1597) and Monteverdi's *Orfeo* (1607) and up to Mozart's Symphony No. 41 in C major (Jupiter) (1788). The more instruments were added to the ensemble, the greater the tension and contrapuntal relations possible between them. Homophony could now rest on a massive symphonic apparatus of sonic diffractions. In a way, the multiplication of orchestral instruments transformed the ensemble itself into a massive keyboard with an incredible tonal range and more textual layers than ever before possible. The composer plays the orchestra like a keyboard.

This new texture is nowhere more dramatically displayed than in the invention of the homophonic art par excellence: opera. The entire Baroque theater apparatus now combined with the orchestra (adding an orchestra pit) to produce an incredible multilayered architectural and phonic image in which layers upon layers overlap and fold over one another in waves of tension and diffraction. The layers of sound mirror the layers of the theater, which in turn mirrors the folds of the audience's Baroque clothing, and the layers of theater masks, in turn mirroring the scalloped stucco ceilings. All of these layers form kinesthetic relations that support a single monodic solo and homophonic melody that rides across the surface. By the eighteenth century, the rise of *opera seria* introduced a new virtuosity into the operatic solo that allowed homophony to reach its absolute tensional limit without breaking from its polyphonic and harmonic support, as it eventually did in modern composition.

In short, the invention of the keyboard gave birth to polyphonic (and homophonic) music that was defined primarily by a sonic tension and contrasting relation (counterpoint) between two or more melodic or harmonic lines. Instead of a single monophonic hierarchy between central tones and their peripheral resolutions, keyboard and polyphonic music multiplied, diffracted, contrasted, and connected divergent tonal lines.

EPISTOLOGRAPHY

The next major relational aesthetic innovation that rose to prominence during the Middle Ages is that of the written image of epistolography: the art of letter-writing. Like most things, the origins of narrative prose letter writing go back to the ancient world, where it was used for governmental communication and occasionally as a literary medium, such as in Ovid's *Heroides* and Cicero's *Epistulae*. However, literary and widespread usage of letters was historically limited in the ancient world by a number of material factors, including the fragile nature of papyrus, the relatively less literate population, the more efficient roadway, the verbal messenger system (in the case of the Romans), and others. With the increasing use of vellum in the fourth and fifth centuries CE and eventually the increasing use of paper from the eleventh and twelfth centuries onward, the Middle Ages became "the golden age of letters."[23] From the fifth to the seventeenth centuries, there was an absolute explosion in the use of letters as a primary medium of communication across distances between all literate parties, from lovers to priests to administrators.

During this time the material structure of the letter defined the kinesthetic distribution of images for several major literary arts. The historical predominance of the letter is first and foremost a kinetic success. The material kinetics of the letter introduced a radical new mobility and durability to the literary recording surface. By making a larger number of recording surfaces compact through folding and enclosing, the letter also increased the mobility of written communication: in speed, type of transport medium, and storage.

For example, one person could carry numerous letters by horseback more quickly and store more of them in one place than any other written medium—scroll, manuscript, or book. Furthermore, the increased durability of the letter made possible by using parchment instead of crumbly papyrus or fragile clay tablets also ensured a longer kinetic duration and life span of the letter. Both these material kinetic features of the letter increased the material and temporal circulation between the author and the reader, but also introduced an increasing tensional relation between senders and receivers.

Since more writing could be in more places faster and farther apart, it could be read by more people over great distances in time and space. It was no longer the exception but, rather, the rule to read a text by an unknown and previously unconnected author. The kinetic tension of epistolography is this: The letter holds together and apart kinetically distant authors and readers with one another. The letter's incredible new power of circulation and mobility made it possible for an author to inscribe it at one point in the circulation and transport it and then further along for the letter to enter into a new circulation of reading and/or reproduction and possible further transportation. The letter thus introduced an increased tension between the points of its circulation and was fixed in the rigid link of the text itself—between multiple readers.

The problem of interpretation, literary indeterminacy, and intermediation is a part of every graphism, but in the Middle Ages it became a primary and defining feature of the whole kinesthetic literary image. The whole medieval apparatus moved through a linked system of letters. Social circulation was coordinated through the introduction and resolution of numerous kinetic limits—lovers moved and coordinated their rendezvous by letter, administrators coordinated their supplies by letter, and priests organized their great institutions by letter.

In the prehistoric world, literary kinesthetics was defined by the gathering of natural sounds into the oral song of the poet and his mythogram.

In the ancient world, it was defined by the centrifugal organization of the oral periphery by the written verse. Now, in the medieval world, literary kinesthetics was defined by the network of simultaneous tensional relations between multiple senders/receivers. The kinetic presupposition of the letter is that there is some tension that binds the sender and receiver, relativizing and coordinating their social motions, but also keeping them at a distance such that the letter is necessary to communicate. The vast majority of medieval letters thus have persuasion and rhetoric at the core of their content—the resolution or negotiating of a spatial-kinetic tension between them. Instead of a simple uni- or bidirectional movement from oral to written (centripetal), or written to oral (centrifugal), medieval epistelography is defined by a series of multiple bidirectionally linked oral-written segments. The letter is written, but the messenger or reader then reads the letter aloud, writes a letter in response, and so on. During this period, the arts of rhetoric and oratory cannot be fully separated because of the kinetic structure of the letter. The relations of transport and connection made by messengers and oral readers here take on a new importance and primacy. The form of the letter with its salutations, brief narration, and requests emerges under the more primary kinetic conditions of its transport, including the authenticity of a nonpresent author, short length, and resolution of indeterminate spatial-kinetic relations.

By tracing its development from these early origins, we can see how the art and aesthetics of epistolography eventually gave birth to the more complex novella, romance, and epistolary novel of the late Middle Ages. More specifically, the rise of the letter occurred in three major historical and material movements.

Ars Dictaminis

The first movement is the *ars dictaminis*, the medieval art of prose composition and the writing of letters (*dictamen*). From the times of the early Christian epistles (first century CE), to its increasingly formal structure first laid out in Alberic of Monte Cassino's *Flores rhetorici dictaminis* (ca. 1080),[24] and fully developed through the twelfth century by Alberic's pupil John of Gaeta, the *ars dictaminis* eventually spread from Bologna to France and Germany, becoming the dominant form of written prose narrative.[25]

According to Montecassino, the formal *ars dictaminis* of the eleventh and twelfth centuries was "the rhetorical division of every speech," both public and private, into five parts: "the *exordium* or *proemium*, the *narratio*, the *argumentatio*, and the *conclusio*" and the "*salutatio*," modeled on the six part divisions of ancient oratory.[26] Around the middle of the century, a *cursus*, or spoken eloquence, was added to enhance the aural effects of persuasion in the letter. Of the teachers and practitioners of prose letter writing at this time (*dictatores*), we have only names.

The kinetic structure of the *ars dictaminis* letter is one directed toward "harmony"[27] and the resolution of tension between readers/speakers who move at a distance from one another. The division of the letter into parts follows the division of readers by the *salutatio* (including the name of the sender) and the *exordium* (which prepares the reader to be receptive to the letter). The letter holds sender and receiver together and apart through the intermediary of the letter itself and its performative *oratio*, the voice of the sender spoken by another.

The Epistolary Novella

The second movement is the rise of the *epistolary novella*, a collection of short narrative prose letters. Beginning around the thirteenth century, a paradigm shift took place in epistolography toward less formal and more expressive long-form letter writing. Starting with the work of Boncompagno da Signa (ca. 1165–1240), the letters of the *dictores* (rhetoric teachers) began to include titles and personal information about themselves. They began to write "model letters" for pedagogical purposes that displayed their literary virtuosity of the letter form, and extolled the universality of the *ars dictaminis* as the "artzum liberalium imperatrix" ("empress of the liberal arts") to which theology and philosophy were simply the handmaidens.[28] Their oratory debates drew huge crowds, and they began to increasingly collect their letters into massive organized collections called *summa*.

Many of these changes were attributed to a return to the *epistoles* of Cicero, which were written in prose on humanistic topics but were not confined to the strict form of the *ars dictaminis*. As one Italian *dictore*, Bono of Lucca (d. 1279), wrote of his colleagues—and himself, "They have drawn all things from Cicero as their source."[29] Indeed, later thirteenth-century *ars dictaminis* has recently, and with good reason, been labeled "the first real revival of Cicero."[30] In particular, Boncompagno, a professor of rhetoric, was already transforming his letters into longer autobiographical stories, fictional narratives, and epistolary sequences from four to seven letters long. Boncompagno was thus the first to invent the epistolary novella as a genre within the literary tradition of the Latin West.[31] Under the influence of Cicero's letters, "the essential nature of *ars dictaminis* invited its practitioners to the development of the epistolary novella."[32] Petrarch's

(1304–1374) discovery of Cicero's letters also allowed him to break free from the formal structure of the *ars dictaminis*, and write his own letter collections.

At the same time that the *ars dictaminis* was being transformed into the epistolary novella, the love letter was being transformed into the serial chivalric romance. In the thirteenth century, the chivalric romance emerged as a series or "cycle" of tales and interwoven stories rather than with a single linear plot or main character. The Lancelot-Grail stories or Vulgate Cycle, for example, are composed of numerous prose stories of the Arthurian legend and tell the story of the quest for the Holy Grail and the romance of Lancelot and Guinevere. As early as the writings of Chrétien de Troyes, the late-twelfth-century French poet, these fragmented stories began to receive a narrative structure—beginning, middle, end—such as in *Yvain, the Knight of the Lion* (ca. 1170).

The chivalric romance emerged from the kinetic conditions of epistolic courtly love. Lancelot spends his time seated on his steed, leaning on his lance, waiting, daydreaming of his beloved, pining for her. He is constantly forgetting names or who is speaking to him. The chivalric romance is a story of mobility, travel, quest, and circulation precisely because the lover is at a distance. Distance requires the epistolary form for communication, and the epistolary form presupposes a distance to be traversed by a message or by a story to be sent. Before the chivalric romance emerged in the late twelfth century, it was the medieval love letter that preceded it and shared its linked kinetic tension between the lover and the beloved, the knight and his grail, and so on.

The consummation of either would destroy the very epistolic and kinetic conditions under which the genre functions—that is, a tension or material distance between lovers.[33] In Chrétien's *Perceval, the Story of the Grail* (ca. 1190), for example, Gawain derives enjoyment from the journey itself and not from his numerous unconsummated conquests. The form of the chivalric romance, and in particular the Lancelot-Grail stories, is a resonating series of fragments, each presenting an interconnected and related story in shared tension with the others, similar to the kinetic structure of the epistolic novella.

The more famous of the novellas, however, are those of the fourteenth through sixteenth centuries, including Boccaccio's *Decameron* (1353), Franco Sacchetti's *Trecentonovelle* (ca. 1390), Marguerite de Navarre's *Heptaméron* (1558), and Giovanni Straparola's *The Nights of Straparola* (ca. 1550). Under the influence of Cicero and Petrarch's letters, Boccaccio, another professional teacher of the *ars dictaminis*, also began writing less structured, longer, and more humanistic letters. Following the lead of Boncompagno's widely distributed epistolic literary works *Rhetorica Antiqua* and *Rota Veneris*, Boccaccio assembled his own serial novella of multiple interrelated stories for travelers escaping the Black Plague. Chaucer, Sacchetti, Navarre, and Straparola were then in turn inspired by Boccaccio.

Although these novellas are not explicitly written as letters, they all share the same kinetic structure as the epistolary novella series. They are a series of interrelated stories told while in transit between origin and destination. They echo not only the kinetic structure of the tensional love letter but also the other fourteenth-century popular medieval genre of travel writing.

The *Decameron*, in particular, makes reference to all this in its beautiful proem. The novella opens with the full title of the *Decameron: Prince Galahalt* in reference to a character in the chivalric romance of "Lancelot du Lac," who functions as a go-between or intermediary between Lancelot and Guinevere and helps kindle their amorous love. *Galahalt* plays the kinetic role of the letter-bearer and oral intermediary between the amorous writing and pining of the lovers. We are thus meant to understand the *Decameron* from the beginning as a kind of intercessor or a series of Ciceroian "pleasant discourses" (or love letters) meant to bring "compassion" to those "afflicted" by love loss or by a distance between lovers. The kinetic precondition of the letter, the novella stories, courtly love, and even travel writing, is a held tension between readers at a distance as they change their motions relative to one another through the triangulated medium of the letter or letter bearer (Prince Galahalt).

The Epistolary Novel

The third movement of epistolography is the rise of the epistolary novel, a collection of short prose letters worked into a single, long-form narrative. Beginning around the fifteenth century, another paradigm shift took place in epistolography toward increasingly structured compilations of prose letters. For example, in Diego de San Pedro's *Cárcel de amor* (*Prison of Love*) (ca. 1485), the first epistolary novel, the prisoner Leriano is held captive by his unrequited love for the princess Laureola and he begs the author of the letters we are reading to write down his pleas of love to Laureola.

Again, the letter writer to whom the prisoner dictates and the letter bearer or Prince Galahalt operate as epistolic intermediaries between a lover and the beloved in linked tension with one another. The lover cannot get closer without the beloved's moving farther away, so he must move at a distance, using the letter as the kinetic medium of relation. The reader intrudes on this exchange and thus adds another polygonal line of relation in the author-reader-lover-beloved network. It is again no coincidence that the themes of linked cellular confinement, travel, and love at a distance occupy so much of the epistolic works, including James Howell's *Familiar Letters* (1645–1650). After its early expressions in the fifteenth century, the epistolary novel reached its peak of dominance in the eighteenth-century works of Samuel Richardson, Montesquieu, Jean-Jacques Rousseau, Goethe, Hölderlin, and others, ultimately falling out of favor by the end of the century in favor of the modern nonepistolic novel.

It is also no coincidence that the picaresque novel emerged from these same kinetic patterns in the sixteenth through eighteenth centuries. The picaresque novel first emerged in Spain with the anonymously published *Lazarillo de Tormes* (1554), and is defined by a series of loosely interrelated stories reported in the first person on the travels and loves of the author. The fact that Boccaccio had already produced similar kinds of stories testifies to the epistolary character of picaresque, written as it is from the epistle-like prose recounting of travel and love letters.

Most famously, however, Miguel de Cervantes wrote the series of novellas *Novelas Ejemplares* (*Exemplary Novels*) (1613), which included a short picaresque novel titled *Rinconete y Cortadillo*. He intended this to be for the Spanish what the novellas of Boccaccio were to the Italians. In fact, some scholars speculate that Cervantes began writing *Don Quixote* (1605–1615) as a series of novellas but instead chose to combine them into a single, long-form narrative.[34]

The epistolic tone of *Don Quixote* is quite clear. The book opens with Cervantes's claim to have not invented the character of Don Quixote but to have learned of him from his archives at La Mancha and from other writings translated from Arabic by the Moorish author Cide Hamete Benengeli into a series of short pamphlets. Cervantes then states that he paid someone to translate the pamphlets and tried to work them into a single narrative, even consulting a friend who advised him to put some Latin quotes of Horace in the margins so that he would look like one of the great *dictores*, or humanist grammarians. The kinetic triangulations of *Don Quixote* abound. It is a combination of archival materials and narrative materials translated into Arabic pamphlets, translated again into Castilian, and then wrapped up into a single narrative. Cervantes becomes just one more letter-bearing intermediary between the reader and historical events.

The epistolary novel thus emerged from the weaving together of epistolary novellas, which emerged from the gathering together of epistles into the novella, which emerged from an opening-up of the epistolography of the *ars dictaminis*—all of which presupposes a fundamental tension of negotiated movements between reader/writer, sender/receiver, and written/oral relays held together and apart by the movement of the letter and its bearer/orator.

CONCLUSION

All the newly dominant arts of the Middle Ages are defined by the kinesthetic pattern of relational or linked tensional motions holding images together and apart. Form no longer appears, as it did in antiquity, as an eternal model of images as copies but, instead, as shapes or patterns that emerge out of a more primary intermediary flow of light, sound, or transport. The brilliant insight of the Middle Ages is that aesthetic form is nothing other than the polygonal intersection of a network of material kinetic relations. Even God appears only in and through a fundamentally kinetic network of light relations or the polyphonic flows of sound that sing his glory.

However, after kinesthetic relations have come to the fore as constitutive of formal images during this period, a whole new kind of kinesthetic regime begins to take shape around the eighteenth century that starts to stretch, bend, and segment these same kinesthetic flows in the experiencing subject of modern art. This is the next and final historical regime of aesthetic motion.

SECTION D

The Differential Image

The fourth and final major kinesthetic field to rise to dominance in the West is the differential field. This kinesthetic field increasingly rose to prevalence over the course of the modern period, roughly from the mid-eighteenth to the twentieth centuries. Alongside the medieval and Baroque arts of the previous centuries, a number of new modern arts emerged, including the photographic image, steel-frame construction, the novel, and musical meter. The following chapters examine each of these aesthetic innovations in turn.

CHAPTER 12
Elastic Difference

The kinetic elements of difference in art are those elements that give sensuous emphasis to the elastic unity and mutability of the flows that constitute all aesthetic fields. In chapters four through eleven, we examined how the functional, formal, and relational fields of images emerge from material kinetic patterns of centripetal, centrifugal, and tensional motion. We now turn to the internally differential flows themselves that give mutability to all these regimes of motion. As we saw in Section C, between aesthetic forms there are networks of relational flows within which forms emerge. However, relational flows, like all flows, are continuities, as we argued in Chapter 1. Since every flow of matter is infinitely continuous and pedetic, then the relational links are capable of not just rigid connections but also more supple connections that can add and subtract folds indefinitely. The profound insight of modern art is to have discovered the seemingly infinite elasticity and mutability of all aesthetic images.

Every aesthetic body can fold and unfold itself indefinitely into a multiplicity of singular and differential images. Every folded image can in turn be folded and unfolded again. While aesthetic relation makes formal differences possible, aesthetic differentiation makes it possible for functional images to become dysfunctional, formal images to become deformable, and relational images to become unrelatable.

The goal of this chapter is to provide a description of the conceptual and kinetic features that define this historical period of the image. The subsequent two chapters then demonstrate the same thesis empirically and historically. The theory of kinesthetic difference is thus defined by three interrelated features: difference, elastic motion, and seriality.

DIFFERENCE

Just like the other kinetic elements of art defined in this book (function, form, and relation), the kinetic definition of difference means something different from its typical nonkinetic definition. We already put forward a theory of kinetic difference in Chapter 2, but we now apply it to the historical material of the modern arts.

Kinesthetic differentiation is the process by which the flow of matter folds itself up. At any point, any flow can begin to curve back over itself, producing an image. This image in turn is nothing other than a flow that can itself be internally folded again to produce subimages, and so on in a multiplicity of images. Thus within the relational flow that rose to dominance in the Middle Ages, it can be further shown that the relational flow of matter itself is full of an indefinite multiplicity of further subfolds that compose the microscopic and macroscopic dimensions of the relational field itself.

Modern art is the quest for the infinite depths of the microcosmic and the infinite surfaces of the macrocosmic—the microtones of Karlheinz Stockhausen's *Xi*, 1. ex Nr. 55 (1986) and macro tones of John Cage's *ORGAN2/ASLSP* (as slow as possible), which is 639 years long and will be completed in the year 2640. This infinite differentiation is possible precisely because the flow of matter itself is absolutely continuous and elastic. Every aesthetic field is made possible by its capacity for kinetic differentiation that makes possible both the self-affectation of matter with itself and the material relation between each fold: the differential image.

We can contrast this kinetic definition with the logical definition of difference as the way that two self-identical things are not the same as one another. However, the problem with this logical definition of difference is that it already presupposes the unities and identities of the things that are being contrasted. Difference here appears only as a negativity, lack, or immaterial absence between two identities. Such singular points can then, by definition, have no relation at all. There is only the pure fragmentation of singularities without any description of their process of unification and consequently relative differentiation. Logical difference is therefore not difference at all but simply a presupposition of identities whose discontinuity is radical and therefore indifferent. Kinesthetic difference, therefore, should not be thought of as a negative lack or gap between two identical images but rather as a kinetic process of continuous differentiation whose different images are only relatively different images in the same kinesthetic field. The degree and distribution of these relative differences produce the aesthetic image, as we have shown in Chapter 3.

Elastic Unity

The most basic operation of kinetic difference is that it holds together an elastic unity of singular images. The work of art is what holds together various images of color, texture, sound, and so on as they are added to or subtracted from a common field that distinguishes their collection from others. However, this assembly is a relatively open unity unlimited by fixed functions, forms, or relations. Modern art emphasizes, above all, these other elements—the elastic or flexible nature of the work of art—to accommodate multiple and even divergent images.

In short, a work of art is more or less differential insofar as it renders sensuous a kind of compositional unity or process of unification of heterogeneous images across space and time. The more apparently heterogeneous, diverse, and singular the images held together by the field, the more striking the process of differentiation that distributes and gathers them as relatively different. This effect can be achieved both by a radical heterogeneity of images or by a radical repetition of images. The more images are repeated, the more the process of kinetic differentiation itself comes into aesthetic focus, like Andy Warhol's *Eight Elvises* (1963) or John Adams's *Shaker Loops* (1978). Kinesthetic differentiation is thus brought to the foreground of modern art by both extreme heterogeneity (atonality, expressionism) and extreme homogeneity (minimalism). They are two sides of the same differential aesthetics.

Mutation

The second basic operation of kinetic difference is its mutability. Kinesthetic difference is defined both by its capacity to assemble heterogeneous elements and by its capacity to disjoin them. Differentiation assembles and disassembles images. Through differentiation, matter can become dysfunctional, deformable, and unrelated. All matter is in motion and thus in some degree of constant mutation. The persistence of certain kinetic patterns that define the functions, forms, and relations of the arts are all the result of this motion. Matter continues to fold itself, hold itself together in images, and compose itself but also disjoin itself.

More than other historical aesthetic regimes, modern art brings this to the foreground of sensation. The images of modern art increasingly become untethered from their historical functions, forms, and relations. Aesthetic difference is not only about demonstrating fragmentation but also about demonstrating a continuous modulation or variation of images without an obvious discontinuity. This is obvious in the cinematic arts but also in music. Schoenberg's String Quartet No. 1 in D minor, Op. 7 (1905), for example, does away with the typical quartet division into movements and division between theme and variation. Instead, Schoenberg produces a movement-less movement of continuous variation on the theme itself.

All works of art require some degree of material-kinetic differentiation, but some render more apparent than others the continual modulation and unification of heterogeneous images in a single field of composition. Aesthetic difference is the active capacity of matter to multiply and modulate its field or regime of circulation without breaking apart entirely or merging into another field. What is typically called "abstraction" or "anti-representationalism" in modern art is a poorly posed definition, since it presupposes a memetic model of art. "Nonrepresentational" art is defined only negatively.

However, abstraction is nothing other than the process of giving to matter and sensation their own freedom to express their microscopic and macroscopic affects and images. As its etymology implies, abstraction is a drawing out, a separation, or an internal differentiation of the images that are the material and constitutive conditions for the possibility of aesthetic function, form, and relation in the first place. Far from being derived from or contrasted with representational art, nonrepresentational art is actually more primary and foundational, both historically and compositionally.[1] Aesthetic abstraction is not something radically "other" than function, form, and relation but, rather, is something constitutive of them. It is the kinetic process of functionalization, formalization, and relationality itself. Kinetic difference is not relational; instead, it is the process of differentiation or multiplication that makes the distinction and relation between two images possible in the first place.

ELASTIC MOTION

Elastic motion is the kinetic capacity to expand and contract an aesthetic field of images without breaking the field apart. This is possible because the material flows that relate and compose all fields of images are not only rigid and tensional but also supple and elastic. This elasticity is what allows material flows to fold themselves up into an indefinite series of smaller and larger folds. The profound realization of modern art is that matter is infinitely foldable, serial, and elastic.

Not only is there a continuous series of indefinite folds, but such continuity also presupposes a connective flow running between the folds: the field. The structure of kinetic seriality is such that at any region in the field, a fold can emerge between two other folds. The rigid link that connects two folds can become elastic and thus foldable. Since an affective fold is nothing other than a folded flow, seriality already presupposes the elasticity of the flow. Ipso facto, this elasticity also presupposes the possibility of seriality. Therefore, there is not only an elasticity of the folds but also an elasticity of the flow of the connective field itself insofar as it allows for an expansion and contraction of the entire sequence.

However, the elasticity of the flow of matter is not limited strictly to a one-dimensional movement; it has a multidimensional elasticity. In this way, the elasticity of the field of circulation not only connects the folds to one another in a series but is also capable of looping back and connecting the last fold in a series to the first fold in a series, creating an elastic unity. Accordingly, the elasticity of this circulatory field also makes possible the expansion and contraction of the field itself to include more or fewer folds differentiated within it. The elasticity of the circulatory field in adding and subtracting its serial folds is what gives the image its proper regime of elastic motion—on a par with the other dominant regimes of centripetal, centrifugal, and tensional motion. These two kinetic operations can be distinguished as those of multiplication and oscillation.

Multiplication

The first kinetic operation of elastic motion is the capacity of any given fold to produce an indefinite number of smaller folds within it and larger folds outside it. These additional flows are not separate or discrete folds that are added to the original fold but rather further foldings of the original fold itself. This multiplication or manifolding of the fold is possible only because the original fold itself is elastic and pliable. This elasticity allows for a given aesthetic regime to generate an increasingly dense field without necessarily exceeding its external limits. The limits of the aesthetic field function as the infinitesimal limit that the multiplication of folds approach indefinitely without ever reaching or exceeding it.

Oscillation

The second kinetic operation of elastic motion is the capacity of the aesthetic field to expand and contract without breaking. The elasticity of a circulatory flow allows a kinesthetic field to oscillate back and forth, here and there, without disjoining the folds of the field. Oscillation is simply the expansion and contraction, back and forth, here and there, of the aesthetic field.

SERIALITY

If folds are indefinitely multiple and fields can expand and contract, this makes possible an indefinite seriality. The kinetic operation of seriality is defined by a basic differentiation between two kinds of flows: folded flows and conjoining flows. Any composite body is made of both a number of material or cellular folds and a pattern of motion that connects them. This is the most minimal kinetic distinction that can be made regarding sensuous matter—folded flow or unfolded flow.

Modern art is characterized by a kinetic process of multiplying and oscillating images into indefinite series. Every image can be broken down into multiple images all the way down and built all the way up in the same way, but without totality and without the fixity of aesthetic form: brio-collage. Form and function are not derived from one another but rather emerge within a more primary elasticity of material flows that can be folded up and rearranged with corresponding changes in form and function. Unformed matters and dysfunctional elements achieve a consistency depending on how their series of elements are combined in the assemblage.

Modern art is thus defined by the construction and deconstruction of a series.[2] Another fold can always be added or subtracted to the series to transform the field of circulation into something else, ad infinitum. The kinesthetic elasticity of the field of modern art makes possible a new malleability of matter and image, but it also relies on and presupposes a more primary elastic continuum within which its folds can be multiplied and oscillated here and there. The so-called freedom and experimentation that characterize much of modern art's drive to pastiche, constructivism, and deconstruction of heterogeneous images are derived from an underlying material kinetic condition of difference and elasticity in matter itself that makes such recombination and experimentation possible.

The serialization of matter into combinatorial images is frequently described as duration or temporality. Chronological time, however, is nothing more than the serialization of matter in motion. Just as modern art breaks space up into an indefinite series of combinatorial elements without essential form, so too it breaks up motion into an indefinite series of present moments following one another in an expanded and contracting aesthetic field. As we will see, one of the primary aims of modern art is to show the seriality of images through the passage of time or subjective mutation in the functions, forms, and relations in the arts.[3]

CONCLUSION

Difference, elasticity, and seriality are the three defining kinesthetic features of modern art. This short chapter provides only a general and conceptual definition of these processes, whose material historicity and emergence require the much longer exposition presented in the two following chapters.

CHAPTER 13

The Modern Image, I

Modern art is defined by the historical rise to dominance of an elastic difference within images. Since all flows are continuous, all kinesthetic images are indefinitely multiple and all fields are indefinitely oscillatory in their expansion and contraction. The degree to which an elastic unity of images actively changes function, form, and relation defines a work of art's degree of "kinetic difference." For example, the material relations of light, brought to prominence in the art of the Middle Ages, are now literally exposed as a multiplicity of different photographic and cinematic images. Like light, every relational material flow is indefinitely containable or divisible in a series of different frames that can in turn be edited into a new series of relatively expanded or contracted functions, forms, or relations: cinema. The great insight of modern art was to see all aesthetic fields as constituted by an internal difference and mutability to expand or contract itself into a compositional unity through continuous modulation.

The period of modern art, beginning around the middle of the eighteenth century to the late twentieth, is one of the most diverse in Western history. Therefore, I again remind the reader that these chapters are not an attempt to synthesize or summarize all major arts or works across the arts. The aim of these historical chapters is much more narrow: to examine only the kinetic patterns of a few major arts in just enough detail to support the broader thesis that modern art is characterized dominantly, but not exclusively, by the kinetic pattern of elasticity and difference. The argument of this chapter and the next is thus that the aesthetic field during the modern period is defined by an elastic and differential regime of motion. Each of these chapters marshals support for this thesis by looking closely at six major arts of the modern age: steel, the photographic image, and the novel in this chapter, and meter, the action arts, and molecular arts in the next chapter. Although empirically quite different and distributed over hundreds of years, each follows a similar kinetic pattern or regime, as we shall see.

STEEL

The first major elastic aesthetic field of the modern period is that of steel. The introduction of iron-alloy building materials, particularly steel, into the architectural arts made possible a whole new elastic regime of aesthetic motion. While the invention and use of iron metal alloys, including steel, goes all the way back to ancient Anatolia (1800 BCE)[1] and was used by the Romans,[2] it was not until the late eighteenth century that new production methods allowed cast iron to be produced cheaply enough and in large enough quantities to be regularly used in large building projects.

Iron alloy and steel made possible the advent of modern architecture by giving it a new pattern of motion. In contrast to wood, stone, cement, and other historical building materials, iron and steel have a significantly higher compression and tensile strength. This strength is made possible in particular because of the elastic quality of the metals. Cast iron and particularly steel can sustain significant tensile stress before they are bent, twisted, or broken. As the weight of a structure naturally shifts in different directions owing to earth movements, wind, temperature, and even its shifting contents, steel stretches elastically, oscillating back and forth, expanding and contracting, without cracking or breaking. The elasticity of iron and steel also make possible a new aesthetic elasticity of the structures that they compose.

The use of iron and steel columns allowed support walls to become even thinner and taller than the thinnest and tallest Baroque wall. The height and thinness of these modern walls also made possible even more and larger windows than in any Baroque church. The strength of the walls further reduced the necessity of their number and thus made possible the largest open spaces and floor plans ever seen, permeated with the natural light of enormous windows. However, by flooding architectural space with light, the Baroque play of tensional luminous relations was de-emphasized because there is no longer a contrast between light and dark to accentuate the play of light off the manifold surfaces.

Furthermore, such play of reflective surfaces was reduced by decreasing interior and exterior ornamentation. Light came in from all angles but passed right through transparent glass, contra the colored, semi-opaque stained glass of the Middle Ages. Interior and exterior were no longer contrasted in a dynamic relation; rather, they became differentiations, frames, or relays of the same continuous light. Light became serialized and minimally reflected. The modern room was now a relay, module, or modulation of the beam of light as it passed through. It is no coincidence, therefore, that the first iron architectural structures were sites of transportation, mobility, passage, and openness, such as the Iron Bridge (1781), Coalbrookdale Railway (1768), and the Commissioner's House of the Royal Naval Dockyard (1820s), and not the fixed play of reflective surfaces inside the darkened resonating and reflective church.

Frame Construction

Iron and steel allowed for a new technique of frame construction. Metal-frame construction creates a "skeleton frame" of vertical columns and horizontal beams in a rectangular grid pattern. The skeletal frame is then used to support floors, walls, and ceilings. In contrast to the medieval practice of cutting stones to fit on site, premade metal frames or modules could be assembled elsewhere or ahead of time and then simply assembled on site. Metal-frame construction is thus based on a kinetic differentiation or fragmentation of the building into exchangeable component parts that can be quickly and cheaply assembled, deconstructed, and reconstructed.

Iron Frame

As early as Joseph Paxton's Crystal Palace (1850; figure 13.1), architectural forms were being broken down into component frame modules. Form was beginning to be differentiated into heterogeneous frames. Matter was being abstracted from its organic relations and reassembled. Architectural function became as open as the empty space in which it occurred. The organic architectural body was becoming internally differentiated in favor of a more tensile, elastic, and serialized frame body. The railway station, for example, was increasingly built in the same serial manner as the railway itself—one metal frame at a time, molded and manufactured elsewhere, transported, and assembled.

The Crystal Palace is one of the early precursors, if not the very first work, of modern architecture. Paxton, originally a gardener who designed greenhouses, submitted his design in a contest to build the hall to house London's Great Exhibition of 1851, which would showcase the works

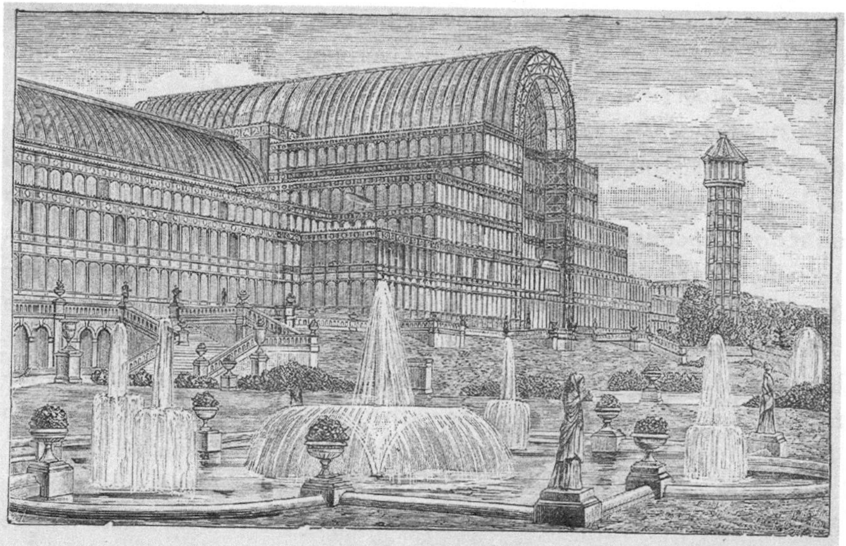

Figure 13.1 Joseph Paxton's Crystal Palace (1850), London
Source: Wikimedia, https://commons.wikimedia.org/wiki/Crystal_Palace#/media/File:CrystalPalaceEngraving.jpg

of industry of all nations. Paxton designed iron frames to fit the largest manufacturable strong cast plate-glass sections, a glassmaking technique invented just two years earlier. The iron-framed glass was prefabricated, shipped to the site, assembled into larger square modules, and then put together. The result was the largest surface area of glass ever seen, and a building that required no interior lighting. The Crystal Palace was cheap, easy and quick to build, and temporary. After the Great Exhibition, the palace was disassembled, moved, and reassembled with some changes at Sydenham Hill and completed in 1854.

The Crystal Palace itself introduced a new elastic and differential pattern of motion into architecture. For example, instead of capturing and reflecting light off an ornamented surface as the Baroque church had done, or concealing a mysterious darkened central core as the ancient temple had done before that, the Crystal Palace allowed light to pass through a series of metal frames. Light became serialized without being blocked or reflected around in dynamic polygonal play. The multiplicity of the Crystal Palace's glass plates transformed the whole building into a kind of photographic montage in which each window took its own snapshot of the flow of light. If the vaulted ceilings of the church became a canvas for painting reflected light, modern structures became cinematic projectors making images as light passed through their transparent frames. We can see this also in the first metal-framed glass-curtain walls by Peter Ellis at Oriel Chambers (1864) and 16 Cook Street (1866) in Liverpool, which transformed their interiors into a cinematic slideshow for the outside world.

Instead of expensive heavy stone, strong, thin, cast-iron frames defined the differential components of the building. The height of the building was no longer limited to the stone arch and vault but, rather, could stand up on its own modular supports in elegant simplicity and tensile elasticity. The whole building became increasingly elastic, mobile in wind and horizontal stresses. By supporting more stress, the floor area of the space could be opened up to all kinds of functions, as was required by the exhibition. Modular frame-style architecture thus breaks form into its deformed components, multiplies them, then expands and contracts them to increase the size and distribution of the open space as needed.

Another early example of modern architecture is Alexandre Gustave Eiffel's Tower (1889), which used a similar iron-frame construction style to achieve a complete transparency of the interior and exterior in a way absolutely new to viewers. Light passed straight through the structure. No longer a contrast of reflective tensions between inside and out, or a mimesis of model forms, the Eiffel Tower explicitly transformed the skeletal framework or formwork that was typically removed after real building materials, like stones, had been set in place, into the work of art itself.

The kinetic inversion is incredible! That which had only been the bare, partially formed, parafunctional, briefly related material conditions of ancient and medieval architecture now were the structure itself. Skeletal framework was the architectural difference that made a difference—the difference *between* form and copy—but the Eiffel Tower stood alone, minimal, unornamented, industrial, prefabricated, itself made from a series of iron molds. The ancient idea of pure aesthetic forms was exposed as already different from itself, composite, partial, fragmented. The form of the tower is nothing other than a composite form of preformed iron columns and beams. The Eiffel Tower robbed aesthetic form of its transcendence by laying bare its multiformed and deformed, differential, and composite skeletal conditions.

At the time of its construction, the Eiffel Tower was the largest building in the world and remained the tallest manmade structure until the Chrysler Building was built in New York City in 1930. Being made of iron, it was also the most elastic and mobile building in the world. The tower was built to sway several inches in the wind. The heat alone causes the top of the tower to oscillate as much as seven inches back and forth and six inches up and down. The relative immobility and permanence of the stone cathedral was thus increasingly replaced with the more elastic, modular, and differentiated metal framework.

Steel Frame

The introduction of steel for architectural purposes was realized by the invention of the Bessemer process in 1855, which made steel production more efficient and cheap, and gave it a dramatically higher tensile and compressive strength. By the 1880s, steel was widely used for construction, allowing a previously impossible new level of elasticity and differential frame composition. Steel-frame construction borrowed heavily from cast-iron construction, but owing to the increased elasticity of its material, a new age of modern skyscrapers was now possible. In response to the shortage of land, the high cost of real estate, and the need for fireproof materials in urban areas, steel-framed skyscrapers were increasingly built in a number of major cities in America and Europe.

The first was the ten-story Home Insurance Building in Chicago, built in 1884 by William Le Baron Jenney. Although it was built from a combination of cast iron, masonry, and steel elements, it was the first large building to utilize steel framing and thus weighed only one-third that of a traditional stone building. In the United States, the first all-steel framed building was the Rand McNally Building in Chicago, erected in 1890. Numerous others followed, building higher and higher, and using the same basic steel-frame construction technique.

Kinesthetically, the steel-framing techniques follow the same iron-frame pattern of elasticity and modular differentiation but to a greater degree. Steel-frame skyscrapers were built to be even more elastic, even more temperature-responsive, and even more lightweight and strong. This meant that buildings became even more vertical, even more mobile in wind and sun, with even taller walls, more open spaces, larger plate-glass curtains with thinner and thinner frames, and even more light. This kind of motion lent itself directly to the needs of economic circulation that were becoming increasingly dominant at the turn of the century.

Cinematic Architecture

While the form of the enclosed church dominated the previous age, the form of the department store, office building, and apartment complex dominate the modern age. One of the great pioneers of capitalist architecture was the great Louis Henry Sullivan (1856–1924). Sullivan famously claimed that form followed function, but what is less often noted is that the connection between form and function is itself neither formal nor functional but rather differential. Sullivan was right that form *follows* function, but we should also observe that function also *follows* form at the same time. Function and form are two dimensions or elements of the same kinesthetic pattern of differential motion made possible by modern steel construction. It is the *following* or *moving between* the two that is crucial.

For example, the expansion and contraction of open space, lighting, ventilation, building size, and window placement, which compose form and function, are possible only under the condition that one has a wide control over them. However, the maximization of open space, light, number of floors, and so on was made possible only recently by the multiplication and oscillation of frame construction. Furthermore, it is no coincidence that Sullivan's adaptation of form and function almost always looked like a maximization of open space, lighting, window size, and so on based on the differential-frame approach.

Sullivan accepted and even celebrated capitalism's increasing demand for the kinetic circulation of commodities, both human and nonhuman. The increasing circulation of modern capital is both the condition for and the effect of an architectural "realism" capable of creating spaces that allow for the maximization of commodity circulation. Large open spaces like those of Sullivan's Guaranty (Prudential) Building in Buffalo, New York (1894–1896), make possible the mobility of furniture and interior walls to adjust to the contingencies of market demands. The multiplication of windows allows maximum light for workers and an increased ventilation system to aid productivity. Workers and landless proletarian renters move between apartments and office-building spaces just as rapidly as the display windows of the department store rotate their displays. It is no coincidence that the apartment, office, and store now begin to look the same under Sullivan's designs, like a series of temporary cubicles or relay points along the continual circulation of human capital—production, consumption, repose, and repeat. The worker on display, the merchandise on display, the tenant on display—all are commodities on display in a well-ventilated society.

Sullivan's problem was no longer how to support an organic function, or to venerate a perfect form, or even how to fix the right relations around them but rather how to circulate them in an indefinite series of modular lifeways always open to mutation. Just as the market expands and contracts, so does the great elastic skyscraper with its flexible steel, ventilated windows, and temporary proletarian populations. For example,

Sullivan explicitly imagined the display windows of his steel-framed Carson, Pirie, Scott Department Store in Chicago (1899–1904) as a series of "pictures" and their steel frames as picture frames. As one circulates between these enormous buildings, one experiences a kind of cinematic architecture. The display-window mannequins on the other side are juxtaposed with the ghostly reflections of onlookers. Instead of intensifying their mirrored refection, the display windows create a series of transparencies, each traversed by a continuous flow of light that brings them together but only as differential repetitions of a modular commodity form—prefabricated and assembled everything.

The Elevator

Skyscrapers are possible only under the kinetic condition of efficient transport mobility through them: the elevator. For the first time, buildings were designed with an enormous transversal shaft running through their floors. The elevator creates an architectural space like no other before, a single transversal room that can only be occupied one module at a time, just as the lens shaft of a film projector is occupied by only one frame at a time. In fact, the windows of many early elevators gave precisely such a cinematic view as they oscillated up and down the great empty shafts. The elevator creates the spatial unity of all the differentiated floors.

By the act of multiplying floors, one also internally differentiates architecture from itself. Each floor is like its own building, and perhaps with its own function and operation that differs from that of other floors. It is a pastiche of different persons, events, and coordinated activities. The elevator is the strange zone of unproductive activity, a vacuole of noncommunication, a dysfunctional module where no work, sleep, or shopping are done. Yet it is the kinetic condition of all architectural function in the building, the differential unity of all form, and the condition of the relation itself between the floors, forms, and functions. It is the continuous differential unity of the different floors. The modern building is thus transformed into a vehicle, a strange kind of vehicle whose destination is completely internal to the vehicle itself. The great oscillating elevator is possible only under the condition of the strong steel-frame core that surrounds the central shaft of the building. The skyscraper and the elevator are mutually constitutive; without the one, the other would be useless.

Reinforced Concrete

Steel also made possible a new fortification and elasticity of the traditionally inelastic building material of concrete. Between 1853 and the end of the century, numerous inventors in France, England, and the United States experimented with the use of concrete containing steel rebar or wire mesh. Rebar is bent into whatever shape is needed, and concrete is poured around it. Steel and concrete experience a similar degree of thermal expansion and contraction, so when the cement hardens around the rough or corrugated steel, it allows tensile stresses from the concrete to be transferred to the steel. The concrete protects the steel from fire, rust, and other damage, while the steel allows the concrete to be molded into any shape that the steel itself can be bent into. Reinforced concrete replaced stone and brick as the primary material for modernist architects.

Again, reinforced concrete follows many of the same kinetic patterns as iron and steel-frame construction but more so. Not only could steel frames be prefabricated, but now truckloads of corrugated-steel rebars could be made thin enough to be easily cut, bent, and assembled on site. This took decomposition, modularization, and differentiation to a radical new level whereby a building could now be deconstructed into its most minimal, exchangeable, modular parts: steel rebar. Architecture became increasingly elastic as more and more steel bars, steel mesh, and steel frames were introduced. Even something as brittle as concrete became elastic. Again, form follows function but only on the condition of the multiplication and deconstruction of matter into its most elastic and differentiated parts such that form can respond to whatever function. However, form and function become responsive and elastic under the more primary kinetic condition of their more general differential plasticity with respect to one another.

Plasticity

In 1917, Dutch architect Theo van Doesburg founded an artistic movement called De Stijl, or "The Style," and also called "neoplasticism" or "the new plastic arts." Doesburg defines neoplastic architecture as follows: "The new architecture is anti-cubic, i.e., it does not strive to contain the different functional space cells in a single closed cube, but it throws the functional space (as well as canopy planes, balcony volumes, etc.) out from the centre of the cube, so that height, width, and depth plus time become a completely new plastic expression in open spaces. . . . The plastic architect . . . has to construct in the new field, time-space."[3] Neoplasticism expresses precisely the elastic and flexible kinetic conditions under which a more general serialization and temporalization of architecture is possible. Reinforced concrete allows for walls to support increasingly larger spaces without the use of columns and pillars that would break up the space. The modular cube can thus be thrown open and allow for a new emphasis on the empty form and open function of the room. Human movement through dysfunctional or openly functional space now emphasizes the seriality of time and temporality produced by movement through space.

In short, the art of open space is a temporal art but only on the condition of a more general destabilization of form and function in motion. De Stijl architecture's focus on abstracting primary colors and simple geometric components in order to recompose them follows the modernist modularization and differentiation of composition into its constitutive parts. For example, De Stijl architect Gerrit Rietveld designed his Schröder House in Utrecht, Netherlands (1924), such that the entirely open second floor is segmented by mobile or sliding partitions that define the rooms, like modern office-cubicle partitions. The flexible elasticity of walls also produces a corresponding expansion and contraction (oscillation) of the time it takes to move through the space. The open plans of Gropius's Bauhaus Shop Block accomplishes a similar elastic pattern of motion for his students. Ludwig Mies van der Rohe's model for a glass skyscraper (1922) relied almost entirely on a solid-steel reinforced concrete core transacting a series of entirely glass-walled floors, thus rendering explicit the elastic and transversal conditions of vertical and horizontal seriality. Just as light traverses horizontally through the floors, so the reinforced concrete shaft traverses them vertically at a perpendicular angle typical of De Stijl cubist composition.

The Dutch De Stijl, German Bauhaus, and Russian constructivist architectural styles all shared a common inspiration from Cubist painting techniques, even if they departed from them in important and different ways. Cubism is an art of fragmentation, modular differentiation, and compositional unity of the heterogeneous. The very idea of a constructivist or productivist style itself already presupposes the prefragmentation or differentiation of matter into compositional images to be constructed into nonorganic forms and relations owing to the new plasticity of steel and concrete. By tearing down the old forms into bits, one can build them up in a new way.

The right angles of cubism and De Stijl even find their way into Le Courbusier's utopian plans for horizontal transport cities and vertical urban cities. The pattern of motion remains the same: one begins with the differentiation of matter and persons into discrete heterogeneous elements, and then constructs them into a horizontal suburban series and a vertical urban series. The two, as architecture critic Lewis Mumford writes, are the mirror image of each other—isolated elements assembled together in a series.[4] Hence, the strange convergence of Le Courbusier's utopian capitalism and utopian socialism are brought together by their shared kinetic celebration of differentiated components stretched elastically in a series along rational Cartesian coordinates: the cubist grid plan.

Even Frank Lloyd Wright's similarly utopian attempt at an "architecture of democracy"[5] at the Robie House (1907–09) and Fallingwater (1936–39) affirms a new organicism only on the condition that the organic body be treated as a series of autonomous individuals held in the open composition of a free space of circulation. However, nature is not an open space like a civic center or capitalist market; it is a densely folded and occupied ecosystem that is paradoxically destroyed and leveled in order to open it up to "free circulation," as Le Courbusier knew well. At Fallingwater, the interior and exterior stretch into each other only by transforming the natural rock into a series of orthogonal bricks stacked at right angles while architectural space is opened around it with the use of reinforced concrete. The natural rock gives the vertical axis of support while lightweight reinforced concrete roofs and porches give the horizontal axis. Wright's attempt at a new organic unity of materials and site is thus paradoxically only possible under a more generalized deconstruction of the site into recomposable orthogonal units.

Deconstructivism

Deconstructivism is the logical architectural conclusion of modernism. It renders explicit and visible what had already been kinetically presupposed from the beginning of modern architecture: that matter is divisible, exchangeable, already broken down into units that can be constructed into something new, or not. Even the deconstruction of form, balance, order, and symmetry is itself a construction. If aesthetic form is already differential and composite, then formation merges with deformation in the same serial recomposition of components. Günter Behnisch's Hysolar Institute (1987) in Germany and Rogers and Piano's Pompidou Centre (1977) both render explicit the prefabricated, differentiated, and heterogeneous nature of the material components of composition that were previously smoothed over by layers of concrete, stucco, and glass. The work of Frank Gehry and Daniel Libeskind shows the act of differential composition in action. This does not make them anti-modern but, rather, all the more explicitly and self-consciously modern. Libeskind's Denver Art Museum (2006), for example, is a composition of heterogeneous three-dimensional elastic metal shapes, including the oblique. Gehry's Guggenheim Bilbao Museo (1997), in Spain, also renders explicit the possibility of curvature and organicism precisely from thousands of prefabricated titanium squares. The continuous, as Wright believed, is here explicitly made possible by the assembly of the discontinuous, the differential, and the elasticity of expanding and contracting metal in the sun. Modern architecture thus fulfills Zeno's paradox of an infinite compositional and differential series that expands and contracts in the open space of an elastic space-time.

THE PHOTOGRAPHIC IMAGE

The second major elastic aesthetic field of the modern period is that of the photographic image, found in photography and cinema. Both introduce a strong emphasis on aesthetic difference and the elasticity of the aesthetic image. With the introduction of the camera, the whole aesthetic field is decomposed into a series of differentiated fragments, frames, and slices of the continuous flow of light infinitely rearrangeable in an infinite series. The flow of networked light relations dominant in the Middle Ages is now divided up into a multiplicity of different shots. Through projection, the resulting images are then elastically expanded or contracted (enlarged or compressed) onto photographic paper or the cinema screen. As Susan Sontag writes in her famous book *On Photography* (1977),

> The photograph is a thin slice of space as well as time. In a world ruled by photographic images, all borders ("framing") seem arbitrary. Anything can be separated, can be made discontinuous, from anything else: all that is necessary is to frame the subject differently. (Conversely, anything can be made adjacent to anything else.) Photography reinforces a nominalist view of social reality as consisting of small units of an apparently infinite number—as the number of photographs that could be taken of anything is unlimited. Through photographs, the world becomes a series of unrelated, freestanding particles; and history, past and present, a set of anecdotes and *faits divers*. The camera makes reality atomic, manageable, and opaque. It is a view of the world which denies interconnectedness, continuity.[6]

This is true not only of photography but of cinema as well, which relies on a similar material kinetic medium: the photographic image. It is true that the photographic image denies or fails to present the infinitely receding continuity of the world that slips through the difference between one frame and the next, but it is also true that the very act of division and framing presupposes the continuity of the material-kinetic substratum that it divides and frames. Photography and cinema express this in different ways.

Photography

The emergence of photography is less an abrupt invention than a gradual transformation by degrees, beginning with the ancient *camera obscura* and culminating in the use of photosensitive silver nitrate to capture a duration of light. The first attempt to capture a photographic image on a chemical-sensitive surface occurred in the photoetchings of the British inventor Thomas Wedgwood, who painted images on glass surfaces and allowed the light passing through to affect the sensitive paper below. However, the images quickly faded. The first successful attempt to capture an image from a *camera obscura* on photosensitive chemicals was made by French inventor Nicéphore Niépce in 1822. By 1826, Niépce was finally able to produce a print of his images, culminating in the earliest surviving photograph from nature, "View from the Window at Le Gras."[7]

These first photographs already reveal the kinesthetic connection between the metal-framed plate-glass window and later photographic cameras. The photographic image is a framed slice of a continuous flow of light, like a view from a window or a glass plate. The aim of the modern transparent glass window is not so much to reflect light as to let light pass through. While the Renaissance picture frame operated as a reflective or mirrored surface to show its network of constitutive luminous relations, the modern picture frame is the window frame that lets light through only to capture a slice of it—to divide its luminous continuity into a series of frames, shots, or blocks of decomposable space-time. The window frame is the first photographic frame, and its plate glass is the first photographic lens. After Niépce died in 1833, his partner Louis Daguerre perfected a process for using silver halides to produce images with shorter exposure (minutes instead of hours), called the daguerreotype process. France made this process public knowledge in 1839.

In 1840, the British inventor William Fox Talbot developed a new calotype process that was similar to Daguerre's but that produced a translucent negative that could be used to print multiple positive copies. Talbot's first photograph using this process is not coincidentally of an oriel window at Lacock Abbey (1835).[8] Again, the window functions to let in a continuous flow of light, but also to frame it, slice it, and divide it up into frozen segments. The negative makes explicit the kinetic reversal at play. The flow of light is captured into that which it is not: the darkness of a relatively fixed chemical residue, but the chemical residue itself keeps moving and fights against the fixity of the image.

The great contribution of John Herschel was to discover a chemical that would stop and fix the light–silver reaction: sodium thiosulphate (1819). From a kinesthetic perspective, the emergence of photography can be seen as one long quest for stasis and fixity in the face of the continuous flow of light and light-driven chemical flows. In a kinetically radical insight, Herschel went even further by exposing the negative image itself on a plate of glass (1839), which could in turn be easily used to produce a multiplicity of positive images employing the same process as used to produce the original image. Thus the photographic image comes full circle on itself as light passes through a glass frame. Glass-plate technology and the negative process remained standard into the twentieth century.

With the advent of Herschel's glass negative, a vertiginous inversion takes place. The original flow of light through the window is now mirrored in the flow of light through the plate of glass that constitutes the "copy." However, since both original and copy are now produced in exactly the same way (light through glass), the whole model/copy relation is exposed at the foundation of photography as actually two aspects of the same, more primary kinetic process. The window and the photograph of the window are frames of glass that show an image. Model and copy are thus simply different images of the same elastic motion of light, and are not part of some aesthetic hierarchy of resemblance. The centrifugal model of antiquity is thus overthrown by rendering sensible this flow of light through multiple frames as nothing but differential images.

Framing

Kinetically, photography is defined by an aesthetic field of differences between discrete slices, frames, or blocks of space-time. Within and between each block is an indefinite multiplicity of differences. Exposing this difference between multiple frames photography thus renders visible what painting cannot—the temporal and aesthetic differences internal to the duration of the act of painting itself. In a series of certain photographs, one sees the differences between the photos in the series, qua series. While a Baroque oil painting might take months or years to be painted and dry, the photograph takes only minutes or seconds. This speed makes possible the photo series.

However, the photograph is not an instant—it is not an empty, instantaneous slice without duration or space—despite a certain prevailing aesthetic fantasy about its perfect stasis, fixity, and realistic representation. All photographs require the exposure of a light-sensitive material for a certain duration. Although the time needed for such exposure is reduced through technological innovation, it asymptotically approaches the "instantaneous" but will never reach it. The photograph will always be a capture of a certain duration of a certain process of light.

In short, because of this duration between every photo in a series, it will always be possible to take one more photo between the others. Photography does not just assume the seriality of motion or "time"; it also produces an aesthetic image of it. At its limit, photography thus assumes the continuity and flow of matter such that the flow can be infinitely divided or folded up into a serial multiplicity of different frame-blocks. Therefore, matter is infinitely differential only on the more primary condition of the pure continuity of its material flow. The kinesthetic flow enters and moves through a series of slices: lenses, glass plates, and copies. It is no wonder that photography lent itself so rapidly and easily to the techniques of discrete scientific quantification in passport photos, medical study, and war photography, where the human body is framed, fixed, and temporally dissected into a frozen body: the cadaver or corpse.

Timothy O'Sullivan's "A Harvest of Death, Gettysburg" (1863) is so striking in part because of the similarity between the immobile dead body of the soldiers and the immobile body of the photographic image itself. As Auguste Rodin writes, "[Photographs] present the odd appearance of a man suddenly stricken with paralysis and petrified in his pose. . . . If, in fact, in instantaneous photographs, the figures, though taken while moving, seem suddenly fixed in mid-air, it is because, all parts of the body being reproduced exactly at the same twentieth or fortieth of a second, there is no progressive development of movement as there is in art. . . . [I]t is the artist who is truthful and it is photography which lies, for in reality time does not stop."[9] Since photography can only be a differential capture of a fixed duration of light, its only chance at indicating its kinetic conditions is the use of multiple exposures, such as Gjon Mili's "Nude Descending Staircase" (1942), a clear reference to the kinetic focus of the futurists and Marcel Duchamp's original *Nude Descending a Staircase No. 2* (1912), and the use of long exposures like Étienne-Jules Marey and Georges Demeny's *Pathological Walk From in Front* (1889; figure 13.2), Frank Gilbreth's *Cyclegraph* (1914), Man Ray's *Space Writing* (1935), and others.

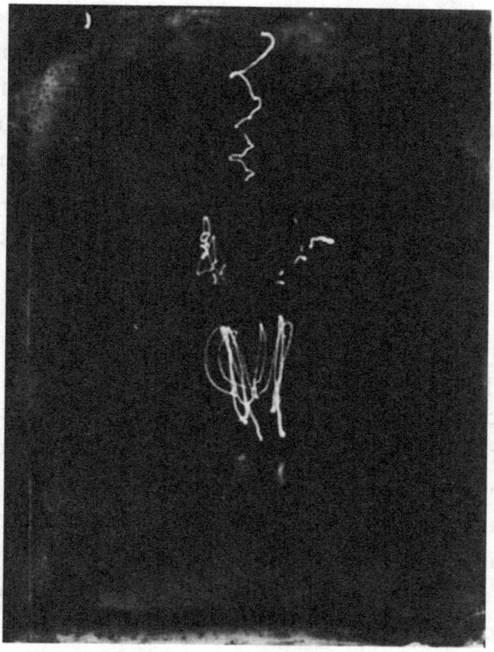

Figure 13.2 Georges Demeny, *Pathological Walk From in Front* (1889)
Source: Wikimedia, https://commons.wikimedia.org/wiki/File:Etienne-Jules_Marey_et_George_Demeny._Pathological_Walk_from_the_Front.jpg#/media/File:Etienne-Jules_Marey_et_George_Demeny._Pathological_Walk_from_the_Front.jpg

Multiple Exposure

By exposing the same plate multiple times, the photographic image reveals the internal differential of the image itself: that every photographic image is already differential, already multiple, already exposed by more than a timeless instant of light. The image is not infinitely thin but rather a block of space-time that can be infinitely exposed, multiplied, and differentiated with respect to itself. This is less a direct representation of motion as its authors may have hoped and more a negative indication that such infinite exposures are possible only under the more primary condition of continuity that makes such infinite division possible.

Long Exposure

Long-exposure photography reaches the same conclusion by other means. Instead of multiplying the differences within the image, it traces the path of light over a given duration and shows the simultaneity of the path. The simultaneity of the process in a single image does not show the procedure but, rather, indicates and gestures to the kinetic conditions of flow and continuum that must have been the case for the capture of the process to be possible. As such, the long-exposure image reveals the fundamental nature of all photographic images as themselves continuous kinetic exposures made visual through short bursts of durational exposures.

Projection

Kinetically, the photographic image is defined by an aesthetic elasticity in the use of projection lenses to expand or compress the light image. Once the original plate or celluloid slide film is exposed, it is mounted on a projector that could make the image larger or smaller as it focuses on a light-sensitive paper for a certain time. This sheet is then soaked in other chemical baths to bring out the image and arrest its process, then set out to dry.

In short, the original process of production is duplicated with a projected light in order to expand or contract a block of light. This elastic enlargement or compression of the image is only possible precisely because light itself is relatively continuous and can be focused or expanded while still retaining the image. The photographic image thus has a new elasticity that painting does not have, which it gains by being continuously stretched and fragmented. By photographing other works of art, photography can then transfer its kinesthetic elasticity to all other visual arts through its duplication and projection of them.

Cinema

Cinema functions according to a similarly differential and elastic kinesthetic field. Already in the 1650s, Christiaan Huygens's magic lantern had begun using painted plates of glass to project images on a screen, moved by hand. The modern invention of manufactured transparent glass made possible a whole new kinesthetics of transparency and seriality. The sensory image was less an eternal form or relation of reflection than a series of semi-transparent layers—differential realities stacked on top of one another without final depth or ultimate surface. Aesthetic relations became laminar, serial, and intercalated.

By increasing the speed of the transition between multiple images, the phenakistoscope (1832) and the praxinoscope (1880s) eventually demonstrated quite dramatically the optical effect of motion produced when the short intervals between frames are moved faster than the eye can replace the previous visual image. The perception of motion is due to a psychological effect called the phi phenomenon. However, cinematic precursors like these are not at all "dissimulations" of motion but are, in fact, highly kinetic apparatuses in their own right. The praxinoscope, for example, was a rotating wheel that could eventually hold hundreds of images. It is precisely the continuity of motion itself that makes possible the appearance of movement in the cinematic image. It is motion that allows for the series of fixed images to give the appearance that multiple images are only a single moving image. This is the great discovery of modern art: that seriality and compositional fragmentation make possible a new composition through *motion*—a continuous fragmentation or fragmented continuity.

Most famously, in the late 1870s, Eadweard Muybridge was the first to combine the discoveries of animation with those of photography. He accomplished this first by setting up a series of cameras (two dozen at most) set to expose in a temporal series, one after another, onto a photographic glass plate. Images made of the photographs could then be placed next to one another and moved in a series by means of his zoopraxiscope. The result was the first animated use of photographic images in *Sallie Gardner at a Gallop* (1878).

Inspired by the combination of these two technologies, there followed a number of film cameras that could take multiple photographs in succession, including a single-lens camera by Louis Le Prince in 1888 that could take a rapid series of photographs onto celluloid film. His first film, *Roundhay Garden Scene* (1888), was shot directly in front of a three-part oriel window that projected out from the nearby house—as if to suggest again the kinesthetic connections between the window, frame, glass, camera, and cinematic image.

In 1891, William Dickson, an employee of Thomas Edison, invented the kinetographic camera, powered by an electric motor and capable of exposing one frame of a single continuous celluloid film strip at a time and then quickly advancing the film in about 1/460th of a second for the next exposure. This new camera was the first technical condition for the practical use of high-speed filming for the next century of cinematography. The Lumière Domitor camera accomplished something similar, using perforated 35mm film.

Framing

Kinetically, cinema is defined by an aesthetic field of differences between discrete slices or blocks of space-time moving at a given frame rate. In addition to the spatial difference between frames achieved in photography, cinema adds a kinetic difference between frames: the frame rate or speed at which each cinematographic image passes across the lens of the camera. The profound kinesthetic discovery of modern cinema is not only that aesthetic forms and their relations are materially decomposable into a series of discrete spatiotemporal blocks or frames of light, as in photography, but also that in between these frames one can locate a more primary spatiotemporal difference that makes the frame and the differences between frames possible.

In other words, the cinematic image discovers a hidden dimension between images: a kinetic gap, a speed or rate of motion that constitutes the photographic unity of a whole series. While photography decomposes aesthetic forms and relations into a series of frames or blocks from which a new frame can be produced, cinema acts on an entire series of photographic images by working between the frames through editing. By further decomposing the continuous flow of light, forms, and relations between frames into time frames, an entirely new spatiotemporal series can be produced through the motion of the projector.

Montage

Montage is the art of framing, deframing, and assembling a series of different frames. While multiple- and long-exposure photography render sensible the vast multiplicity of differences internal to a single frame, cinematic montage and editing render sensible the vast multiplicity of differences between multiple frames and multiple frame series (figure 13.3). In cinema, every frame series can be composed of smaller series and can be part of a larger series, with the minimal limits of the single frame and the maximal limit of the whole film. Between these limits editing occurs.

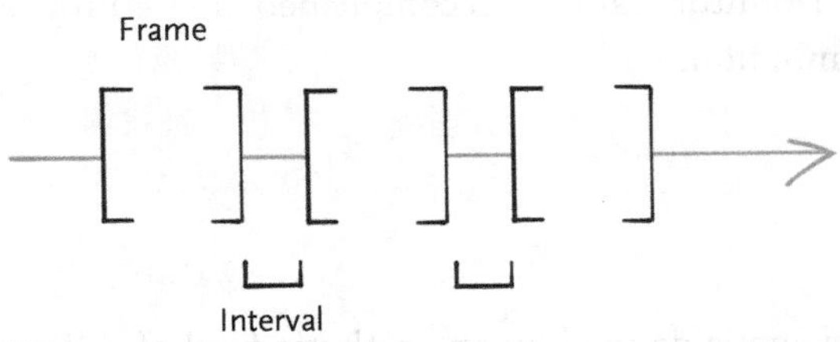

Figure 13.3 Montage and still shot of cinema film

The narrative montage of D.W. Griffith's *Birth of a Nation* (1915) contracts a large composite frame series into a series of smaller ones by shortening each of its composite series and reassembling them. The dialectical montages of Sergei Eisenstein's *October: Ten Days That Shook the World* (1928) contracts each frame series even further, producing a rapid-fire tempo of pure juxtaposition. In Vertov's *Man with a Movie Camera* (1929), these rapid juxtapositions are then themselves assembled into a much larger series: the juxtaposition series. In American and Soviet montage, the frame series is broken down until it reaches its minimal cinematic limit of the difference between rapidly alternating single frames. In French and German montage, however, these fragmented series are reassembled into a new, inorganic whole. In Abel Gance's *Napoléon* (1927), the montage stretches out to include frame series from the entire film (330 minutes) as a whole, including the memorial past and imagined future of Napoleon's life. Cinema fragments the totality of the film in order to recompose it into an inorganic montage of a new *fragmented whole*.

In Lang's *Metropolis* (1927) and Wiene's *The Cabinet of Dr. Caligari* (1920), the frames themselves become split and cut open in the high-contrast chiaroscuro of black and white lines stretching out of the frame. Unlike the quantitative totality of series in Napoleon's life, German montage invokes a qualitative totality of frame series, but one that has gone missing in the unexplainable power of supernatural events and in the darkness of the invisible interval itself only seen through the cracks of the frame and in the montage series as a lack or formless swamp, like the one in F.W. Murnau's *Sunrise* (1927). German montage exposes this expressionistic darkness that limits the frame series on either end of the film but gives it a visibility in the frame itself as darkness—a dark background, a fog, or a shadowed corner. The black, unexposed interval between photographic frames is exposed as the constitutive material kinetic condition for the entire film's own motion. The dark gap between frames is expressed in the frame itself as darkness—the intervalic and invisible darkness that connects the frames and continuously differentiates them.

Still Shot

Cinema also uses long takes of relatively fixed scenes in order to render sensible the interval series that makes the frame series itself possible and mutable. In between every photographic frame is an interval, a difference that makes a difference between frames. While montage renders this interval sensible only indirectly through the decomposition and recomposition of the frame series, the still shot renders this interval sensible as a simple passage of time during which relatively "nothing" is happening in the dramatic action of the film or in the frame itself. Cinematic temporality is a temporality made possible by difference and differentiation in the frame and between the frames. Time passes but only on the condition of a more primary differentiation or division between distinct images made possible by the continuity of the filmstrip and continuous movement of the film projector.

In Chris Marker's *La Jetée* (1962), for example, the stillness of the shots and the use of photographs makes explicit the difference between the photographic image and the cinematic image. When Marker films a photograph, he renders sensible the duration and movement of the film itself as the kinetic condition of the relative immobility of the photo image. This is the great inversion of postwar cinema: Only by filming something that does not move is the movement of the camera itself made sensible. The series of intervals between the frames is exposed as the condition for the frame series itself and the persistence of the immobility of the image.

In Carl Theodor Dreyer's *The Passion of Joan of Arc* (1928), for example, the face becomes frozen, but as frozen the face exposes the implicit movement of the camera itself and the mobility of the dark intervals between frames. This invisible darkness is depicted visually in the black pupils of the eyes. The eyes are the black holes or intervals of the face frame. In Andy Warhol's *Empire* (1964), the duration of the unedited film merges with the real-time duration of someone actually watching the Empire State Building. By merging the duration of the camera and the body, the kinetic condition of both is made sensible. The eye blinks just as the frames pass through intervals.

In Ingmar Bergman's *Persona* (1966), the filmstrip burns and rips off the reel, exposing the intervalic gaps and their material conditions of celluloid, light, and motion that sustain the frame intervals themselves. The continuum of white light is shown as the pure kinetic condition of the film, but only on the condition that the actual film *Persona* remains intact, framed, and not burned at all, even if what it shows is burned frames and white light. It is still a white light divided by the frames of the celluloid film itself. Film cannot escape its own material conditions; it can only reveal them: light, motion, frame, interval. During the duration of this white light in which nothing happens and no one is there, the differential kinetic interval comes to the fore as the condition of duration (figure 13.4).

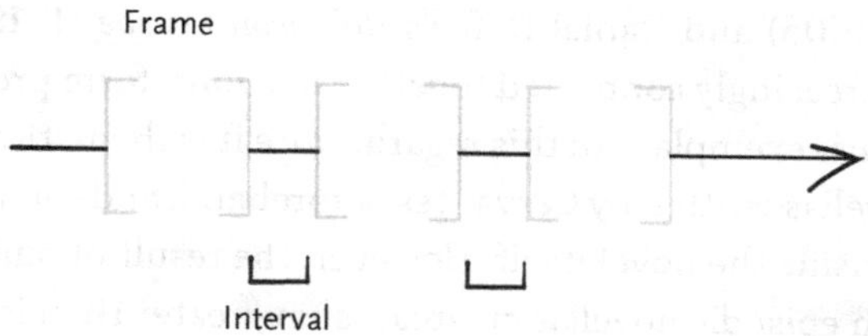

Figure 13.4 Intervalic series in film

Projection

Cinema also introduced a much wider range of elasticity into the image by transforming the film–camera apparatus into the apparatus of projection as well. The introduction of the film projector in the nineteenth century made it increasingly possible to expand a series of images many feet in diameter and thus viewable by large groups of people at the same time. Using the same optical technologies for focusing, the film projector could enlarge or compress the projected image for the duration of the whole film and could accommodate diverse spaces and audience sizes.

Again, the elastic enlargement or compression of the image is only possible precisely because of the continuum of light and of matter in motion that can be stretched to fit the big screen and small screen alike, while still retaining the image ratio. Cinema also allowed for a new oscillation of the image in the forward-and-reverse moving series.

THE NOVEL

The third major elastic field of the modern period is that of the novel. While the tradition of epistolography in the Middle Ages made possible the novella, which gave kinetic primacy to a tensional relation between correspondents at a distance, modern writing introduced a new dominant kinetic pattern of elastic difference internal to the novel itself. The origins of the novel are extremely controversial, and some argue stretch all the way back to antiquity[10] or start as late as the eighteenth century.[11]

However, the kinetic theory of the novel put forward here is defined less by the symbolic content of the work than by the historical material composition of the novel itself. To each origin story, its own criteria. Even in its kinetic definition, however, we find only an incremental transition from the novella to the novel that occurs sometime between Cervantes's

Don Quixote (1605) and Daniel Defoe's *Robinson Crusoe* (1719) as a series of novellas increasingly connected together in a long-form prose narrative.

Don Quixote is exemplary in this regard, since it is thematically related to a series of novellas written by Cervantes beforehand, and even includes two full novellas inside the novel itself. However, the result of building a single narrative from episodic novellas creates a sum greater than its parts. Once a work of literature is made long enough by connecting together several episodes, novellas, or chapters on the same theme, a much larger narrative unity is possible, and arguably necessary to hold the episodes together as dimensions of a single written work.

Just as increasing layers of polyphonic rhythm make possible, but do not necessarily guarantee, the introduction of a dominant homophonic melodic line, multiple layers of episodic novellas make possible, but do not guarantee, a single narrative unity. The struggle of the novel is how to bind the heterogeneous pages of a long book into a single linear unity, a unity of fragments. The kinetic definition of the novel offered here does not give a final definition of "this most pliable of all forms," as Virginia Woolf says, for all novels forever and all time but, rather, adds a new material kinetic dimension to what we consider the modern novel.

The Page

The novel is first and foremost distinguished from the epistle and novella by the material condition of its pagination. By the eighteenth century, the printing press had succeeded in homogenizing the written page in a way that most medieval and Renaissance books had never been. Written manuscripts and books of the early printing press underwent an increasing standardization since the medieval period that finally reached its apex in the eighteenth century. This had two related kinesthetic consequences.

Print Frame

The first consequence is the fragmentation and homogenization of the printed page into the typographical print frame. It is no coincidence that so many great writers describe the modern book as a kind of window frame, kinetically linking it to the architectural frame and the photographic frame. Henry Ward Beecher writes, "Books are the windows through which the soul looks out. A house without books is like a room without windows."[12] The introduction of standardized spacing between letters and words had been underway since the eighth century and eventually became standard practice in seventeenth- and eighteenth-century printing frames.[13] The printing press also introduced an increasingly readable typographical letter image in the vernacular along with page numbers, margins, standardized grammar, syntax, and other elements to aid the reader. Thus, alongside the novel occurred a transformation of the typographical page frame based on the fragmentation of its components into its constitutive typographical elements and their redistribution into a homogeneous page-frame layout.

The standardization of all elements of writing on the printed page of the novel thus defined a new kinetic pattern of ordered differentiation of discrete typographical parts, which was only nascent in medieval and Renaissance novellas. Each element of the page was broken down and abstracted only to be rearranged into a new pattern of typographic elements in the printing press. In short, the page became a page of rearrangeable typographical fragments. The page and book are thus both kinetically defined by the unity of their differential elements.[14]

The second consequence of the page frame is the multiplication of pages into a long-form book. The producibility and homogenization of the page was key to the production and distribution of printed books. The more preformatted the page frame, the faster the assembly of parts. Additionally, this homogenization opened up the possibility not just for more books but also for longer books—novels, in particular. Kinetically standardized printing patterns or page frames made possible longer and longer consistently formatted works. In short, homogeneous page frames made possible long, visually unified, and homogenous books, as required by the typical novel form.

Paper

Longer books were also made possible by the increasing affordability and efficiency of paper production in Europe. In particular, the invention and industrial development of woven paper in Britain in 1757 produced the first smooth printing surface, in contrast to the previous ribbed papers that smeared and distorted printed letters. For the first time, lengthy books could be reasonably produced, distributed, and legibly read by a rising literate population. Thus the novel was not just a series of thematically connected novellas but also a single unified narrative work, divided instead by chapters and unified by a single authorial and often biographical voice, thanks to the long-form serialization of paper pages.

The novel is thus kinetically defined by a seriality of paper pages that makes possible a continuous and sequential long-narrative development. While epistle writings and novellas were often temporally disjoined in content, voice, theme, and even author—befitting their short epistolic and relational material-kinesthetic pattern—the novel differed. The introduction of page frame and paper volume not only gave rise to a single kinetic material operation of the long-form book but also made possible a new sensuous pattern of continuity between large sections of textual images, paper, and paper texture that gave legibility and differential continuity to a single authorial voice constituted in and through a sequential series of different pages and over time. These material-kinetic conditions should not be ignored as unrelated to the emergence of the novel form.

Just as long exposure had in photography and the long take had in cinema, literary art accomplished its own long-form image: the novel. In its more typical long-form style, the novel takes time to read. Thus, one kinetic dimension of the novel is the act of sitting for an extended period of time. This includes both the time during which the author writes and that which the reader reads. Each page of the novel must be turned in sequence. One does not skim or jump around in a novel as one would a newspaper, textbook, or index, or dwell too long on the beauty or sound of poetic verse.

Reading

The novel is a time-lapse image. It makes sensuous to the reader the micro-differentials in the movement of matter itself (time) because of the seriality and necessarily chronological pagination of the long-form printed book. The content of the novel may vary, but the material-kinetic operation of the novel persists in the affection of a unity through temporal and paginated differentiation of the long-form materiality of the book itself. This produces a kind of single author who can maintain kinetic unity over long temporal differences through the act of remembering heterogeneous experiences and unifying them through storytelling. It also produces a reader whose own affections follow the author's, and thus produces a similar unity of memorial and experiential fragments that the novella could not sustain owing to its limited length and nonunified themes, authors, and events.

In the novel, Georg Lukacs writes, "things appear as isolated and yet connected"[15] through time:

> This victory [of unifying the discrete] is rendered possible by time. The unrestricted, uninterrupted flow of time is the unifying principle of the homogeneity that rubs the sharp edges off each heterogeneous fragment and establishes a relationship—albeit an irrational and inexpressible one—between them. Time brings order into the chaos of men's lives and gives it the semblance of a spontaneously flowering, organic entity. . . . Beyond events, beyond psychology, time gives them the essential quality of their existence: however accidental the appearance of a character may be in pragmatic and psychological terms, it emerges from an existent, experienced continuity, and the atmosphere of thus being borne upon the unique and unrepeatable stream of life cancels out the accidental nature of their experiences and the isolated nature of the events recounted.[16]

In contrast to the pre-given organic unity of the epic, the novel thus produces a strange or ironic inorganic unity of parts that are "heterogeneously contingent and discrete" but that are "compositionally united" in the biographical unity of a differential time.[17] Here, differential time and aesthetic difference in general are not a series of instants but, rather, a process of intervalic differentiation or movement that distributes framed differences as different from one another. In other words, serial chronological time is an affective or folded product of a more primary kinetic flow of matter in motion: the continuous materiality of the book itself.

Just as the isolated letters, words, and pages are brought together in the compositional unity of the printing press and long-form book, so "the discretely heterogeneous mass of isolated persons . . . and meaningless events receives [in the content of the novel] a unified articulation by the relating of each separate element to the central character and the problem symbolized by the story of his life."[18] Lukacs thus contrasts the centrifugal motion of the epic, in which the hero is "the luminous centre around which this unfolded totality revolves, the inwardly most immobile point of the world's rhythmic movement,"[19] with the differential nature of the modern novel defined by "the paradoxical fusion of heterogeneous and discrete components into an organic whole which is then abolished over and over again,"[20] like "a surface riddled with holes."[21] This is the perfect image of kinesthetic difference: a surface riddled with holes, a filmstrip riddled with intervals, a building riddled with windows, and so on.

The material-kinetic conditions for writing and reading a necessarily serial long-form novel also produce a strange temporal depth in the images contained in the work itself. The modern novel thus privileges an ironic present, novelty, or newness that is opened up and exposed as a deep interiority of past memory and future expectation. Even when memory and history are evoked in the content of the novel, it is retold from the present. The novel thus exposes the thickness and differentiated nature of the present itself alongside the thickness and differentiated nature of the long-form book that moves with ease between past, present, and future. The narrative present of the novel finds itself rich with memorial depth, feeling, and reflection. There is no longer a "luminous center" of attention but, rather, a nomadic narrative distribution between figures: a chiasmic elasticity that circulates among and between them, panning in and out.[22]

The novel is a technology of interiority. It is, therefore, the persistence of subjective or biographical voice through the temporal differentiation of pages and time that it takes to write and read a novel that exposes the kinetic difference in the time of the novel itself. The novel produces its own differential time series to the extent that its material kinetic condition of long-form pagination makes this possible.

The kinetic structure of the novel is pushed to its limits in such exemplary works as Marcel Proust's *À la recherche du temps perdu* (1913–1927), Leo Tolstoy's *War and Peace* (1869), Victor Hugo's *Les Misérables* (1862), Sir Walter Scott's *Ivanhoe* (1820), Fyodor Dostoyevsky's *The Brothers Karamazov* (1879–1880), and other extremely lengthy long-form novels. These show the full scope of how an enormous amount of heterogeneous material (pages and themes) can be unified by an author and reader, and the incredible time scale of memories and events such a differential unity can be achieved.

The Typewriter

The novel is also distinguished from the epistle and novella by the material condition of its production: the typewriter. The typewriter made possible a new elastic motion that came to define the kinetic structure of the modern novel. The use of the typewriter to write novels introduced a further fragmentation of authorial unity, both indirectly through increasing copyist and dictation practices and directly through the author's use of the machine. The introduction of the linotype typewriter also increasingly liberated the printing of the novel from the slow speed of the letterpress and the corresponding inhibitions imposed on its publishable length.

Circulatory Oscillation

In contrast to the fountain pen and the steam-powered letterpress, the typewriter is defined by a complex series of oscillating switches tapping on moving paper, which in turn moves across a circulating metal plate behind it. The first documented typewriter was invented by an Italian nobleman, Pellegrino Turri, in 1808 in order help the blind countess Carolina Fantoni write. His writing machine had alphabetical keys that oscillated between an off state and an on state, hammering against an ink ribbon to produce legible marks on the paper recording surface.

This first kinetic operation of circulatory oscillation is thus defined by a number of kinetic operations. First, the typewriter is not defined by a single oscillation, or switch between an on and off state, but rather by numerous such oscillations. Furthermore, all the oscillations take place in a continuous series on the recording surface. Each letter oscillates back and forth, but does so in a unique series graphically recorded on the paper. There is thus a binographism of each letter, key, or switch, but also a serial binography that records its sequence. Each new graphic mark expands the inscription process as the paper flows vertically underneath. At the end of each line of text, the platen returns to its starting point. This is the first form of typographical circulation: the continuous circulation back and forth of the platen while the keys oscillate back and forth on it. This horizontal and vertical circulation of paper is the flow within which isolated points of graphic data are inscribed. The text becomes nothing more than a series of discrete, homogenous graphic marks on the continuous flow of paper moving by. The pages of paper are not prebound in the book or fixed in the rigid tensions of the printer's plates but are now allowed to flow into and out of the machine in a loose pile, without beginning or end.

After Turri's typewriter, there was a rapid development of typewriters throughout the nineteenth century. At first the typewriter was slow, but as modifications were made and more people adopted it, it became faster and easier to use. The typewriter became much more what it is: a temporal machine. It shortened the time between each on and off oscillation in the circulation, making it possible to not only type faster than one could write but also increasing the speed at which one could read what was typed. Each finger became a semiautonomous switch in the dactylographic process. As the speed of the keys increased, the time between composition and publication decreased. Between each previous kinetic difference, a new shorter interval was discovered and harnessed. This not only increased the speed of writing and publication but also had an effect on how quickly a long-form novel could be written. The difference between composition and print publication was now the time it took the key to move forward and hit the ink ribbon. The typewriter thus became its own printing press, and therefore typography increasingly liberated itself from the press.[23]

The fact that the earliest writing machines were automatons reveals something about the kinetic structure of the machine itself. The typewriter has always been capable of a higher degree of autonomy than other inscription techniques, either through mechanism, punch card, or teletype, and has always had a higher capacity of automation or autoaffection than any other graphism. The typewriter is the machine that inscribes itself in a series. It is self-circulating in the sense in which it moves its own platen in a series of loops, back and forth, scrolling up and down on a roller. It combines the rolling and unrolling movement of the ancient scroll and the rigid-bar linkage and hinge systems of the medieval book, and gives them all an automatic interconnection and circulation within a single device. The typewriter is the scroll that unrolls itself, the book that turns its own pages as it is written. It produces its own autonomous regime of circulation within which letters appear as so many heterogeneous space-time frames in its flow.

The stylus has always seemed to suggest the authority of a subject who is separate from the recording surface and who uses the stylus as her object. However, the typewriter is the surface, paper, ink, and stylus all at once. It combines and automates them all into a whole new circulation. Just like the novel, the typewriter is a unity of heterogeneous elements. All that is left to be done is to flip the switches, which no longer even technically requires a human being.

Mechanical Elasticity

The second kinetic operation of the typewriter is that of a mechanical elasticity in the movement of extension or stretching that makes possible the continuous oscillation between two positions. The mechanical typewriter makes use of this motion so that the key hammers return to their original position after extension. This is accomplished either through an elastic fluid or oil between the mechanical hinges, or by the introduction of various spring mechanisms in the machine. Coiled metal springs connecting the key lever to the bar linkage system and connecting the universal bar to the u-bar allow the key lever's motions to expand and contract elastically without breaking.[24]

However, the mechanical typewriter is not a purely elastic graphism, since in addition to its elasticity, it relies on a tensional system of rigid linkages and escapement. The two kinetic motions coexist in the mechanical typewriter, but it is the elastic motion that tends to be historically favored as more and more springs are added. There is thus a historical transition from tensional bar-linkage kinetics to increasing elastic-spring kinetics, toward greater and greater mechanical elasticity. It is precisely this increasing elasticity that allows the mechanical typewriter to literally move faster and faster. The more elastic or "springy" the kinetics of the mechanisms, the more easily they can be pushed back and forth, the faster the hammers can retract, and the faster the next hammer can land. The faster the writing, the more can be written, and the greater is the ease of writing long-form novels. The time it takes between one hammer and the next thus contracts, reducing the temporal difference between the two keys down to fractions of a second.

In this sense, the kinetic elasticity of the key-spring mechanism becomes the temporal difference between the keystrokes. Elastic motion is what gives the spatial difference between letters, as well as the temporal difference needed for the hammers to not stick to one another. Elastic movement is therefore the kinesthetic condition for the actuality of the spatiotemporal difference required by typographic writing. The minimal difference in space and time between each letter is what allows each letter to appear as a distinct letter in the inscription series. This difference is therefore nothing other than the movement of the typewriter itself. It is a fundamentally kinetic difference that gives the spatiotemporal difference. The condition of spatiotemporal type and discrete letters is the elastic movement of the typewriter itself.

Modern novelists and poets like Franz Kafka (1883–1924), Johann Wolfgang von Goethe (1749–1832), T. S. Eliot (1888–1965), Stéphane Mallarmé (1842–1898), Guillaume Apollinaire (1880–1918), and almost every other novelist of the nineteenth and twentieth centuries either directly or indirectly relied on the typewriter for the creation and printing of their novels.[25] The automatic writing of Gertrude Stein (1874–1946), the stream-of-consciousness writing of James Joyce (1882–1941), and the short fragmented prose of Samuel Beckett (1906–1989) express most directly a similar kinetic staccato or fragmentation made possible by the differential composition of the keyboard and elastic motion of its automatic apparatus.[26]

Formal innovations in the twentieth-century novel only served to make explicit the kinetic structure that was already implicit in earlier literature by drawing attention to the irony and heterogeneous assembly of letters, words, and pages in the novel itself. However, the stuttering short prose, forgetful characters, narrative loops, and abrupt transitions of Samuel Beckett's *Molloy* (1951) and the nonlinear narrative voice of James Joyce's *A Portrait of the Artist as a Young Man* (1916) are still only made possible on the material kinetic conditions of the seriality of the novel's pages and typographic structure. Instead of pushing the novel to its maximal limits of unification by elongating it to its breaking point, as Proust and Tolstoy had done, the twentieth-century novel pushed it to its minimal limits of differentiation by fragmenting it into smaller and smaller parts until the biographical unity was left, stuttering and confused in the dark, on a drifting boat with no oars, as in the ending of Beckett's *Malone Dies* (1951).

> never there he will never
> never anything
> there
> any more.[27]

The modern novel is less a critique of this kinetic differentiation and more an increasingly explicit affirmation of it as the very kinetic condition of the novel itself. Unity only comes from differentiation, and is thus only a differential or elastic unity.

The kinesthetic rise of steel, the photographic image, and the novel produced a whole new kinesthetics of elastic difference. Something similar occurred in the other modern arts of meter, action art, and the molecular arts, as we will see in the next chapter.

CHAPTER 14

The Modern Image, II

Over the course of the modern period, a kinesthetic shift Heather Pringle, "Quest for Fire Began Earlier as "bulubu bulu bulu bulu from a tensional to a more elastic pattern of motion occurred. The argument of this chapter is that meter, action art, and molecular art are also defined predominately by a distinctly elastic pattern of motion and a differential aesthetics.

METER

The first major differential art we look at in this chapter is meter. The introduction of meter into Western music had its precursors in the transition from the *musica plana* or *musica choralis* of plainchant to the *musica mensurata* (measured music) or *cantus mensurabilis* (measurable song) of vocal polyphonic music in the late thirteenth century through the seventeenth, or what is called "mensural music."

With the addition of multiple voices into vocal chants, a new kinesthetic problem arose with respect to the tensional relations between each voice: exactly when each voice should come in and how it should combine with the others. This first attempt to solve this problem was the invention of musical notation. Early mensural notation thus measured the length or duration of musical pulses such as poetic feet. Mensural music thus introduced a measured ratio or structural relation between each vocal pulsation that marked their points of sonic diffraction. However, the use of different measures or ratios in a single piece of plainchant, Baroque, or Classical music remained highly variable. This was because the measure of various cross-pulsations did not assume a single dominant metric system of hierarchical levels in a single metric but rather only a series of overlapping, differently measured pulsations.

The modern unity of discrete pulse and tone is a historical invention. Medieval polyphonic music, as well as many old folk musical traditions from Europe and Africa, had pulse durations and patterns but none that fit into a single metric system[1] or that always forced a tone such as C major into a discrete pulse duration. It was only after the introduction of the bar and bar line in seventeenth-century European dance music that a single, dominant, and hierarchical system of musical measurement was introduced. Only with a universal system of notation could every tone be given a discrete pulse and every pulse a specific tone. Western polyphonic and contrapuntal music have always been defined by mensural patterns of pulsations, but it was only in the modern period that these patterns began to coalesce into a single notational system of musical measure or meter rigorously defined at larger and smaller differential levels.

Only under these conditions was it possible for music to take on a new regime of elastic and differential motion through metrical syncopation. This is because metrical syncopation emphasizes the difference within and between pulses in a precise way but also makes use of the elasticity of sonic waves *between* multiple metric levels. This occurs in two major stages: The first emphasizes the differential structure of sonic pulse, and the second adds to this an emphasis on the elastic structure of syncopation. We now look at each of these in turn.

Pulse

The first way the metric system of modern music is made possible is by emphasizing the differential structure of pulse. This happens by breaking down sonic pulsations into their shortest, most discrete sonic components. In doing so pulse is internally differentiated accordingly to larger and smaller homogenous and hierarchal pulse units. In music theory kinetic or kinesthetic pulse is referred to as the *tactus*, or physical feeling, of a regular and repetitive back-and-forth between complete oscillations in a sound wave. This kinetic *tactus* not only occurs as a material vibration in the body of the human listener but also in all the surrounding matter that physically responds to this basic binary oscillation.[2]

Before the eighteenth century, most measured musical pulses were largely either *legato* (bound together) or had a significant length or duration. However, from around the eighteenth century to the twentieth century, there was an increasing emphasis on the internal and intervalic differentiation of the kinetic pulsation of the sonic wave. This was possible owing to the accentuation of smaller and smaller pulses contained in the older, longer pulses as well as an accentuation of the interval between pulses with the use of silence (figure 14.1).

Just like its modern sonic compatriots, the pendulum clock and the typewriter, musical pulsation became increasingly defined by a dominant kinetic pattern of oscillation—back and forth, tick and tock, on and off. David Landes calls this rise of oscillatory clock time a "revolution in time,"[3] in which, as Lewis Mumford puts it, "the categories of time and space underwent an extraordinary change and no aspect of life was left untouched by this transformation."[4]

Kinetic pulse is a distinct periodic series of short-durational oscillatory waveforms. This means that discrete pulse is not ontologically or in any other way divided from the continuous material-kinetic movement of the sound wave that produces it. There are not two kinds of things: metric pulse points and an empty medium of metaphysical time or space that contains it. Physically and kinetically speaking, the pulse, beat, or tones that define rhythms, melodies, and harmonies are not ontologically discrete or spatial objects. There are only sound waves in a continuous kinetic modulation, folding and unfolding into different sonic images. There is no need for a metaphysics of time to define pulse or meter.[5]

This definition of pulse thus includes the pulses of both pitched instruments like many chordophones and aerophones, and unpitched instruments like many membranophones and idiophones. During the modern period, the sonic subdivisions of metronomic music became increasingly accentuated in two main ways that can also be divided into two

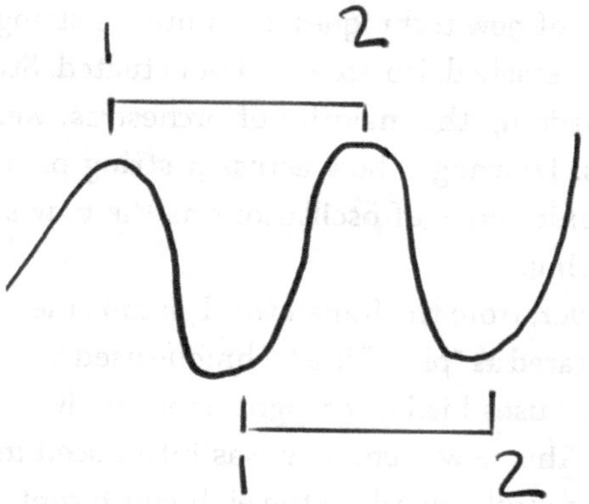

Figure 14.1 Oscillation of kinetic pulse

historical periods: one dominated by increasingly short, fragmentary, and detached synchronous pulsations in the eighteenth and nineteenth centuries; and the other dominated by similarly detached but increasingly isochronous and syncopated pulsations in the twentieth century. The first major differential technique of modern music is the increasing usage of relatively "detached" kinetic pulsations. Pulsations, as we said, cannot be absolutely detached from their sound wave, but they can be *relatively* detached by shortening them and introducing an unpulsed interval between them.

Romantic music of the eighteenth and nineteenth centuries is defined by an increasingly wide range of expressive techniques that push every musical relation of the Baroque and Classical periods (volume, pitch, speed, timbre, and so on) to their relational extreme, including an expressive variation of numerous metered time signatures in a single piece.

However, although Romantic meter remained predominantly synchronous or metronomically variable during this time—as it had since the beginning of Western polyphony—its true kinetic novelty lies in its use of increasingly detached sonic pulses and pulse groups to expose the differential interval within the apparent unity of the pulse itself and between multiple pulses. This occurs with the introduction of two interrelated kinesthetic operations: the sonic detachment or difference that results from a number of new string techniques, including *pizzicato, col legno battuto, staccato*, and *spiccato;* and the sonic unification or repetition that results from the increasing introduction of nationalist elements such as the drum and brass pulses taken from military marches and regional dance music.

Difference

The first kinesthetic operation of synchronous pulsation is one of difference. Through a number of new techniques, the notes of string instruments became increasingly detached, isolated, and punctuated. String instruments, which typically made up the majority of orchestras, were predominantly played with bows. Drawing a bow across a string or a series of strings produces a harmonic series of oscillations as the bow slip-stick action is periodic on the string.

Pizzicato, however, from the Italian word *pizzare*, meaning to pinch, nip, seize, or pluck, notated as "pizz.," is a technique used on string instruments in which the player uses his left or right hand to pluck the notes instead of using the bow. This new technique was introduced in the seventeenth century but came to full flourish in the eighteenth century.[6] It produced a sharper and inharmonic tone of much shorter pulse duration. Just as the name suggests, *pizzicato* therefore pinches or seizes out of what would have been a much larger bound series of tones an isolated pulse fragment of that series, rupturing the harmonic series.

Following the same differential kinetic pattern as the other modern arts, *pizzicato* technique takes the previous historical patterns of motion that defined the function, form, and relation of the aesthetic field and breaks them down into their component parts in order to reassemble them in a new configuration. In *pizzicato*, the harmonic form is thus decomposed into its minimal inharmonic oscillations and timbre components. The same instrument played differently can produce different timbres. Timbre is thus one of the differential conditions of the kinetic waveform. One of the most original efforts in modern music was the exploration of the differential timbres internal to the waveform, extracting and recombining its internally differentiated and even heterogeneous components.

Pizzicato is thus a way to extract or pluck from a tone or series of tones a smaller, discrete pulse. It is a way of showing that any series of tone pulses or wave oscillations is in turn made up of smaller pulses and wave segments. *Pizzicato* brings forth both a latent differential timbre in the instrument and the possibility of a new inharmonic recomposition of the musical form.

All sound is kinetically oscillatory because it is defined by a series of wave patterns. However, one of the kinetic novelties that distinguishes Romantic and modern music from previous traditions is the way these musical traditions break sonic oscillation down into smaller and smaller component micro-scillations and then play between these levels of oscillation (figure 14.2).

The introduction and increasing use of other string techniques, such as *col legno battuto,* using the wooden part of the bow to tap the strings; *spiccato*, or "to separate," tapping the strings vertically with the bow; and *staccato*, the emphasized isolation, or literally "detachment," of a single note from a tone series, all share a similar kinetic pattern: the differentiation of the tone pulse into a smaller series of detached and mobile tone pulses. All these techniques were introduced in the seventeenth century and flourished throughout the eighteenth and nineteenth centuries.

The use of *pizzicato* in Romantic music is almost ubiquitous but used mostly in isolated sections of a work. Some composers, however, dedicated whole movements and acts of their compositions to the use of *pizzicato* and *staccato* techniques, including the ninth movement of J. S. Bach's *Magnificat* (1723–1733); Josef Strauss's *Pizzicato Polka* (1869); Act IV's "Anitra's Dance" of Edvard Grieg's *Peer Gynt* (1874); Act III of Léo Delibes's ballet *Sylvia* (1876); the fourth-movement "March to the Scaffold" in Hector Berlioz's

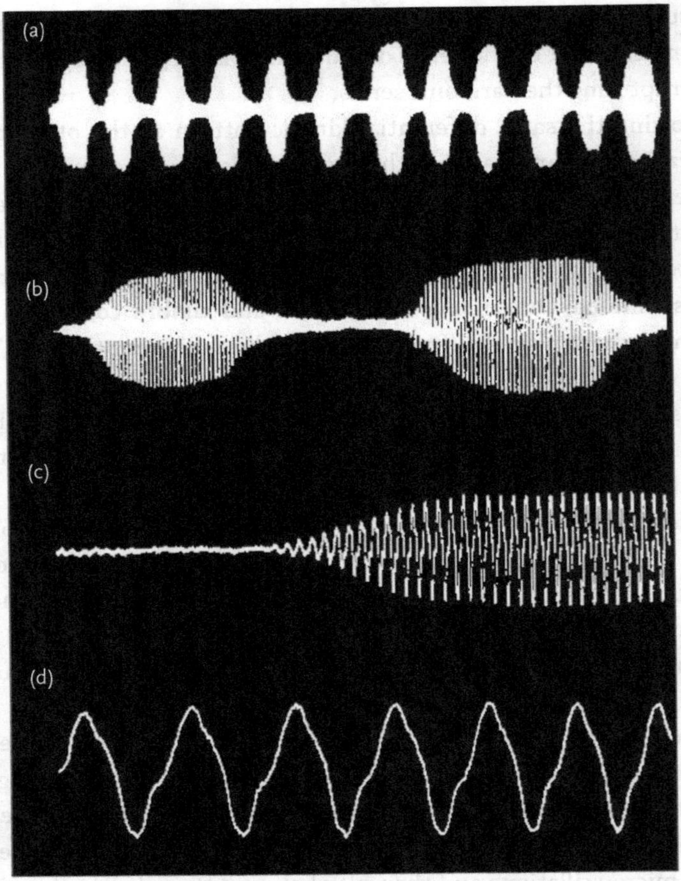

Wave traces of a series of staccato notes (a) on a treble recorder at the rate of six notes per second; in (a) the trace lasts two seconds, (b) a third of a second, (c) 0.1 seconds, (d) 0.014 seconds

Figure 14.2 Wave traces of a series of *staccato* notes
Source: Grove Music Online.

Symphonie fantastique, Op. 14 (1830); and the famous second movement of Maurice Ravel's String Quartet in F (1903). Berlioz's *Symphonie fantastique* makes use of both *staccato* and *col legno battuto*, emphasizing the literal death drive to repeat abstract differences, dramatically related to the death of the beloved and of the murderous lover in his march toward the scaffold. All oscillation is defined by the death grip of this basic binary: the one and the two, tick and tock, love and death. One follows the other.

In the third and fourth movements of Dmitri Shostakovich's Trio No. 2 in E minor, Op. 67 (1944), for example, *pizzicato*, *staccato*, and *spiccato* techniques are all used to produce an oscillation between bowed and detached tones. This produces not only a relational contrast between long and short kinetic patterns but also a differential exposure of the bound *legato* tone series itself as composed of smaller detached fragments. What is living is composed of nonliving fragments. Modern life is ruled by the tick-tock of the clock, the clatter of the typewriter. The discrete moments of time necessarily proceed by *staccato* units toward death—in this case, the death of Shostakovich's good friend Ivan Sollertinsky to whom the piece is dedicated.

The organic whole of biological life and of compositional development is ripped open only to reveal the deeper differentiation and fragmentation of the pulse and interval that support and define all tonal forms and relations. The deconstructive fourth movement ends with an extremely quiet E major *legato* played on the piano while a slow *pizzicato* of the same chord is played on the violin. The *pizzicato* literally rips and detaches the notes from the chord and the inharmonic pulses from the harmonic ones. The violin expresses the true differentially pulsed conditions internal to the *legato* tone, just as the shining unity of the major chord of organic life is both composed of and decomposed into the dead fragments of inorganic matter: minerals and discrete pulses. Shostakovich's "Dance of Death" movement is thus both the dance of dead pulsation and the death of dancing life.

Serialism

Arnold Schoenberg (1874–1951) was the first to extend the kinetic idea of differentiation to tone by devising a twelve-tone serialism. It used all twelve notes of the chromatic scale with equal frequency by ordering them into tone rows without an emphasis on any one note or tone. As Webern stated in his lectures, "All twelve notes came to have equal rights."[7]

This technique is possible only on two kinesthetic conditions: first, that a discrete tone can be sufficiently and clearly differentiated or "detached" or "individuated" from its arbitrary harmonic relations; and second, that these tones can then be ordered into a new, continuous series of temporally and harmonically equal tones. As Schoenberg writes in *Theory of Harmony*, "There are no non-harmonic tones, but only tones alien to the harmonic system.. . . . Possibilities for combining tones are unlimited."[8] Schoenberg and the Second Viennese School achieved these two conditions by relying heavily on the use of the detached pulse techniques of *pizzicato, staccato, spiccato,* and *col legno* to clearly distinguish tones from one another such that they can be clearly and equally serialized in the tone row without extending over one another and creating unintentional, nonserial harmonies or disrupting the linearity of the sequence. Schoenberg used twelve-tone chords in his *Serenade*, Op. 24, which includes plucked mandolin and guitar, and a march and dance movement, to precisely this effect. Anton von Webern also used them in *5 Movements*, Op. 5, and Alban Berg used them in his opera *Lulu* but almost always in either *staccato* or distinct bursts so as not to muddy the distinctness required by the logical ordering of the tonal series and to dramatize the expressiveness of the silent differential interval between tones.

The silent pulse is the difference that makes a difference but that itself must remain silent—heard only *as a difference between tones*. Just as German expressionistic cinema emphasized the constitutive darkness emerging between visual images, so German expressionistic music emphasized the constitutive silence of the sonic image. This is especially true in the case of Webern's incredible octave intervals and the *legatos* in his *Variations*, Op. 27 (1936; figure 14.3), which *impossibly span a rest between two tones*! As Boulez writes of Webern's music, "Each phenomenon [note] is at one and the same time autonomous and interdependent."[9]

In other words, twelve-tone serialism tends not only toward an equality of tone but also toward an equality of *pulse duration*. If tone pulses are played for unequal durations, then this reintroduces a tonal dominance to the piece. Olivier Messiaen knew this well, although Schoenberg did not explicitly grapple with this issue. The differential kinetics of tonal serialism thus spread from this original insight to a more total, integrated, or generalized serialism that included not only tone and duration but also dynamics, timbre, and every other parameter of musical composition. This broader kinesthetic trajectory of serialism can be found in the works of Pierre Boulez, Henri Pousseur, Karlheinz Stockhausen, and others. However, the basic differential kinetic structure of serialism remained the same: "detach" compositional elements from the metric wave and recompose them into a new sequence or series of equally different fragments united only through the act of composition itself.

Repetition

The second kinesthetic operation of synchronous pulsation is one of repetition. The influence of national motifs identified with a specific country, region, or ethnicity, including both militaristic and folk elements, rose to musical dominance in the eighteenth and nineteenth centuries. In their own way, both influences introduced an increasingly repetitive kinetic pattern of differential and detached pulsations.

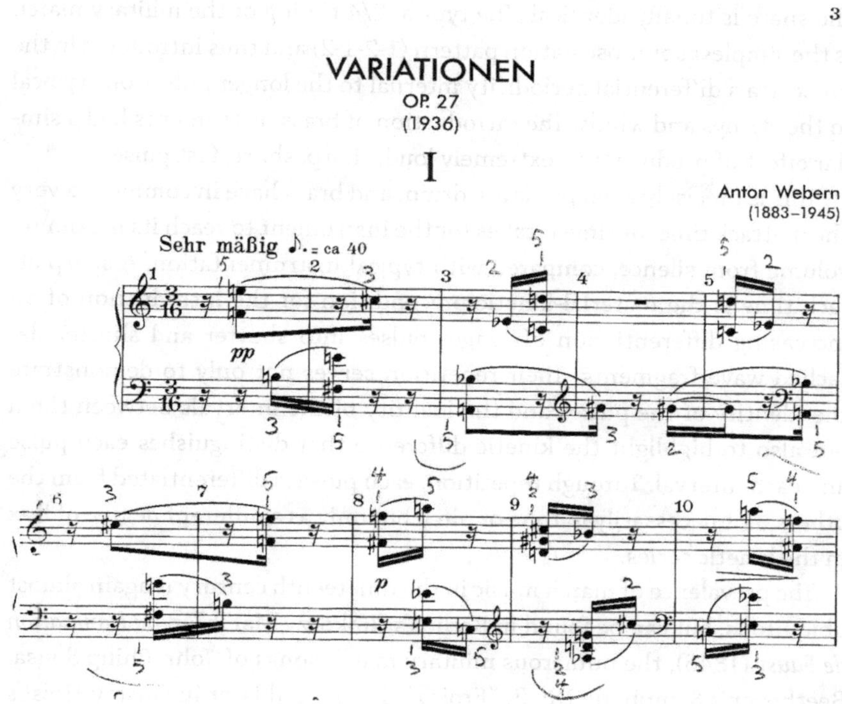

Figure 14.3 Facsimile of the score from Anton Webern, *Variations*, Op. 27, Universal Edition, Vienna, 1979, annotated by Webern for the first performance, in October 1937, by Peter Stadlen. The title page says (in German and English): "Webern's ideas on the work's interpretation set out for the first time by Peter Stadlen with the aid of the facsimile of his working copy containing Webern's instructions for the world première." Note: The score number is UE 16845, whereas the number for the plain score is UE 10881
Source: Anton Webern, Variations, Op. 27, Universal Edition, Vienna, 1979 G. Henle Verlag (January 1, 2016)

The first influence is that of the military march and its traditional instrumentation, drum and horn. The rise of Western nationalism also brought with it the drama of the public musical performance of the march, associated with the power of the national people. March instrumentation has long been driven by the isochronous pulse of the snare drum and the horn. Both had been used in a limited way in earlier Western music but now came to occupy a central role in the musical nationalism of the Romantic period, in part due to the introduction of valved brass instruments in the nineteenth century.

Kinetically, the drum, particularly the snare drum, allows for a more punctuated and detached unpitched pulse that does not necessarily need to follow other instrumentation but rather sets a metric framework of repeated, identical pulses within which the other instruments play. Unlike the variable tones of most other orchestra instrumentation, each pulse of the snare is tonally identical. The typical 2/4 timing of the military march is the simplest sonic oscillation pattern (1-2-1-2) and thus introduces in the orchestra a differential periodicity internal to the longer pulsations typical in the strings and winds. The introduction of brass instruments had a similar effect of producing an extremely loud, sharp, short, fast pulse.

This is precisely what *pizzicato*, drum, and brass have in common: a very short attack time, or time it takes for the instrument to reach its maximum volume from silence, compared with typical instrumentation. A sharp attack time is the material-kinetic precondition for the introduction of an increasing differentiation of longer pulses into shorter and shorter detached wave fragments. Their repetition serves not only to demonstrate the identity of the pulses and the identity of the intervals between them but also to highlight the kinetic difference that distinguishes each pulse and each interval. Through repetition, each pulse is differentiated from the others, not necessarily in tone or pitch but only as a different degree of flux in the kinetic series.

The prevalence of march music in the nineteenth century is again almost ubiquitous, but can be found in Berlioz's "Rokoczy March" in *La Damnation de Faust* (1846), the numerous military march songs of John Philip Sousa, Beethoven's Symphony No. 3, "Eroica" (1804), and later in Gustav Holst's *The Planets* (1914–16) and numerous others. The kinetic connection between military power, authority, attack time, necessity, and the numerous "death marches" (Mendelssohn's and Handel's funeral marches) is that of the simple 1-2 repetitive oscillation pattern. The identical repetition of kinetic pulses exposes not only a difference internal to all larger meters, thus exposing their inorganic and constructed structure, but also the fundamentally open nature of the series itself without essential beginning or end—except for death. The march and the dance, death and love, 1 and 2—all derive their feeling of necessity and determination from simple kinetic oscillation. Adorno is right to locate the origin of modern music in the repetitive attack of the military drumbeat, even if his normative judgments of such music are wrong.[10]

The second influence on repetitive music is that of national folk music—specifically, folk dance music. Most national folk-music traditions, such as the military march, used the more detached short attack patterns of the drum, horn, and *pizzicato* strings and followed more isochronous (homogenous) pulse patterns. This is the case in part because such songs were often dance songs that required the use of regular and discrete pulses for the dancers to move to, at least in the traditional context. The music thus takes on a more repetitive character while the dancers provide the more interesting physical syncopation between the pulses.

The incorporation of national folk dance music such as the German polka, the Polish polonaise and mazurka, the Scandinavian waltz, the French bourrée, and so on were increasingly common in Western art music, which in turn meant introducing increasingly large sections of more isochronous pulsation and more detached tones. All this is evident in dance pieces like Brahms's twenty-one Hungarian Dances (1869); Tchaikovsky's Dance of the Flutes from *The Nutcracker* (1892); Dvorák's sixteen Slavonic Dances inspired by Brahms's dances (1878 and 1886); Bizet's "Seguidilla" in *Carmen* (1875); Johann Strauss II's Neue Pizzicato Polka (1892); Edvard Grieg's four *Norwegian Dances* (1880), and many more, all using dramatic drum, brass, and *pizzicato* to produce a highly *staccato*, kinetically differentiated, and repetitive structure.

Just as architecture, film, and the novel all began to develop a modern long form or time-exposure form that allowed for an increasing internal aesthetic differentiation of continuous matter into deeper and deeper folded-up levels, so music developed its own long form in the symphony, which reached its highest intensity during the eighteenth and nineteenth centuries.

Kinetically, what is exposed by this increasing differentiation in the continuous surface of the waveform or metric wave is that such a wave must have an elastic character so that each successively smaller interval in its surface can be discretely pulsed. Of course, this has always been the case, but only in the modern period was the elasticity of metric multiplicity as emphasized and exploited to this degree, from the contracted micropulses of the snare drum in the waltz and the march to the expanded macro pulses of Wagner's 136-bar-long E-flat prelude to *Das Rheingold* (1876). The modern period is one of a continuous and incremental tendency toward detached pulsation and metric multiplicity.

Syncopation

The second way the metric system of modern music is made possible is by emphasizing the elastic structure of syncopation. This occurs through the increasing use of isochronous pulsation and syncopation, defined by deep metricality or multi-leveled meter.[11] Alongside the rise of musical nationalism's influence on Western art, music was also the rise and spread of national folk and dance music more generally. Most significantly, however, beginning around the late nineteenth century and early twentieth century, a mixture of European and African folk music traditions created by migrants to America developed into a new musical-kinetic pattern.

In particular, it was the mixture of Afro-Latin and Afro-Caribbean folk dances performed in New Orleans's Congo Square that led to this invention. Hundreds of slaves would gather in Congo Square for festivals and celebrations, playing all manner of largely percussive instruments, such as the sheepskin-covered "gumbo box," a frame drum; triangles; and jawbones. Places in Louisiana and a number of other Southeastern states held similar events between 1820 and 1850.[12]

These musical forms were defined by the dominant nonmetric polyrhythmic structure that characterized most slave music of the Caribbean and Sub-Saharan African musical traditions. This particular musical pattern produces what is called a cross-rhythmic or cross-beat overlap between two conflicting pulse patterns and not merely a momentary displacement that leaves the prevailing meter fundamentally unchallenged.[13] The kinetic effect of this polyrhythmic pattern is similar to that of melodic polyphony in the late Middle Ages and the Renaissance, as Afro-musicologist Simha Arom writes. Both were mensural or measured but not strictly metrical, since they were not subordinated to or did not explicitly accentuate the pulses of a single metrical hierarchy:[14] "The arrangement of durations in most African music is still based on the same principle as the medieval *tactus*. No use whatsoever is made of the notion of matrices of regular contrasts of strong and weak beats. African music is thus based, not on measures in the sense of classical musical teaching, but on pulsations, i.e., on a sequence of isochronous temporal units which can be materialised as a beat."[15] As such, both polyrhythm and polyphony emphasize the kinesthetic tension of a convergent and divergent diffraction series, and not the differential multiplicity of a single metrical series.

However, when nonmetered African isochronous polyrhythms mixed with metered European synchronous homophony in late nineteenth-century America, the result was the creation of a new *metered isochronous homophony*. Thus, there was a single isochronous series of metered pulsations in which interlaced rhythmic and melodic lines align, diverge, and overlap—what we call syncopation.

Deep Meter

Syncopation is the play of polyrhythms within the confines of an accentuated metric system. It is "a momentary displacement that leaves the prevailing meter fundamentally unchallenged." What music theorist Mark Abel calls "deep" or "multilevel" meter is thus a kinesthetic regime in which the musical waveform or sonic flow has become increasingly elastic, such that every region of the oscillation can in turn be differentiated or subdivided into yet smaller pulsed oscillations.

> Each off-beat at a particular metrical level becomes an on-beat at the next lowest level, thereby bringing into play new off-beats at half-way intervals which were merely latent at the higher levels. . . . When using a ruler to measure length, if the point to be measured falls exactly on one of the strongest lines of the ruler, then the measurement is a round number of units. If not, then we must look to where it falls in relation to the next strongest lines. Again, if its length does not coincide with one of these, the next longest set of lines come into play, and so on, until a subdivision is found which coincides with the point to be measured.[16]

Syncopation does not just demonstrate the tensional relation between multiple polyrhythmic and polyphonic series; it also sonically accentuates the internal and radially differential nature of any given series with itself. Metric syncopation reveals that each series is already itself indefinitely and elastically multiple at the micro and macro levels. By playing within and between multiple levels of the same metric time signature, syncopation plays between the depths and heights of the micro and macro of a single meter, rendering both explicit and implicit the coexistence of multiple continuous sonic levels at once.

Groove

The differential kinesthetics of detached pulse syncopation define almost all the dominant musical instrumentation and genres of the twentieth century—ragtime, jazz, blues, rock, country, funk, hip-hop, and all their related subgenres. During the twentieth century, "groove" music pushed this kinetic pattern to its limits and even exerted a significant influence over the Western art music tradition in the jazz-infused works of George Gershwin and Aaron Copland, as well as the minimalism of Philip Glass and Terry Riley.[17]

It is no coincidence that the same instrumentation and techniques used to produce the detached pulse of the eighteenth and nineteenth centuries became the most dominant in groove music. The predominance and almost ubiquity of the drum kit, plucked strings (guitar, bass, banjo), and brass instruments in modern groove music is directly related to the need for accentuated sonic detachments to more clearly mark the precisely syncopated metrical patterns. It is no coincidence that brass sharp-attack instruments such as the trumpet and saxophone play such a crucial role in one of groove's most dramatically syncopated genres: jazz bebop. John Coltrane's *Blue Train* (1958), for example, makes use of 4/4 time kept by the sharp pop of the snare and hi-hat, while the plucked walking bass lines pop along in one syncopated rhythm, and the virtuoso brass instruments solo freely up and down the metrical hierarchies. The metric wave of the groove song is not broken but rather exposed for what it is: a deeply differentiated multiplicity of coexisting sonic series that can be elastically stretched across micro and macro levels.

ACTION

The second major art of kinesthetic difference we look at in this chapter is that of action art, which is defined not only by a kinetic differentiation and fragmentation of the work of art into its abstract elements but also by a kinetic differentiation of the artist, audience, and entire aesthetic milieu or field of images itself into fragments. The action arts, therefore, are a synthesis of all the arts into a single aesthetic field and, at the same time, a radical fragmentation of that same field.

Art as action means that a work of art is no longer strictly limited to a canvas or even to the action of the artist himself or herself but, rather, to an event defined only by its spatiotemporal limits and components. For something to be an artistic act, it need only have a delimited time and space in which the act happens. The historical emergence of this newly "expanded aesthetic field" occurs,[18] like modern music, in two distinct but related and continuously developing trajectories: the rise and increasing autonomy of dance in the eighteenth and nineteenth centuries, and the rise of performance art in the twentieth.

Dance

Historically, the kinetic pattern of elastic difference rose slowly to dominance over the course of several hundred years of modern artistic innovation, beginning in the eighteenth century with the increasing aesthetic prominence of dance. This book has limited its discussion of dance to the short discussion of it in the prehistoric period and to modern dance. This is in part because detailed research on the history of dance is scarce owing to poor documentation, but, more importantly, because it was not until the modern period that dance, just like dance music, began to take on a historically enlarged importance and autonomy from the other arts. Humans have always danced, and dance has always played important historical and kinesthetic roles, but in the modern period, two main factors began to give it a radical new importance and differential kinetic character: nationalism and ballet.

Nationalism

The first contribution to the kinetic character of early-modern dance was the rise of nationalism in eighteenth- and nineteenth-century Europe. The rise of nationalism, as we discussed previously, coincided with the rise of folk dance because European nations sought to trace and define the early historical and cultural folk practices that defined them as distinct people.[19] What had previously been a culturally and politically marginal yet pervasive tradition of folk dance in Europe began to be appropriated and transformed by urban bourgeois culture across the continent in the forms of polka, polonaise, mazurka, waltz, bourrée, fandango, jigs, and numerous others. All across Europe, people began gathering in large public areas to dance versions of traditional folk dances. Alongside the rise of public dancing was the rise of national military marches, an admittedly impoverished form of dance, but nonetheless a significant one.

Kinesthetically, the rise of popular national and folk dances and marches is defined by a similar kinetic pattern as the music analyzed in the previous section. The increasing detachment of sonic pulses and the elasticity of the metric wave that allows for syncopation produce a similar pattern in folk dance. National folk dance transformed the movements of the body into a series of moves, steps, or specific actions to be performed and changed in a sequence following the clear and detached sonic pulses of a relatively isochronous beat. The idea that the movements of the body could be spatiotemporally divided into distinct actions was quite old, but the introduction of an isochronous metric beat that legislated the specific and sequential steps of a dance only came much later and reached its zenith of popularity during the dance craze of the early modern period.

Each dance move or distinct action is distinguished from the others only by a spatiotemporal duration or interval, and not by any tangible product in a finished work of art. Dance is thus the "process art" par excellence. Even though it frequently accompanies narrative and music, it does not require them to be considered dance. Simple hand-clapping will suffice. One can still dance without music, keeping time only by a pattern of motion. What historically distinguishes dance from other arts and from everyday movement is nothing but the sequential difference and repetition of actions during the larger spatiotemporal limit of the whole performative act.

However, we should not make the mistake, as Adorno did, of identifying all dance with the military march. It is true that both marches and folk dances follow the same basic kinetic pattern of differential motion, oscillating between action one, action two, and so on, defined by the clearly determined mark of the detached pulse or beat. However, dances and marches differ significantly in their degree of kinetic elasticity. To dance is to draw, pull, or stretch out.[20] Dance is fundamentally an elastic art. Even in the march it is still the body that must continuously stretch out in the infinite sonic gap between the first movement of one foot and the second movement of the other. The pulse of the march can be detached, but the movement of the body is still continuous and thus has a degree of freedom or elasticity between pulses, although certainly less than other dances.

Most folk dances, on the other hand, express a much larger degree of kinetic syncopation between the predetermined series of dance actions. Between pulse A and pulse B there is a kinetic difference or interval to be traversed that can be fulfilled in a number of ways. Thus, alongside the sonic syncopation and freedom in dance music discussed earlier, there is a degree of freedom and bodily syncopation and elasticity in all dancing. Dancing is thus already syncopated before the musical rise of syncopation itself.

Ballet

The second kinetic practice to give new importance and differential kinetic character to dance is the rise of ballet in eighteenth- and nineteenth-century Europe. Although ballet had its historical precursors in the choreographed court performances of the Italian Renaissance courts of the fifteenth and sixteenth centuries, it was not until its national and cultural valorization as an autonomous art form by Louis XIV in the seventeenth century, and ultimately with its radical reform in the eighteenth century by French dancer and ballet master Jean-Georges Noverre, that ballet became sufficiently liberated from its historical subordination to the other arts.

Noverre's *Les Lettres sur la danse et sur les ballets* (1760) called for a revolution in ballet and art dance that he termed "ballet d'action." This new style emphasized the sensitivity of technique to the dancer's anatomy; the importance of the personality of the dancer; the unity of the dance with the plot, music, costume, and entire aesthetic milieu; and the abolition of excessive costume and masks, which inhibited the dancers' movement.[21]

All these changes made important steps toward creating a unified aesthetic milieu of total performance and furthered the aesthetic expression of the dancer.

A similar process of aesthetic synthesis was underway in opera in the seventeenth century, but remained incomplete without the full unification and equality of dance, dance technique, and the dancer into the poetic, theatrical, and musical aspects of opera. Eighteenth-century dance masters such as Jean-Baptiste Lully (1632–1687) and Jean-Philippe Rameau (1683–1764) made important contributions to this aesthetic synthesis, but Noverre's revolution significantly liberated the movement of the dancer's body and opened the door to the *staccato* kinetics of point work and the increasing performative autonomy of dance. Once dance was capable of an equal aesthetic composition and technique of its own in opera, it could then take on a more autonomous expression outside opera, as, for example, in the revolutionary work of the twentieth-century Ballets Russes company and Vaslav Nijinsky (1889–1950).

Kinesthetically, early modern ballet and art dance are defined by the same differentiation of bodily motions into distinct actions syncopated continuously across a relatively isochronous pulse. As the French poet Paul Valéry writes, "dance is an art derived from life itself, since it is nothing more nor less than the action of the whole human body; but an action transposed into a world, into a kind of space-time, which is no longer quite the same as everyday life."[22] Dance works on the pure actions of everyday life, divides them up into a series of space-times, and recombines them into a new world. In other words, dance creates an aesthetic and useful unity from a multitude of fragmented and useless acts. "We can perform a multitude of acts that have no chance of being utilized in the dispensable, or important, operations of life . . . [but in dance they] tend to build up a kind of utility from the useless, a kind of necessity from the arbitrary."[23]

Dance is thus the serial unity of the different, the elastic pull or kinetic stretch across the heterogeneous. Dance is the art of action and of process, and thus it finds its way synthetically into all the other arts. Dance may be defined by the syncopation between different steps or actions, but just as in metered music, aesthetic difference always presupposes a kinetic continuum (meter-wave or action-wave) within which each step or beat is only a fold in its flow. To quote Valéry again, "Events are the foam of things, but what I am interested in is the sea"—that is, the continuous conditions of the discrete folds.

Performance

The second historical trajectory of action art is continuous with the first, but expands and intensifies its differential and elastic patterns of motion far beyond the realm of early modern dance. While dance brings to the historico-aesthetic foreground the differential series of actions or steps in the dance, performance art adds to this an increasing emphasis on the interval or gap between actions or steps in order to expose the elasticity of the act itself.

Like dance, performance art exposes the fact that all art (and life) requires action. Even in the performing arts—music, theater, opera, and so on—it is the kinetic act of performance itself that traverses the interval between one note, word, or pose and another. Between one brushstroke and another is the unpainted action or interval that is missing from the reified work of art. The action that occurs between one stroke or mark and another is the unpainted condition of painting.

Impressionist and Post-Impressionist painters such as Cezanne, Pissarro, Monet, van Gogh, and Seurat dramatize the differential interval between brushstrokes by keeping them distinct instead of blending them like previous painters, thereby drawing attention to this unrepresentable kinetic action of anti-painting at the heart of painting. In modern music, the differential interval of silence, background noise, or bodily action between one note and another is dramatized by Webern through the contrast of octaves and use of multiple rests to highlight the nonmusical interval that is constitutive of music itself. Every art has its hidden kinetic condition or interval that allows for the transition between its explicit differential components. Action art brings to the fore this kinetic interval. This trajectory is composed of several different historical techniques that contribute to the same kinesthetic pattern of action art: futurism, Dadaism, surrealism, and live art.

Futurism

In the early 20th century, a number of Italian artists began creating action-based works of art in every major area of the arts. The first of these was Filippo Marinetti's (1876–1944) publication of the "Futurist Manifesto" on the front page of the major Italian newspaper *Le Figaro* in 1909. The performative act of a public presentation of an aesthetic idea was the first in a long series of performative action manifestos by other futurists that brought to the fore the hidden action of self-promotion that lies as the nonartistic condition of almost every major artist's success. Publishing the manifesto before the existence of a collection of works or group of followers drew direct attention to the promotion itself as an autonomous act or unit of space-time with its own circulation, minus a concrete referent.

In 1910, Marinetti gathered a number of painters together at his home to produce the *Technical Manifesto for Futurist Painting*. Influenced by cubism while living in Paris, Marinetti and the others added to it "a love of speed and danger," producing a kind of kinetic cubism.[24] In 1911, they performed their first show of paintings by lecturing to the crowd and showing how their ideas could be applied to painting. They used performance as a way of holding a crowd but also as a way to change the way that paintings were viewed by the audience. Boccioni wrote, "Painting was no longer an exterior scene, [but] the setting of a theatrical spectacle." Similarly, Soffici wrote that "the spectator [must] live at the center of the painted action." By transforming their paintings into one part of a larger performance milieu or atmosphere, they showed the differential interval of action required of painting by showing themselves painting in a theater or on the street and describing the ideas behind them. The act of painting and the act of viewing were thus exposed as differential parts of the work of art.[25]

One of the core events of futurist performance was the *serate*, or evenings where artists would gather to improvise variety theater composed of rants, marionettes, noise music, and other short experimental performances. There was to be constant invention, no storyline, and an attempt to liberate the audience from being "stupid voyeurs."[26] The audience often became violent and threw things at the performers, thus acting or intervening in the performance with acts of their own—to the delight of the performers. The entire evening became the work of art, with all its variety, fights, and riots included.

Futurist events were often accompanied by noise music created by Luigi Russolo's (1885–1947), with noise instruments (*intonarumori*) that could be manipulated to play any number of different noises. His first piece, *Veglio Di Una Città*, was recorded in 1913 and was preceded by a treatise "The Art of Noises" (1913) and followed by other works titled *Awakening of a City, Meeting of Cars and Airplanes*, and *Dining on the Terrace of the Casino*. His concert *Gran Concerto Futuristico* (1917) angered the crowd to the point of violence and riot—as he had hoped. "Today," he wrote, "music, as it becomes continually more complicated, strives to amalgamate the most dissonant, strange and harsh sounds. In this way we come ever closer to noise-sound."[27] Again, futurism exposes the constitutive intervallic action sounds of modern life—train horns, bus wheels, car crashes, mechanical pulses, crunching, and so on—that compose the background within which music is created and that often occur as one is listening to a piece of music. These kinetic interruptions, silences, dissonances, or differences are not extrinsic to the work of art but, rather, part of its modern kinesthetic milieu.

The strange kinetic connection between early modern ballet and modern performance art has its bridge in so-called futurist ballets. These were inspired in part by Marinetti's love of Nijinsky's ballets. In 1914, Giacomo Balla (1871–1958) privately performed his *Macchina tipografica* (printing press) ballet in which twelve people each performed one repetitive mechanical action and made an onomatopoetic mechanical noise. In 1919, Ivo Pannaggi (1901–1981) covered his dancers in mechanical costumes to "deform the entire figure bringing about machinelike movements." In 1917, he created an actorless "light ballet" of flashing colored light and shapes set to Stravinsky's *Fireworks*. In 1918, Gilbert Clavel and Fortunato Depero created five short performances at the marionette theatre titled *Plastic Dances,* using one big marionette—"the great savage"—as the stage within which smaller savages danced in a rain of cigarettes and flashing lights. In 1927, Enrico Prampolini (1894–1956) and Franco Casavola (1891–1955) performed *The Merchant of Hearts* using only life-size marionettes of human figures and their shadows.

Kinesthetically, futurist ballet not only renders explicit the fragmentation and differentiation of the human body into discrete actions and sounds by modern society, including those of dance but also reveals through live performance the inhuman motions of repetition that produce the habituation of those motions, which we experience as natural. The mechanical printing press, the typewriter, cars, human action, and the performance itself are all rendered visible as performative actions. The abolition of the actor pushes the performative action even further, showing a performer's actions themselves to be nothing more than the habit of repetitive machinelike motions of a puppet. The use of shadow likewise aims to render visible the typically hidden shadows of artistic performance. Shadow is the abject interval or difference required by the use of theatrical lighting, but it is not typically considered as a desired visual effect of action. Shadow theater is an "other" theater included in theater and performance but visually ignored. Futurism again brings out this difference. The *Plastic Dances* are elastic dances—the strings that stretch and pull the bodies of the dancers between their discrete positions, steps, or poses.

The futurists achieved what opera could not—a total or synthetic theater. Futurist theater showed the action of the entire performative atmosphere. In Depero's *Colours*, his characters were cardboard objects—gray (plastic, ovoid), red (triangular, dynamic), white (long-lined, sharp-pointed), and black (multi-globed)—and were moved by invisible strings in an empty blue cubic space. Offstage, performers provided sound effects or *parolibero* such as "bulubu bulu bulu bulu bulu bulu," which supposedly corresponded to the various colors. All the differential intervals of kinesthetic action were brought from the depths to the surface: the inhuman abstract shapes in the human, the vowel fragments in the words, the shadows in the light, the noise in the music, and so on. Synthetic theater was composed of brief, succinct, spontaneous, and even simultaneous performances, such as Marinetti's *Communicating Vases*, in which multiple performances happened at once. Speed and action were ways of uniting the "fragments of interconnected events" encountered in everyday life.[28]

Dada

The Dada movement of the early twentieth century followed a similar trajectory of emphasizing the primacy of the performative act. Dadaists followed the futurist's lead of having aesthetic space-times, *serate*, evenings, or soirées wherein everything that happened was an art act and part of the performative milieu.[29]

The continuing explicit connection between performance art and dance made by Marinetti also was made by Ball, who described dance as the closest and most direct aesthetic material: "It is very close to the art of tattooing and to all primitive representative efforts that aim at personification; it often merged into them."[30] In fact, one of the most famous Dada events was the ballet *Parade* (1917), the collaborative work of Erik Satie, Pablo Picasso, Jean Cocteau, and Léonide Massine. The ballet included noise music, circus acts, and cubist set design and costumes, and caused a riot at the theater. Like futurism, Dadaist performance extended outside the theatre or soirée in street performance. George Grosz and others walked around in costumes, creating new names for themselves and promoting their art movement. Grosz even performed an art event by urinating on expressionistic paintings (1918).

Surrealism

Out of the decline of Dadaism in 1921, surrealism emerged with similar avant-garde performance-art practices, soirées, and all the rest. However, surrealism also emphasized a new element of performance art: that the irrational and unconscious actions of the body and mind are themselves also performances and aesthetic acts. The total atmosphere of the futurist art action pushed as far as one could go from the theater to the streets to the sky in *aeropittura* (aeropainting), but the surrealists went deeper into the fragmentary acts of the presubjective interior. These performance arts of the unconscious were expressed in absurdist ballets such as Jean Cocteau's *Les Mariés de la Tour Eiffel* (*The Wedding on the Eiffel Tower*) (1921). Here, inanimate objects come to life, a telegraph office comes into existence out of nowhere, a lion eats one of the guests, and a "child of the future" then kills everyone.

Another aesthetic expression of the prereflective aesthetics was André Breton's concept of automatism, described in the *First Surrealist Manifesto* (1924). "Surrealism," he writes, is a "pure psychic automatism, by which an attempt is made to express, either verbally, in writing, or in any other manner, the true functioning of thought . . . the higher reality of certain hitherto neglected forms of association, in the omnipotence of the dream, in the disinterested play of thought." By 1921, Breton and Soupault had written the first "automatic" surrealist poem, "Les Champs magnétiques" ["Magnetic Fields"]. The field is precisely that background intervallic difference between objects that bears them and differentiates them. The aim of automatic writing, drawing, painting, and performance was precisely to not think about what one was doing. Thus, even the idea of subjective artistic intention was stripped as a requirement of performance art.

Action art becomes limited only by its spatiotemporal difference and serial unity as process—nothing else. Prereflective action can be motivated by dreams, drugs, or madness. All of these are used in works like Strindberg's *A Dream Play* (1902), Antonin Artaud's *Jet of Blood* (1925), Luis Buñuel's film *Age of Gold* (1930), and Tristan Tzara's *The First Heavenly Adventure of Mr. Aspirin* (1916). In Salvador Dali's film *Impressions of Upper Mongolia* (1975), for example, Dali urinates on a metal pen every day until the acid from his urine creates an image on the pen. At close range this becomes an imaginary landscape through which Dali quests for a giant hallucinogenic mushroom.

Action Painting

Action painting developed out of this tradition of avant-garde modernism. From the 1940s through 1960s, a number of Western artists began emphasizing the action and process—both conscious and unconscious—of painting by spontaneously dribbling, splashing, or smearing paint onto the canvas. In this way, painting—one of the most visual of arts—came to focus on the nonvisible and often unconscious actions of the matter of the body, paint, and canvas. Instead of a finished work, the painting was just a record of the more primary process of the act of painting. In 1952, the American art critic Harold Rosenberg defined action painting by the same criterion as earlier avant-garde modernism: as "an arena in which to act."

The milieu thus became a performative unity of fragments and series of actions, none more primary than the other. Everything in the material process of painting, sculpture, and other arts itself became an action: the act of the paint's own texture as it is topologically built up in Arshile Gorky's *Tracking Down Guiltless Doves* (1936), the act of taking shape as it flies through the air before hitting the canvas in Jackson Pollock's *Autumn Rhythm* (1950), how it reacts chemically to fire in Joan Miro's *Burnt Canvases* (1973), or how the canvas reacts to drying, cracking layers of paint in Franz Kline's *Orange and Black Wall* (1959) or his *Nijinsky* (1940). Again, the primacy of dance, and Nijinsky in particular, persists in the modernist action arts. Wolfgang Paalen's influential book *Totem Art* (1943) makes this explicit by connecting indigenous art, dance, ritual, expression, and action painting.

Performance Art

From the 1960s through the 1980s, action art continued to develop previous performance techniques like burning art, body and costume art, and sound art, and to synthesize them into similar space-time unities like the first *serate*, now called "happenings" or "events," that included body motion, recorded sounds, written and spoken texts, found art sculptures, and even smells. What all happenings, conceptual art, performance art, installation art, and earth art have in common is that they push modernist action art to its absolute limits in the kinds of acts that can be art.

All acts with a determinate and differential space-time can be art acts. Since every action is a certain space-time, and each space-time can be divided further into smaller ones, the domain of action art is unlimited. In the case of earth art such as Robert Smithson's *Spiral Jetty* (1970) and Andrew Rogers's *Time and Space* (2011), it is the performative milieu not only of construction but also of deconstruction as the natural environment slowly changes and affects the work of art in a particular place over time. While the happenings of the 1960s exposed the daily, human, and bodily micro level of space-time actions, earth art exposed the geological, natural, and macro level of space-time action.

What all these movements have in common is the radical discovery of an elastic and intervallic act within and between all other arts. In this sense, modern art explodes the very category of art as a specific kind of thing—and reinvents all art as process, performance, and differential action.

MOLECULE

The third and final major aesthetic practice of differentiation we look at in this chapter is that of the molecularization of food and medicine in the eighteenth through twentieth centuries. Aesthetic images of taste, touch, and smell are not universal, and so they, too, follow a similar historical periodization as do the other arts. During the modern period, this took the form of a molecular differentiation of the gustatory, tactile, and olfactory arts. The molecular turn began with Robert Boyle's 1661 treatise *The Sceptical Chymist*, in which he posited that matter was composed of clusters of particles or "corpuscles" whose rearrangement produces chemical changes in larger matters. The details of this thesis were worked out over the coming centuries and had important consequences for the transformation of the culinary, medicinal arts, and fashion.

Nutrition

The discovery of nutritional vitamins, or "small quantities of unknown substances essential to life,"[31] as the Russian surgeon Nikolai Lunin put it, occurred in the eighteenth century, beginning with the Scottish surgeon James Lind's discovery that vitamin C deficiency causes scurvy. Over the course of the nineteenth century, scientists isolated macronutrients—carbohydrates, proteins, sugars, and the chemical elements carbon, nitrogen, hydrogen, and oxygen—that make up all food, and determined how to measure their proportions. The idea that all foods are composed of invisible molecules and nutritional compounds redefined culinary aesthetics around nutrition and a balance of these tiny molecules. In contrast to previous diets, modern diets and balanced meals are defined not by their macro-level appearance but by their molecular composition. Each component is itself not a food, but when they are combined, they produce a chemical compound we call "food." In short, molecules are the kinetic differentiation of food. The "good," "balanced," or even beautiful meal is increasingly defined by its combination of macro and micro differential elements.

Furthermore, cooking food became nothing other than the rearrangement and molecular elasticity of these components to form new culinary compounds. One very simple change produced by this was the introduction of canned and processed foods made possible by altering the microbacterial composition of the food substance through heat or pasteurization. Canned and processed foods entered into people's diets because of our understanding of the molecular and bacterial composition of those foods and thus their recomposition into something other and more differential than they were before. How far can meat proteins be stretched and mixed into soybean "filler" before we no longer consider it "meat"?

Microbiology

In medicine, a similar kinetic move occurred in the nineteenth and twentieth centuries, away from the study of infectious disease epidemiology and toward bacteriology and virology. Instead of distilling a substance and extracting its ethereal essence, modern medicine began to look at the population of component microscopic elements of a disease or of the body and its cells. The question was no longer of essential individuals and their relation but, rather, of the differential components that make up them and their relations as a population of heterogeneous elements. Vaccination is the treatment of populations within both the human ecology and the social ecology. Microbiology is the differential biology of the organism and the elasticity of its composition to include more or less of certain bacteria, viruses, and other micro elements and still maintain homeostasis. The question of modern medicine was no longer how to return to an original state of health, as it was in antiquity but rather how to achieve an elastic equilibrium between a multiplicity of competing internal populations of bacteria and micronutrients. Just like the aim of capitalist economic equilibrium, there is an aim for nutritional and bacterial equilibrium in which the doctor struggles to add and subtract discrete microscopic substances in the proper exchange ratio.

Fashion

In another sense, the modern arts also produced a social molecularization of classes, castes, and groups into *individuals*. We have not discussed the fashion arts so far in this book, for the same reasons we have not dealt with dance—it came to its most dominant aesthetic expression only in the modern period. Before the modern period, people wore clothing to keep warm (function), to model or represent the divine (form), or to express a social hierarchy (relation)—although usually all three were in some specific historical mixture. In all these, however, there was not yet a personalized and wide social expression of a vestiary form with a fraction of the breadth displayed in the eighteenth and nineteenth centuries, made possible with industrial textile production and sufficient media circulation to promote clothing fashions.

Kinetically, modern fashion transforms the human body into an individual or socially molecular body—that is, an individual whose clothing expresses him- or herself as an individual within a larger cultural population of fashion trends. The existence of rapidly changing fashion trends indicates the existence of a constantly fluctuating population composed of changing individuals. In this way, it is the tiny details and differences of the clothing that distinguish the individual—an extra ruffle, a single feather, an added inch of fabric. Different fashion components of dress are added and subtracted in an elastic equilibrium—pushing a style one step ahead or back without going too far and appearing ridiculous. Fashion is dress liberated from the strict confines of class, caste, and God. It is a strange art that spreads horizontally among slightly differentiated, mobile individuals. It takes the tiny microscopic fragments of everyday life—the differences that make a difference in our appearance—and reshapes them into something new: fashion.

Modern fashion reached its zenith in the aesthetics of dandyism, which is an aesthetics of difference because it works on the individual at the level of preindividual vestiary elements. The molecular individual is itself composed of further vestiary elements that produce its function, form, and relation. To change and take control of one's own appearance is to take control of one's own function, form, and relation, but since these changes occur in collective waves, they also become collectively crafted aesthetic patterns that appear at the macro level of populations as well. The microscopic differences of individuals thus produce emergent patterns defined by the collective addition and subtraction, or an oscillation between vestiary components. "These beings have no other status, but that of cultivating the idea of beauty in their own persons, of satisfying their passions, of feeling and thinking," as the French poet and dandy Charles Baudelaire writes. As the avant-garde made explicit at the turn of the twentieth century, the nineteenth-century dandy was the first to take his or her own living, moving body as an aesthetic object par excellence.[32] It is worth quoting Albert Camus at length on this point.

> The dandy is, by occupation, always in opposition. He can only exist by defiance. Up to now, man derived his coherence from the Creator. But from the moment

> that he consecrates his rupture from Him, he finds himself delivered over to the fleeting moment, to the passing days, and to wasted sensibility. Therefore he must take himself in hand. The dandy rallies his forces and creates a unity for himself by the very violence of his refusal. Profligate, like all people without a rule of life, he is only coherent as an actor. But an actor implies a public; the dandy can only play a part by setting himself up in opposition. He can only be sure of his own existence by finding it in the expression of others' faces. Other people are his mirror. A mirror that quickly becomes clouded, it's true, since human capacity for attention is limited. It must be ceaselessly stimulated, spurred on by provocation. The dandy, therefore, is always compelled to astonish. Singularity is his vocation, excess his way to perfection. Perpetually incomplete, always on the fringe of things, he compels others to create him, while denying their values. He plays at life because he is unable to live it.[33]

The constant proliferation of singularities of micro differences and little novelties compose the molecular individual and its vestiary elasticity. Every change must be surpassed by yet another, pushing the series ahead in an infinite expansion or contraction of retro fashions. Fashion in this sense is akin to the action arts. The wearer of fashion becomes an actor in his or her own life. The individual's whole life and all social ritual are exposed as the performance of micro differences in flare, ruffle, pant length, and so on.

CONCLUSION

All the newly dominant arts of the modern period are defined by the kinesthetic pattern of differential and elastic motions within and between images. Relation no longer appears, as it did in the Middle Ages, as the only fundamental aesthetic medium holding all forms together and apart. Now, the materiality of relation itself and all its forms are composed of more primary and differential fragments. The genius of modern art was to have discovered this constitutive difference and to have begun rebuilding these fragments into something entirely new.

This chapter concludes our historical analysis of the moving image and gives us a material and historical foundation for interpreting the hybrid kinetic structure of contemporary arts. Art history is not dead. It all returns and resurfaces in the present, mixing in hybrid combinations. Although their name and appearance might have changed, the same regimes are still at work today.

We turn now in Part III to the kinetic analysis of the contemporary image and the coexistence of these four major historical patterns.

PART III

The Contemporary Image

Unlike the previous historical aesthetic regimes, the contemporary image is not defined by a single kinetic regime or field. Instead, it is defined by the hybridity of all the previous regimes. This is precisely why we had to demonstrate in Part II the concrete historical emergence and definition of each of these regimes: because they compose the present.

If these regimes seem to take on a different organization for us today, it is not because the material-kinetic patterns described in Part II are no longer accurate; rather, it is because they have been mixed in new ways and new ratios. However, new hybrids are still hybrids of older kinetic patterns, even if they are masked by the fetishism of "new" combinations. In order to understand our contemporary kinesthetic milieu, one cannot ignore the historical and material regimes that not only condition its emergence but also compose its concrete hybridity. This is just as true in politics, ontology, and science as it is in art—"the past reappears because it is a hidden present,"[1] as Octavio Paz writes.

In particular, it is the historical dominance of the electrical or digital image that today brings to the foreground of sensation several hidden dimensions present to varying degrees in all previous kinesthetic fields: hybridity, pedesis, and kinetic feedback. These newly emphasized and expanded dimensions do not apply only to the latest images. Now that the digital image has passed over a certain threshold of historical dominance, we cannot help but begin to see that these kinetic dimensions are integral parts of all historical images—although emphasized in varying degrees—since they are also tendencies that kinesthetic regimes are often trying to actively suppress, redirect, or capture for their own purposes.

The argument of Part III is that the contemporary image is defined by the digital image but that the digital image has two sides or dimensions: a hybrid image and a generative image. While the hybrid image makes possible a radical "remediation" and mixture of all previous aesthetic fields through digitalization, the generative image is defined by a new electrical creativity.

CHAPTER 15

The Digital Image

The discovery of the electromagnetic field defines the electrical and digital image. It also reveals something not only about contemporary "digital media" but also about all sensible matter and its capacity for self-affection or image-making. All sensible material images are composed of electromagnetic fields that have folded over and interacted with themselves in the form of discrete sensible images. The manipulation of electromagnetic flows into various interactions, folds, and patterns of circulation that define the digital image are therefore not radically new aesthetic practices. Every distribution of sensibility and images presupposes the kinetic manipulation of the electromagnetic field. Every body carries an electrical current.[1] All sensible matter circulates charged electrons.

Today we tend to think of electrical flow and its manipulation as something unique to our age, but there is no absolute historical break or ontological division in the continuous modulation of electrical current. Rather, there are thresholds, like water coming to a boil. Today we have certainly crossed such a threshold in our capacity to manipulate electrical flows more directly, more minutely, and in more ways than ever before. Philosophically and kinetically speaking, however, this does not change the fact that all organic and nonorganic beings have already been doing this basic activity for much longer. To treat the digital image as a category that exists only after this threshold is an arbitrary historical, disciplinary, and conceptual bias.[2] It is like saying that steam is something categorically different than water.[3]

In this chapter, we look at the material and kinetic structure of the electromagnetic field itself and consider the way in which it has hybridized contemporary aesthetics. In the next chapter, we turn to an analysis of the generative image in its ordered, disordered, and contemporary varieties. The aim of these last two chapters is to show convincingly that the image has always been digital and generative, and that all art has always been kinetic. It's something we are only now beginning to appreciate.

THE ELECTROMAGNETIC FIELD

The electrical or digital image is defined by the material-kinetic structure of the electromagnetic field. Just like all the other historical kinesthetic fields, the electromagnetic field is not limited to a single historical period, even though it crosses a certain historical threshold of kinesthetic dominance around the late twentieth century. It coexists with the function, form, relation, and difference that defines all aesthetic fields. In fact, it is the material kinetic continuum in which and through which all aesthetic functions, forms, relations, and differences flow, fold, and circulate. The entire history of the image takes place as series of bifurcations, knots, and weavings in this field. However, the electromagnetic field, like the kinetic flow, is not a sensible image on its own. It is only the material kinetic condition of self-affection that makes possible the centripetal, centrifugal, tensional, and elastic patterns of circulated images in the other fields.

However, just because the electromagnetic field is invisible, that does not mean it is immaterial or immobile. It has both energy and momentum. Contrary to new-media hype, electricity is not unreal, virtual, immaterial, hyperreal, pure simulation, or any other such nonsense.[4] The electromagnetic field is real, kinetic, and material. It is continuous and extends infinitely in all directions like a vast ocean. When a region of the electromagnetic field affects itself, folds, or "interacts," as physicists say, it produces a sensible image: a photon wave. Like waves in the ocean, the movement of these photons allows for a transfer of energy between charged electrons, themselves wave folds in an electron field. These charged electrons absorb and emit photons, causing other electrons to move randomly between atoms. When a wave of photons is strong enough to move all the free electrons in the same direction—Einstein's photoelectric effect—it produces an electrical current that releases surplus photons as energy/light/heat as electrons move between the orbits of the different atoms in the conductor material. Negatively charged electrons flow from negative to positive in an electrical circuit. The kinetic transfer and release of energy flows occurs throughout nature and follows the second law of thermodynamics: entropy.

The electrical or digital image is, therefore, defined by a fundamentally continuous, material, and kinetic process by which all material self-affection is possible. This kinetic process gives the digital image its three features: hybridity, kinetic feedback, and pedesis.

Hybridity

The first kinetic feature of the digital image is its hybridity. The digital image is capable of a high degree of hybridity because it is possible to turn the photoelectric effect on or off and thus introduce a binary code into the electromagnetic field. If all images are made from this primary electrokinetic field, then all images can be similarly coded and transcoded from one medium to another. Any medium can be coded, manipulated, and combined with others, allowing for a dramatic diversity of hybrid media and remediation.[5] This degree of hybridity is possible because an electromagnetic flow can be released or not, on or off, one or zero. Electrical or digital binarization pushes the kinesthetic regime of elastic difference to its most radical conclusion but only on the condition of an absolute continuity and flow of the electromagnetic field itself.

Historically, the exploration of the direct control over this process of differential binarization or "electrical elasticity," as Maxwell called it,[6] began in the nineteenth century. In 1800, Alessandro Volta built the first electrochemical circuit consisting of two electrodes, zinc and copper, in an electrolyte bath of sulfuric acid mixed with water or a saltwater brine.[7] This voltaic pile or electric column was the first electrical circuit capable of producing a steady electrical current and thus making possible a new regime of radically elastic electrical circulation. Volta's invention in turn made possible the first forms of electrical typography or telegraphy (from the Greek τῆλε, *têle*, "at a distance").

In 1804 and 1809, the first electrochemical telegraphs connected up to thirty-five wires, one for each Latin letter and numeral, between two locations. On the sending end, an electrical current, made possible by Volta's battery, was sent; and on the receiving end, the wires were immersed in separate tubes of acid. When a current was sent, the acid released hydrogen bubbles.[8] Between 1809 and 1832, a number of similar telegraphic systems were invented, but by 1833, the number of telegraphic wires had been reduced—first down to sixteen, then to eight by Pavel Schilling, then finally to one by Carl Friedrich Gauss. In 1832, Schilling was one of the first to start using a binary system of communication, but in 1833, Gauss and Wilhelm Weber began using a single wire to communicate an entire alphabetic language, using purely electrical means. An electrical current at one end would be given either positive or negative voltage pulses by moving an induction coil up and down over a magnet. Wires were then run all over the German town of Göttingen.

In 1837, Samuel Morse independently developed a similar electrical telegraph and binary-coded alphabetic system in the United States. Based on these developments, several telegraphic printing systems were developed in the 1850s that were capable of inscribing electrical signals directly onto paper. These early telegraphic systems were used primarily as stock-price-ticker systems and used the same binary method of inscription. Based on these binary codes, one of the earliest practical stock-ticker machines, the Universal Stock Ticker developed by Thomas Edison in 1869, was able to automatically decrypt the codes and print in alphanumeric characters with a printing speed of approximately one character per second.[9]

In the twentieth century, the usage of binary-coded flows reached its quantum limit with the invention of the transistor. A transistor is a composite of two semiconductor materials that when electrically (not mechanically) energized releases a flow of electrons from within the materials themselves. A simple transistor is composed of three pieces of silicon. The two on the ends are made from silicon doped with phosphorus, and the piece in the middle is made from silicon doped with boron. Since silicon has four electrons in its outer shell, phosphorus has five, and boron has three, the electrons from the phosphorus side of the transistor move toward the boron side of the transistor to fill the electron holes in the silicon lattice. The phosphorus side is the "source," and the boron side is the "drain." In between the two is an electrical contact called the gate, which applies an electrical charge across the two sides, causing the negatively charged electrons of phosphorus (source) to flow into the positively charged holes in the boron, then to phosphorus on the other side (drain), and then through a connective circuit back around to the source again. In short, with the application of a very small voltage, a circulation of electrons between negative and positive charges is produced.

A nonmechanical electrical circuit is thus produced by mobilizing the subatomic and quantum properties of crystalline semiconductors. Semiconductivity is a quantum-mechanical phenomenon because in classical physics there is no way for an electron to move through "close-packed" atoms. Furthermore, in classical physics there is no such thing as an electron hole or differently massed electrons.[10] The transistor is therefore a binary quantum switch that can be opened or closed simply by modulating an electrical voltage. The difference between open and closed is no longer mechanical but rather quantum-mechanical. Thus, when a computer key is pressed it introduces a modulation or fluctuation in a long and complex chain of electron flows beyond the simple oscillation of its initial impact. Instead of each key marking a one or a zero, each key introduces a ripple or modulation into the constant flow of electrons in the transistor series, which produces an enormous amount of other oscillations. It is as if each switch of the key were attached to thousands (1970s) or billions (2010s) of other related switches.

In this way, the electromagnetic flow of electrons is reduced to its most simple binary difference: on or off, open or closed. Anything that can be coded can be transcoded and thus hybridized, but only on the condition that the hybrid product is also defined by the aesthetic difference of binarization. Digital sounds, digital sights, digital textures, and even genetic coding all presuppose a more primary material continuum of which they are only its differential fragments.

This feature of the digital image—its capacity for radical differentiation and hybridization through binarization—is often used to define the entire image. However, this would be a critical mistake. The digital image is not identical to its binary code, precisely because the digital image is defined by the flow and fold of the electromagnetic field, which precedes the code as its basic material condition. The forgetting of the material-kinetic nature of the digital image has led many scholars to false affirmations of "immaterial information"[11] and false rejections of its apparently "discrete" nature.[12] The digital image not only captures material kinetic novelties into a binary synchronization that advances certain capitalist valorization schemas; it also escapes them at the same time. It is thus too simplistic to merely valorize or demonize the digital image. What is needed is a more bi-valent and kinetic approach.

Like everything else only more so, the digital image can be divided into expanding and contracting differentiated series of bits (electron folds in the EM field). In this way, the digital image appears to be the kinesthetic pinnacle of a fragmented and fragmenting modernism with respect to its hybrid image. However, the situation is complicated by the fact that the digital image is also defined by two other kinetic features that are not reducible to this differentiation and in fact even undermine it entirely, as we will see in the sections that follow.

Pedesis

The second kinetic feature of the digital image is its pedesis. The electromagnetic field that defines the digital image is, like all fields at the quantum level, fundamentally stochastic and indeterminate. Quantum indeterminacy is the chaotic movement of fluctuations in fields. At the level of quantum fields there are no discrete particles (such as photons or electrons) until there is a fold or interaction in the field. The equations of quantum-field theory are successful not because they are able to predict these particles and trajectories but because they set out ranges of probability for the existence of a particle or trajectory, based on an initial measurement and domain.[13] However, in the end all fields retain a true degree of chaos or pedesis of which the existent particle and its path are only shells washed up on its great shoreline. The existence and trajectory of photons and electrons in the electromagnetic field, for example, are both defined by this pedesis even if a certain degree of control and prediction of their motion is possible through certain probabilistic equations.

The quantum indeterminacy, or pedesis, of the electromagnetic field has real effects on the kinetic structure of the digital image. The open flow of charged electron waves in an electrical circuit is only a tendency to flow in a single direction, not a universal motion. Furthermore, the closed state of electron waves is similarly just a tendency to stop flowing but is by no means universal. What this means is that the electron wave is simply less likely to move through a thicker transistor gate than through a thinner one. Since all regions of the electron wave also include those that go through the transistor-gate oxide, this means that sometimes the flow of electrons does not stop but actually moves through the gate oxide. Physicists call this "quantum tunneling." The pedesis of the field causes "soft errors," "single event upsets," or "noise" in the digital image. In the past few years, scientists and engineers have been discovering that this is a more prevalent problem than thought and will only become more of a problem as transistor size continues to decrease.[14]

The flow of the electron wave always leaks at the quantum level. Electron waves do not all stop when the current is removed. At the quantum level, all electrons are in constant pedetic motion, even when a current is not flowing. Thus in addition to their electrical circulation, they have a quantum circulation that does not follow the same rules: they move pedetically, and they leak. As transistors get smaller and smaller, the oxide gate between the source and the drain becomes thinner and thinner, inducing more charge in the channel, boosting the current, and making the transistor faster. But the thinner the oxide on the gate, the more electrons are likely to tunnel through it and introduce "noise" into the system. However, oxide levels cannot be lowered much less than one nanometer, which is about where they are today, or too many electrons will flow across the channel. Engineers are already at work on alternative conductor materials and protocols for working with instead of against the tunneling effects through more quantum probabilistic equations, but the future limits remain uncertain.[15] However, none of these engineering efforts will change the indeterminate and pedetic nature of the electrical and quantum field. When a digital command is issued in a digital binary system, there is only a probability that it will be executed, not a certainty. The dream of downloading human consciousness into machines or creating a perfectly controllable artificial super intelligence will inevitably run up against the pedetic nature of matter and the material agency of the digital image itself.

Quantum noise occurs in both natural and technological electrical flows. However, because its differential structure binary code is particularly affected by these "noisy gates," it can become destabilized more easily than, for instance, an organism whose internal processes are much less hierarchical and centralized. Bioelectrical processes do not function by complex binary-code systems and so single-upset error is built into the functioning in a vast network of bioelectric feedback loops.

The pedesis of the electron field is a key defining feature of the digital image, not just a defect. It is the material kinetic condition for both the possibility and the impossibility of the digital image. It is precisely because of the pedesis of free electrons that they can be mobilized from atomic shell to shell, creating electricity, but it is this same pedesis that allows them to tunnel and deviate, producing errors, events, and novelties in the flow.

Feedback

The third kinetic feature of the digital image is feedback. An electromagnetic field is defined by the movement of charged particles from negative to positive polarities. The electromagnetic field is thus defined by a kinetic pattern of circulation. This circulation or flow of matter follows the same structure as defined in Chapter 3 on the aesthetic field. All kinesthetic fields are defined not only by the circulation of matter and image but also by the recirculation of matter back into that same field. The movement of circulation and recirculation is a type of kinetic feedback in which the flow of matter returns back to itself and modulates itself again and again. Each material flow introduces unique perturbations, particle excitations, and entropic and pedetic deviations that return back through the circuit and remodulate the others. The electromagnetic field is, therefore, a fundamentally relational field in the sense that it interacts with itself and changes itself through intra-actional feedback. When a field interacts with or affects itself, it produces an image or sensation that appears in the form of a particle, a minimum of sensible matter. All digital images, and to some degree all affective images and their kinesthetic fields, are therefore defined by this material kinetic interaction or feedback loop of intra-action. The contemporary digital image simply raises this process to the level of sensation in a way that emphasizes, expands, and diversifies this process to a much larger degree than any other kinesthetic field or pattern.

Digital media are often defined by their "interactivity," or dynamic relation between user and media.[16] Analog media, Lev Manovich writes, are things we passively consume or interact with only mentally, while digital media have keyboards, software, and hyperlinks that we interact with physically.[17] Arguably, all the supposedly new features of contemporary digital media can be seen as subcategories of this interactivity. "Digitality" is electromagnetic interactivity in a binary electrical circuit; "hypertexuality" is an interaction with links; "virtuality" is an interactive extraction of information or immersion in a digital image; "networked" images are defined by the mediated interaction with other people, and so on.

However, kinetic interaction is not just the structure of contemporary media; it is the structure of *all fields of images*. The image is by definition, as we argued in Chapter 2, self-affection. When matter interacts with itself, it produces sensory images, qualities, and pleats in being. Historical images all presuppose a degree of interaction and mutual transformation both between matter and itself and between matter and human sensation.

In order to see a painted canvas, waves of light must be capable of materially transforming the human retina and of reflecting back from our bodies onto the canvas, coloring it in turn—in addition to a million other material interactions of the milieu. The human body, just like other material bodies, is not passive. As one looks at a painting, one's whole body actively sways and one's eyes follow the line and color of the painting, dilating and blinking, along with a million other active micro motions, perhaps even more actively than one clicks on a hyperlink. Feedback and interactivity, therefore, do not divide evenly along fascicle mind/body, active/passive, or culture/nature lines.

Accordingly, the idea of "cyborgs" ("cybernetic" plus "organisms"), popular in the 1990s, is an inappropriate portmanteau. Organisms have always been bioelectric and cybernetic. That said, computer software does allow for a greater degree and range of aesthetic transformation than what is typically possible in the viewing of a painting in a museum. However, this is simply a matter of degree made possible by new technical proficiencies and historical experimentations.

Marshall McLuhan gets closer than Manovich to the kinetic structure of this interactivity when he writes: "Electromagnetism seems to be in its technological manifestations an extension of our nerves and becomes mainly an information system. It is above all a feedback or looped system. The idea of feedback, of being involved in one's own participation, in one's own audience participation, is a natural product of circuitry. Everything under electric conditions is looped. You become folded over into yourself. Your image of yourself changes completely."[18] What defines the digital image is not a category of reified historical media objects but, rather, a type of kinesthetic process defined by the feedback loops of the electromagnetic field itself present in both human and nonhuman bodies. Of course, McLuhan is incorrect in thinking that electromagnetism only occurs in modern technologies and thus wrong in his whole developmentalist history of technology.

What makes both previous and contemporary kinesthetic fields interactive digital images is not whether someone pushes a key or touches a screen but, rather, whether and to what degree a mutual transformation of sensuous images occurs in the electromagnetic field. Again, it must be admitted that the twenty-first century is defined by the pedetic digital image in the same way that antiquity was defined by the centrifugal formal image.

There are, however, different degrees of kinesthetic feedback corresponding to an emphasis either on the hybridity or on the pedesis of the digital image. A more limited feedback loop is a field of aesthetic circulation with a relatively short iteration process that terminates in a predetermined output. For example, a word-processing computer program is set up to produce the letter "a" whenever a certain key, usually the "a" key, is pushed. The depression of the key causes a certain binary configuration of transistors to open and close—"01100001"—to be precise. After the operation is complete, the letter "a" is produced on the screen. User input is thus captured, visualized, and can be responded to with more input. When bugs or soft errors are detected, a notification appears and is forwarded to the software developers in order to improve the functioning of the software build and its predetermined algorithmic function. This feedback system is limited by the same constraints that give it such incredible hybridity and utility: the limited and highly ordered nature of its algorithms. The more functional the utility, the more order is required and the less randomization or pedesis is incorporated.

On the other hand, it is also possible to create less limited feedback loops that incorporate more disorder in their algorithmic structure. For example, a computer program can run a randomized or complex algorithmic operation that will produce much more unpredictable and disordered products. This is possible because instead of running only a small series of highly ordered operations, it inputs the products of its operations into its operations indefinitely. This is not just a simple programming loop or iteration but, rather, a loop with a twist or algorithmic pedesis that takes advantage of the pedetic nature of the *electrical flow itself* and its capacity for ordered disorder.

However, at the very limit of algorithmic disorder there is always the quantum disorder of the flow itself. Within the kinesthetic field of the digital image one can try and order it more or less, but one is always just along for the ride down a turbulent river of quantum perturbations. The question is simply to what degree one captures them in a highly ordered system or attempts to follow them by trying to reproduce the same interactional and pedetic motions.

THE HYBRID IMAGE

These two degrees of feedback manifest themselves in two different but related kinds of digital images in the twenty-first century: the hybrid image and the generative image. Both images are defined by a degree of hybridity, pedesis, and feedback but in different ratios. In the remainder of this chapter, we analyze the hybrid image, and in the next chapter we analyze the generative image.

The first image is the hybrid image and is defined by two kinds of hybridity: a transcoded hybridity and a kinesthetic hybridity. Both kinds of hybridity work precisely because of the incredible level of highly ordered electromagnetic operations possible in twenty-first-century digital aesthetics.

Transcoded Hybridity

The hybrid image is made possible because the electromagnetic field can be binarized. Since all sensuous material has an electromagnetic field, it can be similarly binarized and transcoded into other media objects. These media objects thus become transcoded hybrids capable of producing sensuous images of multiple media objects. Hybrid media abound in the twenty-first century, but the transcoded device par excellence of our century is the mobile device—laptop, smartphone, tablet, and so on.

The mobile device is a specific kind of body that is capable of expressing all the aesthetic functions previously expressed by an increasing number of other bodies. For example, a mobile device can function as a television, phone, stereo, address book, calendar, clock, notebook, photo album, ruler, thermometer, calculator, game machine, mailbox, map, canvas, musical instrument, and many other things. Because of the mobility and mutability of the electrical flow itself, all previous arts can be digitalized and expressed through the mobile device.

Kinesthetic Hybridity

In addition to this transcoded hybridity, there is a second "kinesthetic" hybridity of the contemporary aesthetic field. Because the mobile device is itself a material kinetic object, it is defined, like all other contemporary aesthetic fields, by an admixture of the kinetic patterns that have emerged historically. There is thus a kind of mutual hybridity between the mobile device that transcodes and modulates all previous aesthetic fields and the fields themselves that compose the material kinetic condition of the aesthetic object that can effect such a remediation.

However, since a complete analysis of the exact kinesthetically hybrid ratios of all modern aesthetic objects (televisions, cell phones, electric garage doors, dishwashers, and so on) would be too long and repetitive, let's consider just one of the icons of our age: the computer. After this analysis, I hope the reader can see, based on Part II of this book, that we could conduct a very similar kinesthetic analysis for any other aesthetic field.

The kinesthetic conditions of the most seemingly immaterial processes of the computer can be found in the most material fields in the history of the image. Let us consider each one in turn.

Centripetal Function

The first condition for the production of the computerized digital image is the centripetal motion of the functional kinesthetic field. Computers are made of certain materials and not others. Silicon is chosen because of the functional aspects of its atomic and subatomic structure, as a superconductor of electricity. Plastics are chosen because of their functional relation to the computer as a whole: light weight, hardness, nonconductivity, and the affordability of their petrochemical source. Rare-earth metals such as neodymium, yttrium, terbium, and europium are extracted through strip mining and refined through toxic processes in order to build LED screens, hard drives, and other parts of the computer. Metal ores (iron, nickel, aluminum, and others) are also dug up from strip mines deep in the earth from around the world, and refined through toxic processes owing to their functional relation with the aesthetic field of the object: durability, malleability, and so on.

All these materials are dug up from the vast global periphery, refined, purified, and shipped to centralized processing and manufacturing plants where they are consolidated and assembled into massive accumulations based on their interdependent functional cycle in the final computer. Additionally, since computers do not function without an electrical source, they require a vast peripheral network of wires, themselves made of centripetal accumulations of metal and plastics. These power lines are then run all over the earth in order to be centripetally collected into billions of electrical outlets pooled and consumed regionally.

These various material-kinetic process are not usually considered part of the aesthetic field of the computer, but they are very much in the same basic way that the Neolithic fire was made possible by a centripetal accumulation of fuel from the periphery to the center. The computer is itself a work of art created first by gathering raw materials from the centripetal periphery and then assembling them into a single aesthetic image. There are, of course, differences between fire and the computer, but both have a degree of centripetal aesthetic functionality that defines the material conditions of their production.

Centrifugal Form

The second condition for the production of the computerized digital image is the centrifugal motion of the formal kinesthetic field. All the functional raw materials that compose a computer are formed by molds. Once the raw materials are transported to a centralized location, they are then pooled, melted down, purified, extracted, and injected into molds of all kinds. Every tiny part of the computer is molded into its own form. Raw material enters the hollow center of the mold, is formed or stamped, and is ejected from this central mold to the periphery to cool and be prepared for assembly.

Each transistor is cut from a mold or form that allows thousands of them to fit together inside the body of the computer, which is itself formed of molded plastic and metals. The form of the keys, the width of the keyboard, and the shape of the track pad are all formed by the constraints of the human user. Matter is extracted, liquefied, and formed into a computer body within which all the functional matters fit into their own places, like a jigsaw puzzle. Even the screws are molded by the ancient art of metallurgy.

Tensional Relation

The third condition for the production of the computerized digital image is the tensional motion of the relational kinesthetic field. The functional materials and molded forms are related to one another based on the circulation of the photoelectrical current that moves through them. The materials that compose the computer are only functionally related as conductors or nonconductors in relation to the more primary flow of electricity that moves through them. Each of the parts functions only in electrical relation to the others to produce a total functional kinetic use-value.

Each component of the computer also receives the form it does because of the electrical relation that must move through the formed space of the computer body and fit together. The jigsaw forms of the computer are thus determined with respect to one another so they fit along a continuous circuit of electrical flow. If a laptop battery expands, then other areas must shrink in give-and-take functional and formal relationships between battery life, processor speed, number of ports, hard-drive space, speaker quality, and so on. The parts of the computer thus change in tension with one another.

Elastic Difference

The fourth condition for the production of the computerized digital image is the elastic motion of the differential kinesthetic field. Each of the related forms in the computer can in turn be divided into component parts, making it possible to continually multiply processor speed by shrinking the size of the transistors. Smaller and smaller transistors subdivide the electrical flow into more and more possible gates for the production of binary code. The electrical flow thus expands and contracts between a relatively small number of transistors while the computer is at rest and between a larger number when a large program or series of applications is used. A larger degree of differentiation allows for a wider range of elastic expansion and contraction of the total set of transistors in the computer. The more internally differentiated the computer is by its transistors and components, the more it can produce increasingly faithful transcoded hybrids of other media.

Pedetic Feedback

The fifth condition for the production of the computerized digital image is the pedetic motion of the interactive kinesthetic field. Function, form, relation, and difference are all brought together in the computer in an electrical circuit or feedback loop between the positive and negative poles of the electromagnetic spectrum. Like all other works of art, computers are interactive insofar as they affect and are affected by other matters. However, they are also defined by an internal electromagnetic feedback pattern of circulating electrons, without which none of its internal components would have the function, form, and relation that it does. The internal modulation of each component occurs in this electrical circuit, continuously modified by the pedetic flow of current. Since the whole system itself is defined by this electrical feedback, the system can respond to itself, executing programs autonomously, as well as responding to external input and interaction by users.

Differentiation occurs only as a differentiation of the pedetic electromagnetic flow. The closer one gets to this flow, the higher the noise-to-signal ratio. The more one approaches the pedetic electrical flux, the more chaos begins to creep in and the more complex and autonomous the electrical feedback becomes. This is where the generative image comes in, as we will see in the next chapter.

CHAPTER 16

The Generative Image

As mentioned in Chapter 15, the digital image has two sides: the hybrid image and the generative image. While the hybrid image is defined by a high degree of ordered differentiation, binarization, operationalism, and ordered electromagnetic flow, the generative image is defined by a high degree of pedesis and feedback in the electromagnetic flow. Digital hybridity serves to maximize utility by transcoding multiple functions into a single, all-powerful digital hybrid utility: "There's an app for that." Digital generativity, on the other hand, introduces higher levels of pedesis and feedback that cannot be predicted and tend to have a much less useful outcome.

Rather than only hybridizing and cannibalizing older aesthetic fields into the highly ordered language of computer code, the generative image unleashes the creative, pedetic, and interactive aspects of the electrical flow itself. Instead of making programs under the complete control of the programmer, one can make programs that introduce a higher degree of unpredictability, disorder, and feedback into the computational and aesthetic process.

In this way, the generative image increasingly privileges the process of kinesthetic interaction above the subject (creator) and object (product). By doing so, the generative image discovers and makes intensely sensible the kinetic nature hidden in all previous aesthetic fields. As we showed in Part II, art has always been kinetic, interactional, and pedetic, but never in history have these features defined an entire historical field of images to the degree that they do today. It is only because of the historical dominance and prevalence of the digital image that we can now see that the image has always been fundamentally kinetic. In fact, this discovery is what allows us to justify a full reinterpretation of the history of art in this light—as a kinesthetic history.

However, the electromagnetic field is not just something we find today and in art history; it is something at work in all matters. In the sense that all sensible matter is made up of elementary particles that have a spin, there are magnetic fields for all matter. All of sensible matter is thus defined by the same movement, pedesis, and interactivity that characterize the electromagnetic field itself.

All the quantum fields that make up reality are defined by, among other things, the same fundamental features that define the electromagnetic field and the digital image: granularity (hybridity), indeterminacy (pedesis), and interaction (feedback).[1] Therefore, the kinesthetic exploration of the electromagnetic field is, in a way, also an exploration of quantum-field patterns more generally, and thus of nature. The primacy of motion in the flux and feedback of the electromagnetic field thus reveals to us the primacy of motion and flux in all of sensible nature.

All this makes the generative image seem quite diverse and encompassing, which it is. Generativity (pedesis and interaction) is a dimension or aspect of all materiality. It brings together nature and aesthetics in the same procedural kinetic field. However, among different generative images, there is a range in the relative degrees of disorder and feedback that can be distinguished. We start first with a description of the more ordered generative images and move on to the more disordered ones; but first, a note on randomness and pedesis.

RANDOMNESS AND PEDESIS

One of the defining features of the generative image is pedesis, which as I argued in Chapter 1, is not random. In fact, there is no such thing as absolute or ontological randomness. Absolute randomness would require a motion to be completely unaffected and thus unrelated to any prior motion that would determine its subsequent action in any way. Randomness and affection are thus incompatible. All matter is self-affective, sensuous, and thus nonrandom. The idea of randomness is a logical category, a pure mathematical idealism, as if we could abstract matter from its motion and affectation.

Pedesis, on the other hand, is unpredictable precisely because it is relational, affective, and sensuous. At each moment, movement is connected to what came before and indeterminate with respect to what it will do next.

It is a relational stochasticism. Turbulence, for example, is matter in motion with a very high degree of unpredictability for humans owing to the number of disordered variables involved in the process. Yet turbulence has relatively ordered patterns, spirals, swirls, vortices, and so on that begin to emerge from the initial disorder. Matter is pedetic not only at the quantum level of indeterminacy but also at the subatomic, atomic, molecular, visible, meteorological, and cosmic levels. Pedetic motions crystallize into relatively ordered patterns at every level of nature, even if they take millions of years to emerge. There is no pedesis that does not generate at least a little bit of order, no matter how minimal or over how long a time period, at some level of reality. Since matter in motion is always related to other matter in motion, and so on indefinitely, there is never a case of absolute determination or absolute randomness, although pedesis gives rise to more or less stable configurations like the electron patterns of the atomic elements, the speed of light (299,792,458 m/s), or the rate of acceleration at which things fall on the surface of the earth (9.81 m/s^2).

There are thus two factors barring absolute randomness and absolute determination—and both have to do with motion. The first is that we do not and cannot have an absolute knowledge of all the material variables in the universe, because the universe is constantly expanding and in process, and thus there is no totality of the universe. There is no Laplacian demon. The second is that since all matter is kinetic and relational, it cannot be unaffected and thus is not entirely random. If all of matter in motion were truly random, there would be no universe and nothing would hold together for very long. Now, on to the generative image.

THE ORDERED GENERATIVE IMAGE

The first kind of generative image is the more ordered type. Ordered generative images are those that have a high degree of relatively discernible order and pattern, and a minimal amount of feedback or interaction. In a broad sense, all kinesthetic images are generative to some degree, because all matter is pedetic and interactive. All works of art require a material medium through which an artist or creator creates and through which matters interact with one another. Since all matter is in motion and all media are material, all aesthetic media are defined by the kinetic process and thus have some degree of pedesis and feedback.

The question, then, is to what degree this kinetic process itself is emphasized, highlighted, explored, or liberated in art. The history of art and aesthetic practice can thus be reinterpreted from the perspective of the present as the history of the subordination and capture of the pedetic feedback process into the ordered patterns of centripetal, centrifugal, tensional, and elastic motion. All the previous patterns of art discussed in Part II of this book are examples of how relatively disordered motions can be ordered into distinct kinetic patterns that define aesthetic function, form, relation, and difference.

The Arts

Ordered aesthetic fields always confront the internal obstacle of pedesis. In painting, the disordered texture and color of the fabric, canvas, or grained-wood panel is overcome with the use of gesso or a neutral paint. The artist also confronts the relatively disordered texture of the paint itself—the way it drips or falls sometimes from the brush if it is too liquid; the unpredictable way in which lighting changes the colors of the canvas; the mistakes made by a gust of wind, an unsteady hand, a bumped elbow, spilled paint, the pedetic movements of the model, or the unpredictable light of the sun as it moves through the sky and behind clouds; and so on. In sculpture and architecture, the artist confronts the decay, erosion, and decomposition of the stone or plaster over time, the birds that take up residence under the eaves of the church, the cracks in the walls from earthquakes, the mistakes of laborers, the patina from oily hands that touch the floors of the building or the feet of the sculpture, periodic fires, and so on. In music, the artist confronts the pedesis of improperly tuned instruments, the background noise of the music hall, the scuffling of feet, the murmur of the audience, the acoustics of the room, the whim of the conductor, and so on. The history of Western art has almost always tried to overcome and crush this disorder.

The history of aesthetics is predicated on an ordering of interactivity, as well. All aesthetic matter must be interactive in order to be affected by the artist and in turn affect the artist and other matters. However, the sculpting of this interaction into desired kinds of interactions always runs up against the interaction of nondesirable elements of the milieu: weather patterns, earthquakes, the emotional state of the artist or viewer, the temperature of the room, gusts of wind, and so on. The interactivity of light on canvas affecting a viewer is emphasized and privileged, while the interactivity of light to fade, melt, or discolor the canvas is de-emphasized. The glare of the varnish, for example, is not supposed to be part of the painting, but nonetheless affects the work of art as a whole.

The vast majority of art history has been the erection of the most elaborate mechanisms for overcoming the pedetic tendencies in matter and the interactive relations traversing the whole milieu. Art has thus always worked with pedesis but largely with the aim of overcoming it in favor of its own kinetic regime of order.

Nature

Nature, too, produces ordered generative images. For example, solids are made of pedetic flows of vibrating particles moving very slowly and within a fairly limited range. This gives them a relatively more ordered structure compared with their less ordered fluid and gaseous states, whose atoms and molecules move much more rapidly and dynamically. Nature also produces relatively stable physical patterns, such as the six-sided symmetries of honeycombs and snowflakes; the radial symmetries of flowers and sea anemones; and so on. It also produces spiral patterns and vortices in the leaf arrangements (phyllotaxis) of sunflowers, pineapples, seed heads, the nautilus, and various mollusk shells. Even weather patterns use spiral shapes to lower the pressure of increasingly disordered and weather systems. Spots and stripes on animal skins and the tessellated scale patterns on snakes and certain fruits are also distinct ordered patterns in nature. Cellular division and bifurcation produce a repeating dendritic pattern similar to algae and most plant growth. Electrical discharge also tends to produce similar treelike branching patterns.

Many of these processes can be closely approximated by highly ordered mathematical equations like self-similar fractal patterns, L-system models of oscillating bifurcation, dendritic Lichtenberg figures, Fermat's spiral ratios, Fibonacci numbering of phyllotaxis, Weaire–Phelan soap-film bubble angles, and others. The fact that these mathematically ordered models give such close approximation of natural patterns indicates a profound order to many natural processes. Environmental feedback can thus be overcome to some significant degree by metastable patterns that resist change and outside influence. Snakes shed their skins, and mammals shed their fur, but it always regrows in the same kind of pattern.

However, the fact that abstract mathematical equations are never completely accurate and can never reproduce a physical system exactly also indicates that natural processes are not as highly ordered as equations and are sensitive to the initial conditions of material pedesis and interaction with their milieu.[2] No snowflake is identical to another.

Digital Image

The modern kinesthetic image thus pushes this ordering process to its absolute limit with the binarization, hybridization, and operationalization of almost all material processes. With only ones and zeroes, binary logic has tried to reproduce the whole of nature. Computer programing in particular is the most radical attempt so far to purify and control the kinesthetic process of affectation at the smallest possible level to generate a highly predictable and controlled output or product. This has made possible electrically controlled heating and air conditioning, pressure-sensitive toothbrushes, weight-shifting skyscrapers that automatically adjust to wind speed and load, time-sensitive home lighting, digital door locks, location-sensitive smartphone notifications, and metabolic-sensitive exercise alerts, as well as biometric finger, face, and eye recognition; autocorrect software; and on and on.

There are all kinds of drawing, graphics, and design programs aimed at transcoding images and even generating images created by direct user input on the screen. Images can be manipulated, touched up, and so on. Most early software art, ASCII text art, and so on followed a highly ordered programming language. Most of these images have and are meant to have very little disorder. Many video games, especially older ones, also have a relatively high level of order and not much more feedback than a normal piece of software.

In all this, a user is still interacting with a material process (a computer program) to generate an ordered output. Pedesis and feedback can be statistically reduced but not eliminated. Today we are on the cusp of reaching the historical and material limits of this reduction as we approach the quantum level.

THE DISORDERED GENERATIVE IMAGE

The second kind of generative image is the relatively more disordered type. Disordered generative images are those that have a higher degree of relative disorder, unpredictability, pedesis, and feedback. This is always a question of degree. There is no absolutely disordered image, only a range of disorder relative to another image. The more disorder generated in the image, the less frequent and more insensible the pattern. The flow of matter is what supports the affective fold and aesthetic field, but pedesis is what disrupts it, redirects it, or mutates it. The recirculation of the kinesthetic field is what stabilizes it, patterns it, distributes it, and orders it, but this same feedback mechanism is also what amplifies mutations and reorders the distribution of folds.

Just as it is possible to reinterpret the history of art as a subordination of pedesis and feedback to the dominant kinetic patterns of motion, it is also possible to give voice to a concurrent minor art history that has escaped this capture and has aimed instead, to one degree or another, to render disorder sensible.

The Arts

There is a long, albeit minor, tradition in Western art of emphasizing pedesis and feedback to varying degrees. In *A Deluge, with a Falling Mountain and Collapsing Town* (1515; figure 16.1), for example, Leonardo da Vinci states that he used the appearance of humidity and condensation on windows and walls as an inspiration for painting landscapes, rocks, and rivers or unstable phenomena like fluids, smoke, or clouds.[3]

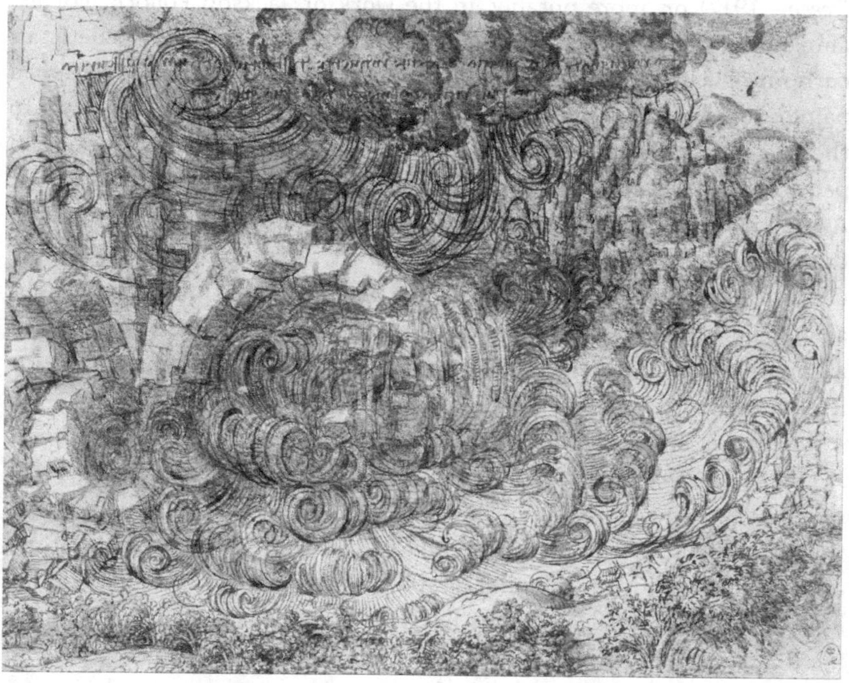

Figure 16.1 Leonardo da Vinci, *A Deluge, with a Falling Mountain and Collapsing Town* (1515)
Source: Wikimedia, https://commons.wikimedia.org/wiki/File:Leonardo_da_Vinci_-_A_deluge_-_Google_Art_Project.jpg#/media/File:Leonardo_da_Vinci_-_A_deluge_-_Google_Art_Project.jpg

The pedesis of the hairs on the brush as they are mashed on the surface of the paper can also give rise to disorderly patterns on which disorderly images of the sky and earth can be figured, as in Alexander Cozens's *Streaky Clouds at the Bottom of the Sky* (1786). Other techniques such *frottage*, or rubbing, use a pencil on paper over an uneven surface that produces a pedetic pattern to be elaborated on, as in Max Ernst's *Le Foret pétrifié* (1929). The pedetic element here is the arbitrary nature of the rubbed object and the stochastic bouncing of the pencil over the surface.

The pedesis of the air itself has also been used as important way to increase the pedesis of the work of art. Marcel Duchamp's *Trois Stoppages Étalon* (1913; figure 16.2), for example, uses the aleatory fall of a single meter-long thread to remeasure a meter. He writes, "If a thread one meter long falls straight from a height of one meter onto a horizontal plane, it twists as it pleases and creates a new image of the unit of length."[4] Since the thread is so light, the slightest bit of turbulent and stochastic air movement will cause it to fall in a slightly different configuration each time.

Painting can benefit from a similar aeropedetic method, like John Cage's *Strings 1-62 No 5* (1980) inspired by Duchamp, or John Arp's method of dropping cut-up pieces of colored paper from a height onto paper, *Grand Dessin* (1917) or, more notably, in the work of Jackson Pollock. For example, Jackson Pollock's *No 31* (1950) relies on the effect of pedetic air currents on a liquid medium (paint) to pedetically reshape flung or dripped paint onto the canvas. Prereflective arm movements scatter the paint into the air, where the real painting is done not by the hand but by the air and paint left to itself in the air, and then the canvas as it shapes the splatter. Aerodynamic and fluid-dynamic processes can be used together to introduce pedesis and material generativity into art. For example, Andy Warhol's *Oxidation Painting* series (1978) uses the pedetic flow of urine through the air to oxidize copper paint on canvas. The flow of liquid in air is subject to all kinds of stochastic turbulence and splatter, and the chemical reaction itself is subject to unpredictable shapes and speeds of oxidation.

In Francis Bacon's work, we find paint splatter and rubbing combined. *Figure in Movement* (1978), for example, begins with splattering paint randomly on the canvas, painting with it, and scrubbing it out in a continual feedback loop of formation and deformation of the figure. Bacon's aim is not to reproduce the photographic movement of the body (inspired by Eadweard Muybridge) but, as he says, "the opposite of natural movement."[5] "I work much better in chaos. Chaos for me breeds images."[6] And "The way I work is totally, now, accidental, and becomes more and more accidental, and doesn't seem to behave, as it were, unless it is accidental, how can I recreate an accident? It's almost an impossible thing to do. . . . [An

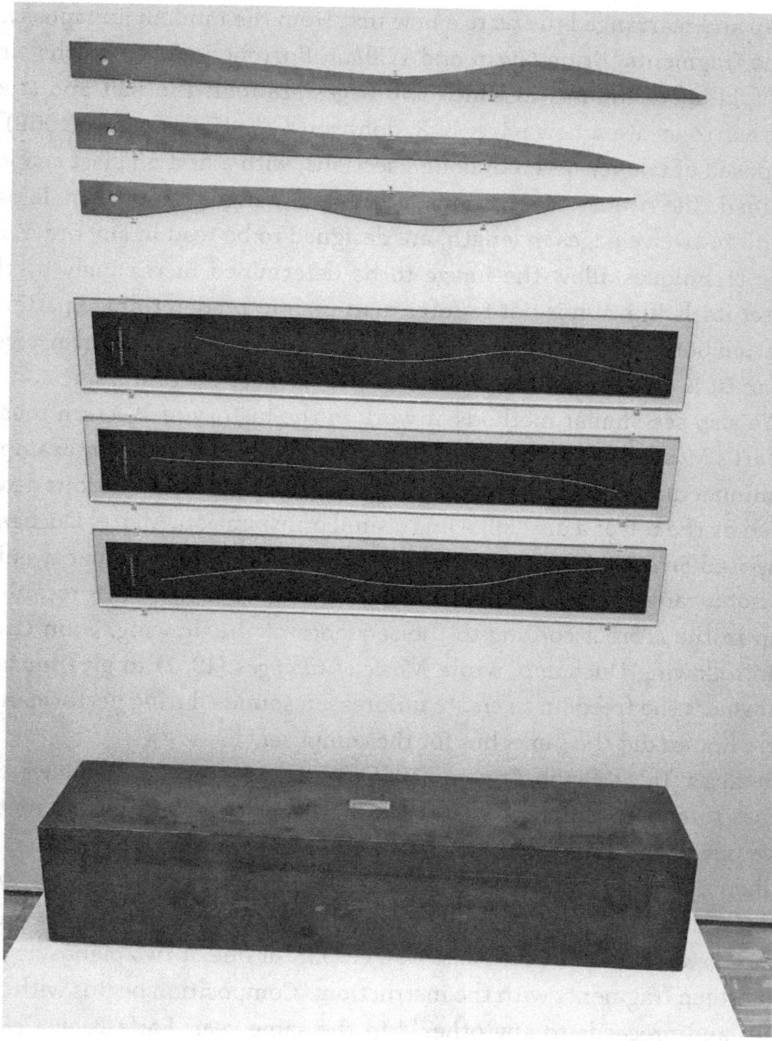

Figure 16.2 Marcel Duchamp, *Trois Stoppages E'talon* (1913)

accident] out of which [the bodies] could move as though out of pools of flesh rose the images."[7] Images, for Bacon, are not something that pre-exist the material-kinetic process of their generation. In contrast to the classical image of the eternal unchanging god, which the artist copies, inspired by the muse, Bacon's work shows how images emerge from the bottom up, through material-pedetic genesis.

Not only in painting but in literature as well, pedetic methods have been used to give agency back to the matters themselves. Tristan Tzara, for example, popularized the *découpé*, or "cut up" technique, in which a text is cut up and rearranged to create a new text from the random juxtaposition of the fragments. Brion Gysin and William Burroughs invented the similar "fold-in" technique, in which two pages are folded in half and stuck together to create a new page. B. S. Johnson's *The Unfortunates* (1969) is composed of twenty-seven unbound sections, with a first and last chapter specified. The twenty-five sections in between, ranging from a single paragraph to twelve pages in length, are designed to be read in any order. All these techniques allow the image to be determined increasingly by the matter itself. The aim is not to force matters into predetermined patterns of action but rather to allow the matters to unfold and express themselves. The artist is there only to facilitate, not to dominate the process.

We also see similar methods at work in the history of Western music. Mozart's *Musikalisches Würfelspiel* (*Musical Dice Game*) (1792), for example, is a minuet made by cutting and pasting together prewritten sections determined by the roll of a die. Following a similar inspiration, Marcel Duchamp composed *Erratum Musical* (1913) by randomly picking from a hat twenty-five notes ranging from F below middle C and up to high F, then recording them in the score according to the sequence of the drawing. John Cage, again following Duchamp, wrote *Music of Changes* (1951) to give musical performers the freedom to create unforeseen sounds during performance. Pierre Boulez did the same, but for the composer.

Boulez's Third Piano Sonata (1955–57/63), for example, allows the pianist to choose different routes through the score, and in one of the movements has the option of omitting certain passages altogether. In Karlheinz Stockhausen's *Klavierstück XI* (1956), however, pedesis is given to the ordering of the sequences of musical fragments. Even more radically, Morton Feldman's *Intermission 6* (1953) for one or two pianos begins with fifteen fragments with the instruction, "Composition begins with any sound and proceeds to any other."[8] In the same year, Earle Brown composed *Twenty-five Pages* (1953) for one to twenty-five pianists, in which the pages are to be arranged in a sequence chosen by the performer(s), and each page may be performed either side up; events within each two-line system may be read as either treble or bass clef.[9] In all these cases, the composers have attempted to introduce a degree of pedesis and interactivity into the musical work of art.

Sculpture, in turn, has invented its own attempts at pedesis. Alexander Calder's *Mobile* (ca. 1932), for example, balances various shapes that hang in the air to be moved by pedetic currents of air. In doing so, sculpture is given pedetic motions depending on the temperature and viewers in the environment. In fact, it was Calder's *Mobile* that inspired the aleatory music of Brown and Feldman. All manner of kinetic sculptures have since

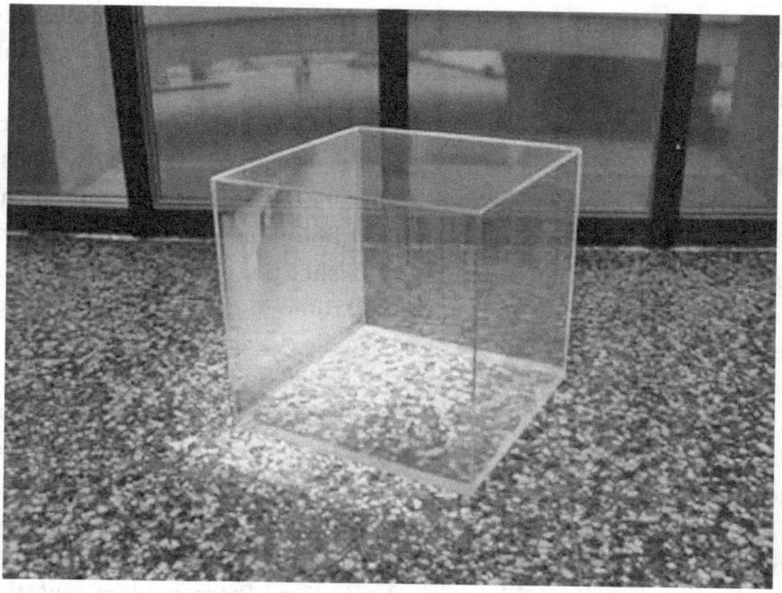

Figure 16.3 Hans Haacke, *Condensation Cube* (1963–1965)
Source: Wikimedia

been created that use aleatory wind patterns to influence their motion, including various musical sculptures such as wind chimes and fabrics such as Christo and Jeanne-Claude's *Valley Curtain* (1972), *The Gates* (2005), and *Floating Piers* (2014–2016). One of the most interesting analog pedetic methods, however, is Hans Haacke's *Condensation Cube* (1963–1965; figure 16.3). Haacke took seriously Leonardo's advice about humidity on the walls, but instead left the condensation to move on its own, heating, cooling, dripping ever new in its own enclosed cube. The stochastic patterns of water molecules give rise to unpredictable patterns of accumulation and dissolution on the sides of the cube in direct feedback with the season, temperature, and viewers in the room.

Nature

Nature also produces more disordered generative images. For example, fluids and gases are much more stochastic because atomic and molecular movement in fluids and gases is pedetic. This tends to give rise more easily to chaotic changes in pressure and flow velocity, or turbulence. Turbulence appears in nature as waves crashing on the shore, as fast-flowing river rapids, as storm clouds churning, or as smoke dissipating. The lower the viscosity of the fluid, the more easily turbulence emerges. The pattern that raindrops make as they fall on a surface is subject to innumerable variables of turbulent flows of wind and water, and is thus highly disordered.

Nature is also noisy. Atmospheric noise is the name given to the highly disordered changes in radio-wave frequency primarily caused by the lightning discharges of thunderstorms. Typically more than 2,000 thunderstorms are active throughout the world at a given moment, producing on the order of 100 flashes per second. Thermal noise (Johnson–Nyquist noise) is the electronic noise generated by the thermal agitation of charge carriers, usually the electrons. Thermal noise is also related to the photoelectric noise of highly disordered photon waves moving pedetically through an electrical current. This is in turn related to the quantum indeterminacy or pedesis of all quantum-field fluctuations.

Another example of pedesis in nature is genetic mutation. In biology, modern evolutionary theory attributes the diversity of life to pedetic genetic mutations followed by natural selection. These random mutations then interact in a biological and bioelectric feedback loop with the environment, which is itself composed of randomly mutating and evolving organisms.[10]

Yet another example of pedesis in nature is radioactive decay. The rate at which an unstable atomic nucleus loses energy by emitting radiation in the form of various particles is pedetic. Owing to the quantum nature of the fields that define these particles, it is impossible to tell when it will begin to decay, no matter how long an atom has existed.

Even a simple coin flip, dice throw, spin of a roulette wheel, or lottery-ball machine has far too many variables to discern any ordered pattern in their outcomes, owing to the material nature of air patterns and the micro movements of repulsion. All of this produces a certain aesthetic beauty. As Gert Eilenberger, a German physicist of nonlinear science, writes,

> Why is it that the silhouette of a storm-bent leafless tree against an evening sky in winter is perceived as beautiful, but the corresponding silhouette of any multi-purpose university building is not, in spite of all efforts of the architect? The answer seems to me, even if somewhat speculative, to follow from the new insights into dynamical systems. Our feeling for beauty is inspired by the harmonious arrangement of order and disorder as it occurs in natural objects—in clouds, trees, mountain ranges, or snow crystals. The shapes of all these are dynamical processes jelled into physical forms, and particular combinations of order and disorder are typical for them.[11]

Digital Images

Images have always been generative to one degree or another. What the historical dominance of the digital image and its pedetic electromagnetic field does, however, is draw our attention to the glaring fact that the generative image has always been with us, like underground volcanic activity popping up here and there to create new land.

Today, for the first time in history, we have found a way to directly engage and manipulate the most pedetic and highly interactive media in nature: the quantum field. One of the fundamental reasons that nature is capable of producing the highly disordered and pedetic images that it has is because matter itself (as quantum fields) is pedetic and interactive.

Beyond dice throws, drip painting, condensation patterns, and so on, the combined powers of the digital image—hybridity, pedesis, and interaction—are able to transcode almost any disordered material process, filter, and add degrees of material pedesis, expanding its interactivity immensely. No other aesthetic medium, including others in the natural world, are or have ever been capable of this level of direct modulation of hybridity, pedesis, and interactivity of a single image. The aim of digital generative art is, therefore, not to mimic or represent the products of nature's processes (as Leonardo dreamed), or even to simply mimic natural processes themselves (as Fermat dreamed) but rather, to *invent new kinesthetic processes* with the new tools we have.

Algorithms

One way this is accomplished is through the manipulation of digital algorithms. An algorithm is nothing other than an operational feedback loop. As such, it is not necessarily digital. Euclid, for example, invented an algorithm for calculating the greatest common divisor of two numbers, *a* and *b*. However, digital processing is capable of executing vastly more complex and vastly more iterated algorithms at dramatically faster speeds, introducing a qualitative change in the aesthetic process and product. Bioelectric feedback and electromagnetic fields have always been present in Western art. Even the simplest organisms are capable of extremely high levels of pedesis and interactive feedback. However, never have such processes been the *direct medium* of such radical feedback (in the electrical circuit) as they are in the digital algorithm.

Just as there are degrees of order and disorder in the generative image, so are there degrees of order and disorder in algorithms. Since we have already discussed the basic aim of the ordered computer-programming algorithm—to control output—we focus in this section on the more disordered types of algorithms. There are, for example, three ways to introduce disorder into the algorithmic feedback process: into the input, into the process, and into the output.

Pedetic Input: Because of the hybrid nature of the digital image, one can transcode natural, social, artistic, atmospheric, and thermal processes into binary code and input that code into the algorithmic process. For example, a computer can gather pedetic information from websites that share their atmospheric noise, thermal noise, or radioactive decay data. It can also draw on large-scale data-mining information. For example, Luke Dubois's *Hard Data* (2009) for string quartet is a data-mining sonification project that uses statistics from American military actions in Afghanistan and Iraq as source material for an interactive audiovisual composition based on an open-source "score" of events. Since computers only function through determinate commands, introducing disordered input offers a degree of disorder that a typical program alone could not achieve. More important, however, the introduction of disordered input is not simply an issue of hybrid representation but, additionally, of modulation and creativity. For example, one can record the bioelectrical signals emitted by various plants using sensors plugged into a computer and then correlate these patterns with various sound volumes, tones, and tone durations, as in Mileece Abson's *Quartet* (2012). One can record the sound made by falling rain and loop it, overlap the loops, slow them down, speed them up, distort them, introduce audio feedback, and do a thousand other things to them to produce a uniquely pedetic sound process.

Pedetic Process: Digital algorithms can also draw more directly on the disordered state of the computer hardware itself to modify its own algorithms. For example, a computer can use the recent number of keystrokes from the computer; the speed of the hard disk; the temperature of the central processing unit; and the day, time, or other arbitrary information from the computer itself to generate variously disordered algorithms. Now Intel even makes computers with a built-in Bull Mountain "randomization process" that actually uses the computer's own quantum thermal pedesis of its photoelectrical wave to generate "random" numbers. What used to be a quantum tunneling "error" can also be used to introduce high degrees of pedesis into the algorithmic process itself, when needed. In this way, the computer uses its own internal quantum process as a method for introducing pedesis.

Pedetic Output: Digital algorithms can also use algorithms that are only relatively disordered, but that begin to display patterns well beyond the scale of human sensibility. These are often called "pseudorandom generators" simply because they tend to produce random numbers according to fixed patterns whose output only appears random to us. They thus produce a relatively pedetic output even if the input and process are themselves not pedetic.

All three of these types of pedesis become radically amplified in the feedback operations of the digital algorithm. Together, the extreme hybridity, pedesis, and feedback of the digital image make possible new experiments in art using the kinetic agency of matter itself. The contemporary image thus also makes possible a new aesthetics never before possible and one in which the human is not the origin and end of sensation.

CONTEMPORARY GENERATIVE ART

We live in the age of the digital image not only because of its powerful hybridity and power to order material reality but also because of its powerful pedesis, interactivity, and power to disorder reality—to create new kinesthetic processes. The beauty of contemporary generative art lies not in its "random" number generators and the sublime affirmation of chaos against the orderliness of contemporary reality. Rather, it is in its capacity to create new kinesthetic processes that play in the complex region between highly ordered and highly disordered images. It gives a high degree of kinetic agency to the matters at work.

The primary question for contemporary generative art is thus how to harness a degree of pedesis in whatever way it can, enter it into an interactive feedback loop, and see where it goes. Humans are just along for the ride. In contemporary generative art, the kinesthetic process itself becomes primary. Subject and object, input and output are folded back over themselves in an interactive feedback loop to be modulated as a whole, continuous process. This has always been the case in all art to varying degrees, even though most arts have tried to block it and confine it. Today pedesis and interactivity have become a primary and dominant focus of the most cutting-edge aesthetic experiments.

Generative Visual Arts

In the visual arts, pedetic computer algorithms can be used to produce thousands of iterations with numerous parameters, like color, line length, width, thickness, rotation, texture, distortion, noise, brushstroke, and so on. The artist selects parameters, type of algorithm, and degree or type of pedesis—Perlin noise, loops, iterative variance, and so on. Pedesis can be introduced from the input, process, or output. An incredible variety of stochastic naturalistic processes can be animated, with different results each time.

In a rapid series of such animations, Maxime Causeret's *Order From Chaos* (2016) shows the pedetic patterns of raindrops hitting a surface and spreading, pedetic branching patterns of plants, swarming behaviors of insects, soap-bubble patterns, cellular bifurcations, coral meandering patterns, and more. The images generated are not meant to be copies of natural products but rather their own visual expressions of how stochastic algorithms can produce ordered patterns just like nature can but this time with new resulting organisms.

More disordered still is Maurizio Bolognini's *Programmed Machines* (1988–), composed of enclosed computers generating flows of continuously iterated pedetic images. In the 1990s, Bolognini programmed hundreds of these computers and left them to run ad infinitum. Most of them are still working today. Of these works he says,

> I do not consider myself an artist who creates certain images, and I am not merely a conceptual artist. I am one whose machines have actually traced more lines than anyone else, covering boundless surfaces. I am not interested in the quality of the images produced by my installations but rather in their flow, their limitlessness in space and time, and the possibility of creating parallel universes of information made up of kilometers of images and infinite trajectories. My installations serve to generate out-of-control infinities.[12]

In another work, *Collective Intelligence* (2000), Bolognini used similar machines to project random lines of light onto public surfaces and allowed mobile telephones to interact with them, changing the patterns in real time and creating "generative, interactive and public art."[13] Bolognini thus introduces pedesis and feedback at every level of the aesthetic process. The input is interactive and collective from the population, and the computer processing then randomizes the input, resulting in a highly pedetic and interactive output.

Radicalizing this idea even further, Scott Draves's *Electric Sheep* (1999–) is a computer screensaver that runs iterative fractal flame patterns with a number of different animated parameters. The screensaver is what your computer dreams of while it is asleep, a reference to Philip K. Dick's novel *Do Androids Dream of Electric Sheep?* Users can interact with the process by liking or disliking various iterations. This input then reprograms the genetic algorithm as the process mutates to become more interesting to the viewers. Users can also program and upload their own fractal processes whereby they "breed" or mix with the others to produce more iterations. There are currently about 500,000 active users a month.[14] Again, pedesis and feedback are incorporated at every level with the aim of finding the most beautiful middle ground of complexity between too much order and too much disorder in the image.

Generative Literary Arts

Contemporary generative literary works go beyond the cut-up and fold-in methods of the Dadaists to produce much more pedetic and interactive works than previously possible. Philip M. Parker, originally an affiliate of the Fluxus group, used a mathematical algorithm named "Eve" to produce digital poetry based on graphic theoretical relations between words in the dictionary. He has produced more than 1.3 million poems in this manner. He has even used similar algorithms to produce entire books—200,000 of them.[15]

More recently, Jason Nelson has used generative methods to create digital and interactive hyperpoetry. His famous "Game, Game, Game And Again Game" (2007) uses flash media to create an audiovisual mashup of text fragments, sounds, and video in an interactive video game format. "I made this. You play this. We are Enemies" (2009) develops the same idea. His "Uncontrollable Semantics" (2006) creates a series of words on the four corners of the screen, each with its own sound and image. As one clicks on the different words, new word–image combinations are created. Poetry becomes a series of continually modulated feedback loops. A similar feedback loop of interactive options occurs in Neil Hennessy's "JABBER: The Jabberwocky Engine" (2000), in which randomly floating letters are connected to form new combinations of neologisms that produce pronounceable English words, but with no dictionary definition. These are then incorporated into poetic works.

Jean-Pierre Balpe has even produced stochastic and interactive novels such as *Trajectoires* (2000) and *Fictions d'Issy* (2005) by using algorithmic and interactive methods. The stories are continuously generated sentence by sentence, and readers can shape the outcome by using their phone's keypad. Balpe's work and many others are contained in the first volume of the *Electronic Literature Collection* (2006) and they represent an amazing diversity of generative literary works.[16] All these give the materiality of words a maximum of pedesis and interactivity by tying them to the kinetics of the digital process.

Generative Plastic Arts

With the advent of 3D printing, generative algorithms can now be modeled directly into plastic media. Although the technology is still in its infancy, some of the initial creations are incredible. Among the most amazing examples are the sculptures and architectural columns made by Michael Hansmeyer (figure 16.4), who is an architect and programmer who uses algorithms and computation to generate unique architectural forms using a simple feedback algorithm of topological folding. Hansmeyer begins his designs with a single cube and then begins to stretch and bend the cube, applying his folding algorithm to different parameters such as depth, curve, and line. The results are incredible—forms so complex that the "artist" could not possibly have "an idea" of them. The whole matter-form distinction collapses onto itself as matter becomes morphogenetic and semi-autonomous. According to Hansmeyer, 99 percent of the algorithms end up producing noise. Only those with certain modulated parameters produce the most complex forms. In addition to the Doric, Ionic, Corinthian, and undulating orders of columns, Hansmeyer has produced an entirely new architectural order: *the generative order*.

Nervous System, a generative design studio, uses algorithmic and stochastic code to create unique sculpture, jewelry, light fixtures, and even

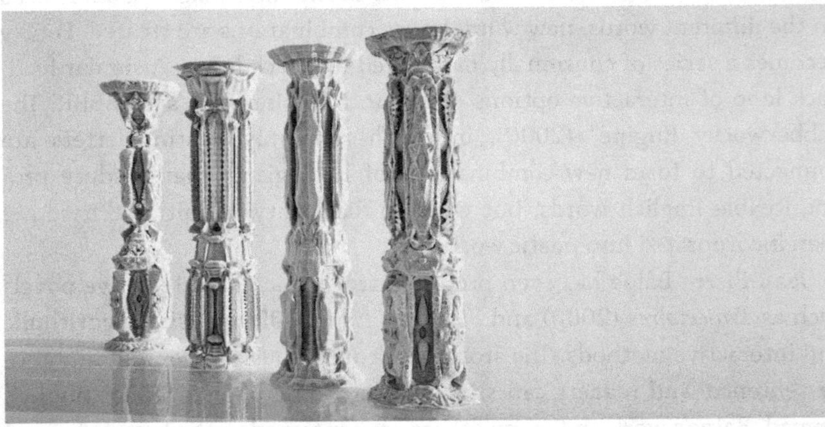

Figure 16.4 Michael Hansmeyer, *Columns* (2010)
Source: From artist's website, © Michael Hansmeyer, http://www.michael-hansmeyer.com/projects/columns.html?screenSize=1&color=1#1.

clothing using 3D printing. Their *Floraform* sculptures are similar to the biomechanics of growing leaves and blooming flowers. Their *Xylem* (2D) and *Hyphae* (3D) sculptures use algorithms that produce structures similar to those found in the veins of leaves. These patterns are used to generate jewelry, lamps, sculpture, and even architecture. Their *Kinematics* sculptures add a fourth dimension to 3D printing by creating a design system of hinged panels with a simulation strategy of folding and compression to produce customized designs that can be fabricated efficiently by 3D printing. The structure is printed as one part but has thousands of interconnected pieces that require no assembly. The result is kinetic dresses, lampshades, jewelry, and more.

Additionally, their website includes interactive software that allows anyone to design his or her own sculptures and print them. Nervous System's designs thus use hybridity to physically transcode binary code into 3D and 4D sculptures. They use pedesis in their stochastic algorithms, and they use interactivity in the user interface which is sensitive to its initial and continuous conditions. The purpose is not simply to maximize noise or feedback or to copy natural patterns but also to produce new patterns through the modulated use of noise and feedback. The purpose it to give the electromagnetic field its maximal kinetic and material generative agency.

Generative Sonic Arts

It was Brian Eno in the 1970s who first coined the term "generative music," but the scene has expanded dramatically since then. Today, generative music has vastly outstripped Mozart's dice throws, futurist noise music, and even the later modernist aleatory music of Cage, Feldman, Boulez, and others. These earlier works relied on comparatively simple pedetic parameters and limited feedback systems, and they remain but modest precursors to the much more hybrid, pedetic, and interactive works of generative music today.

Some of the first works to introduce a higher degree of pedesis and feedback were Stockhausen's *Kontakte* (1958–1960), Terry Riley's *The Gift* (1963), Brian Eno and Robert Fripp's *No Pussyfooting* (1973), and Eno's *Discreet Music* (1975), the latter which used a new tape-loop feedback system combined with an echo unit and a continuously modulated graphic equalizer to change the timbre of the sounds. This allowed sound to turn back over itself in an ever-expanding and interactive modulated feedback pattern of sonic images. Similar modulated tape-loop feedback systems continue to be used today by various ambient music artists, such as Ous

Mal, Taylor Dupree, Tape Loop Orchestra, and William Basinski. For contemporary musicians, the tape-loop process also introduces a new focus on the pedetic sound of the tape noise itself.

Although present in Eno's and Riley's early work, and emphasized in works like Steve Reich's amazing *Pendulum Music* (1968), which swings microphones over speakers generating patterned yet chaotic feedback, contemporary artists have turned increasingly toward the stochastic noise, feedback echo, and hiss of the tape itself—amplifying it, looping it, and dramatizing the noise of the electromagnetic field. This is part of a much wider trend by contemporary generative musicians to seek out pedetic sounds such as tape hiss, noise, vinyl-record crackle, CD-skipping sounds, microphone feedback, FM radio static, and other irregular, pedetic and traditionally undesirable musical sounds created by the pedesis of the EM field. The aim is not simply to reproduce these sound images but also to work with them and use their stochastic patterns as the basis of new feedback loops and patterns of their own.

In the Caretaker's *An Empty Bliss Beyond this World* (2011), for example, vinyl crackle is amplified and echoed to the point where it equals the volume of the looped vinyl melodies. In a slightly different vein, Burial's *Burial* (2006) uses the static crackles and pops reminiscent of those that occur in maxed-out speakers and loose or old audio cables, or the static electricity pops from the audio mixing equipment and microphone itself. These crackles become the sonic milieu of his hyperdub loops. Glitch albums like Oval's *OvalDNA* (2011) combine various melodic audio feedback tones with CD skipping noises, as if one had taken a knife to a CD's surface and then stuck the CD back in the player. The use of FM static in Olli Aarni's *Pohjoisen Kesä* (2012), or his use of field recordings of underwater insects in *Vesiä* (2017), or Mileece's interactive bioelectrical feedback sounds gathered from plants all accomplish the similar aim of introducing pedesis into the audio feedback loops for sonic modulation—to give noise "a life of its own," to paraphrase Pollock.

Even more dramatic, however, is the use of numerous types of digital pedals, oscillators, tone generators, and computer software to produce highly diverse and numerous loops of sound that can all be modulated in medias res and with more technical precision than any tape-loop audio noise. The famous Japanese noise musician Masami Akita "Merzbow" has produced particularly pedetic and abrasive albums such as *Pulse Vegan* (2014), using both granular synthesis software and numerous digital sound boxes or pedals. In his most recent work, the software transforms his sounds into "clouds" or flows of micro sounds that can then be modulated continuously and generatively as a whole, according to a number of different parameters and computer algorithms.

Curtis Roads, a media-arts professor and composer of *Point Line Cloud* (2005), describes the process in fluid dynamic terms:

> Beneath the level of the note lies the realm of sound particles. Each particle is a pinpoint of sound. Recent advances let us probe and manipulate this micro acoustical world. Sound particles dissolve the rigid bricks of musical composition—the notes and their intervals—into more fluid and supple materials. The sensations of point, pulse (series of points), line (tone), and surface (texture) emerge as the density of particles increases. Sparse emissions produce rhythmic figures. By lining up the particles in rapid succession, one can induce an illusion of tone continuity or pitch. As the particles meander, they flow into liquid-like streams and rivulets. Dense agglomerations of particles form clouds of sound whose shapes evolve over time.[17]

Granular or pulse software thus introduces into music a new fluid dynamics of flows to the sonic image, letting it pedetically meander into periodic densities or folds that are then woven into a larger sonic texture like a fabric. However, the term "grains" of sound is misleading because each micro 1-50 millisecond sound sample or "grain" is buffered by an amplitude modulation or "envelope" that connects the grains in a sonic continuum. Wave-scanning techniques also can eliminate the need for the envelopes by having the grain boundaries always meet at the zero-crossing point of the respective signals. The resulting composition is thus sonically *continuous* and has a highly fluid character to it like the sound of rushing water, crashing waves, or a turbulent dripping faucet. Barry Truax's *Riverrun* (1986), for example, is a direct statement on the fluid dynamic nature of micro-sonic generative image composition. "From the smallest rivulet to the fullest force of its mass, a river is formed from a collection of countless droplets and sources. So, too, with the sound in this composition which bases itself on the smallest possible 'unit' of sound in order to create larger textures and masses. The title is the first word in James Joyce's *Finnegan's Wake*."[18]

Such modulation was impossible with the instruments and techniques available before the late twentieth century. For the first time ever, it is possible to modulate noise-pitch-rhythm as the complete sonic continuum that it is, at the smallest possible audible levels of the waveform, thus introducing an incredible new range of pedesis. Recent works integrating granular synthesis also include Ian William Craig's *Centres* (2016), Kaitlyn Aurelia Smith's *Ears* (2016), and Multicast Dynamics's *Scandinavia* (2016).

Mixing various higher degrees of pedesis into the digital input, process, and output produces numerous genres and subgenres of electronic, electroacoustic, and experimental music: glitch, drone, ambient, postclassical, noise, tape music, field recordings, found music, circuit bending, sound sculpture, vaporwave, chopped and screwed hip hop, and many more.

Additionally, contemporary generative music introduces a new level of hybridity and feedback never before possible in music. Brian Eno's latest album *Reflection* (2017), for example, is a brilliant mixture of pedesis, hybridity, and feedback. The album uses stochastic algorithms to determine the parameters of the sounds.

> Because everything in the pieces is probabilistic and because the probabilities pile up it can take a very long time to get an idea of all the variations that might occur in the piece. One rule might say "raise 1 out of every 100 notes by 5 semitones" and another might say "raise one out of every 50 notes by 7 semitones." If those two instructions are operating on the same data stream, sometimes—very rarely—they will both operate on the same note . . . so something like 1 in every 5000 notes will be raised by 12 semitones. You won't know which of those 5000 notes it's going to be. Since there are a lot of these types of operations going on together, on different but parallel data streams, the end result is a complex and unpredictable web.[19]

Second, the album uses an interactive process of modulation as Eno "tweaks" the parameters during playback over and over again.

> Pieces like this have another name: they're GENERATIVE. By that I mean they make themselves. My job as a composer is to set in place a group of sounds and phrases, and then some rules which decide what happens to them. I then set the whole system playing and see what it does, adjusting the sounds and the phrases and the rules until I get something I'm happy with. Because those rules are probabilistic (—often taking the form "perform operation x, y percent of the time") the piece unfolds differently every time it is activated. What you have here is a recording of one of those unfoldings.[20]

Third, the album uses a hybrid transcoding of the music into an audio-visual-haptic software application that allows users to touch a colored screen and modulate the endlessly looped stochastic patterns for themselves.

> REFLECTION is the most recent of my Ambient experiments and represents the most sophisticated of them so far. My original intention with Ambient music was to make endless music, music that would be there as long as you wanted

> it to be. I wanted also that this music would unfold differently all the time—"like sitting by a river": it's always the same river, but it's always changing. But recordings—whether vinyl, cassette or CD—are limited in length, and replay identically each time you listen to them. So in the past I was limited to making the systems which make the music, but then recording 30 minutes or an hour and releasing that. REFLECTION in its album form—on vinyl or CD—is like this. But the app by which REFLECTION is produced is not restricted: it creates an endless and endlessly changing version of the piece of music.[21]

Reflection is thus an attempt at mimesis of neither natural products nor natural processes but, rather, a way of becoming what it is: matter in motion—pedetic, hybrid, and interactive. Just as the flow of matter has no beginning and no end, neither does *Reflection*. The three creative stages Eno describes for this work match up directly with the kinetic ones laid out in this book: (1) pedetic material flows intersect at a constellation, (2) fold into a distribution of affective loops, and (3) are continuously modulated as a whole woven field of sound. Eno writes,

> The creation of a piece of music like this falls into three stages: the first is the selection of sonic materials and a musical mode—a constellation of musical relationships. These are then patterned and explored by a system of algorithms which vary and permutate the initial elements I feed into them, resulting in a constantly morphing stream (or river) of music. The third stage is listening. Once I have the system up and running I spend a long time—many days and weeks in fact—seeing what it does and fine-tuning the materials and sets of rules that run the algorithms. It's a lot like gardening: you plant the seeds and then you keep tending to them until you get a garden you like.[22]

Numerous other efforts to increase the interactivity and hybridity of music abound. Media artist Scott Snibbe, for example, has created a number of such interactive music album applications, such as Bjork's *Biophilia* (2011) and Metric's *Synthetica* (2013). Snibbe's app Motionphone (2012) integrates sound, kinetic motion, and visual animation. As users move their fingers across the screen, their movement is animated and looped. These can then be shared and interact with other users' kinetic sculptures online.

CONCLUSION

This final chapter concludes our study of the generative image, but the field of generative art is growing exponentially. To some degree, generative elements are at work in an increasing number of popular and consumer media. The rise of "smart" media coincides with the rise of the digital image, and with it the potential to release and recover a new generative image that was hidden in all previous works of art. The argument of this chapter is not that the generative arts are the only or best forms of art. On the contrary, the function of contemporary generative art has been to blaze a trail that shows us what all the arts are capable of, to one degree or another. The generative arts have invented a new artistic pedesis and laid the groundwork for a new materialist aesthetics of the image. The challenge now is to unfold the consequences of this discovery in all the arts. The aim of this chapter is not to simply valorize the digital or even the generative image but to draw our attention to what the contemporary image shows us about the image more generally and what it is capable of today.

Conclusion

The Mobile Image

We live in the age of the mobile image. Today, more than ever before, we are surrounded by hybrid images of all kinds that circulate freely and mix with contemporary images. This incredible mobilization and proliferation of images forces us to rethink the basic structure and definition of the image itself—as something fundamentally *kinetic*. The advent of the digital image, defined by a continuous flow of electricity, forces us to see that the image is not and never has been a representation of a static model. Images have always had a material agency. Movement, and not representation, has always been central to the image, making possible a new materialist aesthetics.

This book thus has made three main contributions to the philosophy of art and aesthetics.

THE KINETIC THEORY OF THE IMAGE

Its first contribution is to offer an original kinetic theory of the image. Traditionally, the image has been viewed as either objectively or subjectively derived from something else. A relatively static object, subject, or human structure was assumed as primary and the image was what moved in between them. Even when the image has not been treated explicitly as a representation, it has typically been thought of an expression or production of *something else*. Even contemporary theories of images as a copy of copies or copies without originals, still miss the point. The image is not a copy, and there was never a model to have gone missing. In contrast to these previous theories, this book proposes a new definition of the image as a reflection, a duplication, or a fold in moving generative matters. All images are sensuous, and all sensations are images. Images both sense and are sensed. The image is thus not something strictly visible. There are images of sight and sound, just as there are images of taste, smell, and touch. The image is also not unique to humans or to organic life.

The original contribution of Part I, then, is to have provided a kinetic and materialist theory of the image defined by the flow, fold, and field of sensitive matters. As such, it reorients the central problem of aesthetics and art history, moving it away from the question of representation and anthropocentric constructivism, whether linguistic, social, psychological, or otherwise, and toward the distribution and analysis of regimes of moving images with their own material agency and generativity.

THE HISTORY OF THE IMAGE

The second contribution of this book is that it offers an original conceptual and historical methodology for the study of art and art history. If the study of the image is not a question of representation but rather of kinetic distribution, then we need to understand what kinds of distributions have been invented and to what degree and with what mixture they persist in the present. Part II of this book thus presented neither a universal ontology of affect nor a merely empirical history of works of art but, rather, a study of the kinesthetic patterns or historical regimes of aesthetic motion.

Unlike merely empirical art histories, kinesthetic regimes of motion prefigure, persist, and mix well beyond their initial empirical manifestation, making their analysis much more broadly applicable to the study of art, art history, and sensation widely construed. Thus, the kinetic method of this book makes no attempt at an ahistorical ontology of sensation, affect, or image; rather, it offers a regional ontology from the perspective of the early twenty-first century. Based on the apparent primacy of mobility revealed in the digital image, it proposes an answer to the simple question: What must images at least be like for them to be capable of this kind of motion? In doing so, it thus discovers a previously hidden dimension of all hitherto existing images: the primacy of their motion.

THE CONTEMPORARY IMAGE

The third major contribution of this book is its offer of an original theory of the digital image defined by its materiality and mobility. In contrast to the first wave of new-media scholarship that defined the digital image as largely immaterial and virtual, this book provides an analysis of the material and kinetic dimensions of the digital image and its conditions of circulation. While more recent new-media scholarship seems to be taking the material dimension of the digital image more seriously, this book adds to this literature a complete conceptual and analytic framework that connects the study of the digital image with the rest of art history and the structure of affection more broadly.

The electrical flow that defines the digital image is historically novel in some ways, but not in others. The digital image thus allows an incredible degree of hybrid mobile images, but in a more general sense, electrical flows also pervade all material images. The digital image is not just about hybridity and remediation; it is also about the creative pedesis and feedback of the electrical flow itself: its generative power. This includes both contemporary digital and historical nondigital generativity. The digital image thus presents the twenty-first century with an incredible aesthetic decision: how and to what degree to treat the digital image as an instrumental tool for merely replicating images or as a means for releasing a more generative flow in all matters, thus generating completely new images.

LIMITATIONS AND FUTURE WORK

The beginning of this book set out four limitations for the project. First, the book is limited in its historical scope to human art history. Second, within this historical scope, it is limited to a strictly kinetic analysis of the moving image in this history. Third, this kinetic analysis is additionally limited to the study of only the most dominant kinesthetic regimes. And fourth, it is limited geographically to the near East and Western world.

These limitations have not been imposed in principle but rather in practice in order to make this book manageable to write and to serve as a launchpad for further additions and elaborations by others. Limitation, however, is not the same as error, and so it is the author's hope that this book will be judged by what it set out to do within these limitations, and not by what was outside its practical scope at the time.

Following these limitations, there are a number of areas where future work is needed. First, the historical scope could easily be expanded beyond human art history by looking at the structure of sensation and image distribution made by living and nonliving beings more broadly, since the creation of images is not limited to human beings.

Second, this kinesthetic approach can be used alongside or as a supplementary dimension to existing methodologies in the study of sensation and the arts. For example, temporal theories of cinema could easily be supplemented by adding a material and kinetic dimension like the movement of the camera and projector, and thus providing a more robust analysis of *kinetic* time.

Third, one can supplement the dominant history given here with a number of minor histories that coexisted within the dominant ones but that follow a different mixture of kinesthetic regimes. For example, generative or aleatory arts have always been around and have posed certain challenges to the dominant aesthetic regimes of their age. This is one possible direction to explore.

Fourth, one could expand the geographical scope of this kinesthetic method to the Eastern and colonial worlds. It is hard to imagine that Western art would be what it was without the influence and exploitation of the Eastern and colonial worlds. While the history of Western art in this book is not, it is hoped, technically inaccurate, it is by no means a complete story—none ever is. Eastern and colonial worlds have their own major and minor historical periods and arts that proceeded alongside that of the West's, both influencing and being influenced by it.

Fifth, this book was often forced to sacrifice empirical depth for historical scope. One cannot have both without producing a much larger book or imposing stricter limitations. As such, it leaves out many great artists and works of art, and even those it includes are treated often unevenly as short illustrations or examples to show a larger historical pattern. This is completely intentional and part of the method and scope of this project. Therefore, almost any of these illustrations could be more fully expanded, elaborated, supplemented, and even challenged using the kinesthetic method of this book.

It is my hope that this book will not be an end in itself but rather only the beginning of a new materialist and kinesthetic theory and history of the image.

NOTES

INTRODUCTION

1. Today, 77 percent of developed countries and 40 percent of the entire world use the internet. It has become the single-largest mechanism for the production, mobilization, and consumption of sensory media. Statistics, as per International Telecommunication Union, http://www.itu.int/en/ITU-D/Statistics/Pages/stat/default.aspx.
2. Luca Turin, *The Secret of Scent: Adventures in Perfume and the Science of Smell* (New York: Harper Perennial, 2007).
3. See Mark Hansen, *New Philosophy for New Media* (Cambridge, MA: MIT Press, 2004), for an excellent literature review of this growing field of study. Mark Hansen, *Bodies in Code: Interfaces with Digital Media* (London: Routledge, 2006); Erin Manning, *Relationscapes: Movement, Art, Philosophy* (Cambridge, MA: MIT Press, 2012); Brian Massumi, *Parables for the Virtual: Movement, Affect, Sensation* (Durham, NC: Duke University Press, 2007); Ossi Naukkarinen, "Aesthetics and Mobility: A Short Introduction into a Moving Field," *Contemporary Aesthetics* 1 (2005): ?; Simon O'Sullivan, "The Aesthetics of Affect: Thinking Art Beyond Representation," *Angelaki* 6, no. 3 (2001): 125–35; Melissa Gregg and Gregory J. Seigworth, eds., *The Affect Theory Reader* (Durham, NC: Duke University Press, 2010).
4. Early nineteenth-century German psycho-physicists such as Gustav Fechner, Wilhelm Wundt, and others were among the first to pay attention to the real effects of movement in aesthetic perception. However, the focus of their studies remained one of human psychology and human affect, and not the mutual transformation of the entire entangled affective structure or pattern of motion. More recent efforts in sociology, history, and anthropology have done well to focus on the activity and movement of images, but have largely treated the image as a socially and anthropocentrically constructed one that is reducible to a fundamentally *social object* or *thing*. See Arjun Appadurai, *The Social Life of Things: Commodities in Cultural Perspective* (Cambridge: Cambridge University Press, [1986] 2016); Richard Davis, *Lives of Indian Images* (Princeton, NJ: Princeton University Press, 1999). Alfred Gell's *Art and Agency: An Anthropological Theory* (Oxford: Clarendon Press, 1998), has an especially relational theory of material agency. He contrasts his own "action-centered approach" (6) with the typical "semiotic approach" in which humans project linguist features onto objects. This is a move against the "linguistic turn" in art theory. However, in attempting to avoid a linguist constructivism, Gell falls

back into an anthropocentric and social constructivism in which "the art object is a function of the social-relational matrix" and only "mediates social agency," reducing "art objects as extensions of persons" (9). George Kubler's *The Shape of Time: Remarks on the History of Things* (New Haven, CT: Yale University Press, [1962] 2017), falls prey to a similar anthropocentric and social formalism as Gell.

More broadly, in the recent turn toward "materiality" studies across the disciplines, the focus is almost exclusively on the ways *things* and *society* co-produce each other. See W. Keane, "Semiotics and the Social Analysis of Material Things," *Language and Communication* 23 (2003): 409–25; L. Meskell, ed., *Archaeologies of Materiality* (Oxford: Wiley-Blackwell, 2005); D. Miller, "Materiality: An Introduction," in *Materiality*, ed. D. Miller (Durham, NC: Duke University Press, 2005), 1–50; P. Pels, "The Spirit of Matter: On Fetish, Rarity, Fact, and Fancy," in *Border Fetishisms: Material Objects in Unstable Places*, ed. P. Spyer (New York: Routledge, 1998), 91–121. Anthropologist Nicholas Thomas looks at the agency of material objects in colonialism and empire in Nicholas Thomas, *Entangled Objects: Exchange, Material Culture and Colonialism in the Pacific* (Cambridge, MA: Harvard University Press, 1991). Bill Brown's "thing theory" examines how things are given new meanings in late nineteenth-century literature; Bill Brown, *A Sense of Things: The Object Matter of American Literature* (Chicago: University of Chicago Press, 2003). All these works importantly look at the agency of *things* (not images), and only with respect to their social construction and social functions.

One of the most interesting attempts at a theory of the image that begins with *the primacy of the material agency of the image itself* is that of the German art historian Horst Bredekamp, *Theorie des Bildakts* (*Theory of the Image-Act*) (Frankfurt: Suhrkamp, 2013). In contrast to Alfred Gell's social-relational reduction of the image and Hans Belting's pyscho-physical reduction of the image in Hans Belting, *An Anthropology of Images: Picture, Medium, Body* (Princeton, NJ: Princeton University Press, 2014), Bredekamp gives us a new theory of the image, irreducible to its "extension of the human body or society." His theory of the "image-act" is to make images themselves actors possessed of sovereign agency separable from their handling or their perception by people (51). Unfortunately, by rooting the "right to life" of images in Aby Warburg's interesting, but also socially and anthropocentrically limited, idea of the "pathos of images" (see especially 298–99), Bredekamp is drawn back into Warburg's *Distanzbildung*, where the purpose of images is simply to create space for human beings to regard themselves—to fulfill their own potential in the recognition that humankind. Images, then, have agency, but only for human reaction, will, desire, and perception. "The 'I' becomes stronger when it relativizes itself against the activity of the image" (328).

Another strong attempt to overcome the social reductionism and anthropocentrism of matieriality studies is Ian Hodder's *Entangled: An Archaeology of the Relationships between Humans and Things* (Malden: Wiley-Blackwell, 2013). For all its merits, however, it does not provide a theory of the image, but reduces the affective image to an object or thing. This is also a problem in the other "object-oriented" approaches to aesthetics, from which this book draws its inspiration. *Theory of the Image* shares some things in common with these attemepts. For these varied scholars images are not static copies, or linguistic/semiotic constructions. However, this book is also distinct from previous approaches insofar as it choses to answer the "decisive question" (51), as Bredekamp puts it (about human image relations), by figuring humans and society themselves *as images*, not objects, socially constructed or otherwise. Images are related to but not strictly identical to things or objects; see part I. In this book, the division between images and humans is dissolved from the start and the "decisive question" is refigured instead around regimes of affective motion within which humans and society are only particular knots; again, see part I for the full development. See also Thomas Nail, "Marx: The Birth of Value," unpublished manuscript.

5. The increasing mobility of the image is not an "epochal concept" of "our era" in a univocal or exclusive sense. It is only one of the most widespread and powerful features of the present, among others. See Gabriel Rockhill, *Interventions in Contemporary Thought: History, Politics, and Aesthetics* (Edinburgh: Edinburgh University Press, 2016), 51–52.
6. This is quite different from most of the literature in digital media studies, which tends to focus on contemporary digital media. The work of media archeology scholars and Mark Hansen are often an exception to this. Jussi Parikka, *What Is Media Archaeology?* (Cambridge: Polity, 2012). Erkki Huhtamo, *Media Archaeology: Approaches, Applications, and Implications* (Berkeley: University of California Press, 2011). Siegfried Zielinski, *Deep Time of the Media: Toward an Archaeology of Hearing and Seeing by Technical Means* (Cambridge, MA: MIT Press, 2006). See Hansen, *New Philosophy for New Media*.
7. Theory, however, also has its own material kinetic process of inscription. The study and of this process of inscription warrants its own independent investigation. See Thomas Nail, "Being and Motion," unpublished manuscript. See Levi Bryant, *Onto-cartography: An Ontology of Machines and Media* (Edinburgh: Edinburgh University Press, 2014).
8. Henri Bergson, *Matter and Memory*, trans. Nancy M. Paul and William S. Palmer (New York: Zone, 2005), 19.
9. For more on the methodology of kinetic philosophy, see Nail, "Being and Motion."
10. See Nail, "Being and Motion" for a full description of the historical method.
11. Plato, *Plato's Timaeus*, trans. Peter Kalkavage (Newburyport, MA: Focus Classical Library, 2001), 37c–e.
12. We can see a later expression of a similar idea in Aby Warburg's interesting, but also socially and anthropocentrically limited, idea of the "pathos of images"; and in Bredekamp's theory of the image-act in which images have agency, but only for human reaction, will, desire, and perception. "The 'I' becomes stronger when it relativizes itself against the activity of the image"; Bredekamp, *Theorie des Bildakts*, 328.
13. Tom Sparrow, *The End of Phenomenology* (Edinburgh: Edinburgh University Press, 2014); Karen Barad, *Meeting the Universe Halfway: Quantum Physics and the Entanglement of Matter and Meaning* (Durham, NC: Duke University Press, 2007); Hodder, *Entangled*; Bryant, *Onto-cartography*; Manuel Delanda, *Assemblage Theory* (Edinburgh: Edinburgh University Press, 2016); Diana Coole and Samantha Frost, *New Materialisms: Ontology, Agency, and Politics* (Durham, NC: Duke University Press, 2010); Nail, "Being and Motion."
14. Merleau-Ponty's essay "Eye and Mind" (1961), for example, makes great strides toward overcoming the anthropocentrism and constructivism of earlier phenomenology, including his own. In "Eye and Mind," Merleau-Ponty aims to give back historicity to the image itself as a continuous fold, fabric, or pleat

in being: "the world is made of the same stuff as the body" (163) because it is "visible and mobile: a thing among things" (163). While the emphasis of the text remains largely on the human body, at the same time it also aims to break down the division between image and body.

15. While they remain anthropocentric humanists, they also allow for radical historical changes in existing social and aesthetic structures. See Theodor Adorno, *History and Freedom: Lectures 1964–1965*, trans. Rolf Tiedemann (Oxford: Wiley, 2014).
16. Massumi, *Parables for the Virtual*, 23–45.
17. Delanda, *Assemblage Theory*; Massumi, *Parables for the Virtual*; Manning, *Relationscapes*; Jane Bennett, *Vibrant Matter: A Political Ecology of Things* (Durham, NC: Duke University Press, 2010); E. William Connolly, *A World of Becoming* (Durham, NC: Duke University Press, 2011); Alfred North Whitehead, *Process a Reality* (New York: Macmillan, 2014), 73.
18. Jean Baudrillard, *Simulacra and Simulation*, trans. Sheila F. Glaser (Ann Arbor: University of Michigan Press, 2010); and Gilles Deleuze, *Logic of Sense*, trans. Mark Lester with Charles Stivale (New York: Columbia University Press, 1990), 253–65.
19. According to the *Oxford English Dictionary*. Oxford: Oxford University Press, 2004. Print.
20. See Michel Serres, *The Five Senses: A Philosophy of Mingled Bodies* (New York: Bloomsbury Academic, 2016).
21. Bergson, *Matter and Memory*, 9.
22. An inversion of Bergson's claim that "Matter, in our view, is an aggregate of 'images'"; Bergson, *Matter and Memory*, 9.
23. Karl Marx, *Captial Volume 1* (New York, Penguin, 1976), 102.

CHAPTER 1

1. Serres, *The Five Senses*, 5. "The phrase, which became the motto of empiricist philosophy, seems to have no one identifiable author; it is often assumed to have been said by Aristotle, though the phrase does not appear in his work, while others assumed that it was first used by Thomas Aquinas and John Locke. The earliest use of the phrase detected so far appears to be from the thirteenth century."
2. Lucretius and Marx follow a similar trans-empirical methodology. See Thomas Nail, *Lucretius I: An Ontology of Motion* (Edinburgh: Edinburgh University Press, 2018); and Nail, "Marx: The Birth of Value."
3. Again, this is only the case given the ontological primacy of motion in the present. On the question of whether motion and matter will always be primary or whether something else will emerge we must remain agnostic, any such speculation is necessarily metaphysical.
4. "The way to see the energy/momentum of a field is to arrange some clever experiment in which a series of 'microscopic' movements of energy and momentum in the field kick off a chain reaction of larger-scale movements of energy/momentum until a 'macroscopic' thing is affected in a way that we can see. This is basically what designing an experiment is all about." Personal correspondence with Brian Skinner, a researcher in theoretical condensed matter physics at MIT.
5. See Sean Carroll, *The Particle at the End of the Universe: How the Hunt for the Higgs Boson Leads Us to the Edge of a New World* (New York: Dutton, 2012).

6. Karl Marx, *Grundrisse: Foundations of the Critique of Political Economy*, trans. Martin Nicolaus (New York: Penguin Classics, 2012), 687. Furthermore, quantum fields can only be observed through the visible effects they create and not in themselves. In order to generate mass and particles, quantum fields by necessity must have energy and momentum. Since, as Einstein showed, mass and energy are convertible, particles are born from and return to their quantum fields. Quantum field energy becomes particle mass, becomes field energy, in a continuous momentum or movement. Therefore a quantum field is just as material as particles are—even if the field itself is not empirically visible—because *particles are nothing other than folds or excitations in flow of fields*. Because quantum matter is always continuously fluctuating at every level there is always already a flow of matter. This flow then folds up into a particle.
7. Bergson, *Matter and Memory*, 9.
8. Quantum fields are observable and measurable to some degree, but not reducibly so, owing to superposition, mobility, and nonlocality.
9. For a full-length treatment of the concept of kinetic materialism, Nail, *Lucretius I*.
10. Karl Marx, *Early Writings* (New York: Penguin, 1992), 423, "Ninth Thesis on Fuerbach."
11. See also Bennett, *Vibrant Matter*. Elizabeth Grosz, *The Incorporeal: Ontology, Ethics, and the Limits of Materialism* (New York: Columbia University Press, 2017). See also the vitalist process ontology of *Teotl* (power/energy/vital force) in Aztec metaphysics, discussed in James Maffie, *Aztec Philosophy: Understanding a World in Motion* (Boulder: University Press of Colorado, 2014). Maffie explicitly compares Aztec vitalism with Spinoza's pantheistic philosophy.
12. Terms like "agential" or "performative" or "diffracted" materialism by Karen Barad, or even "animate" materialism by Mel Chen, are better because they do not rely on the language of life and forces. See Mel Chen, *Animacies: Biopolitics, Racial Mattering, and Queer Affect* (Durham, NC: Duke University Press, 2012).
13. For a critique of this vital biopolitical fetishism, see: Elizabeth Povinelli, *Geontologies: A Requiem to Late Liberalism* (Durham, NC: Duke University Press, 2016). For a Marxist critique of vitalist new materialism, see also Jennifer Cotter, "New Materialism and the Labour Theory of Value," *Minnesota Review* 87 (2016): 171–81. See also Simon Choat, "Science, Agency and Ontology: A Historical-Materialist Response to New Materialism," *Political Studies* 66, no. 4 (2018): 1027–42, for an even-handed assessment of the problems with vitalist new materialism and the contributions of Marx to new materialism.
14. See Chen, *Animacies*, 11. "My purpose is not to invest certain materialities with life."
15. For a full typology and critique of various new materialisms, see Thomas Nail, Chris Gamble, and Josh Hannan, "What Is New Materialism?," unpublished manuscript.
16. For a full criqitue of Deleuzian neo-vitalism, see Thomas Nail, "The Ontology of Motion," *Qui Parle: Critical Humanities and Social Sciences* 27, no. 1 (2018): 47–76.
17. "This then is one account of nature, namely that it is the primary underlying matter of things which have in themselves a principle of motion or change"; Aristotle, *Physics* II. 193a28–29. "This is Motion. This becoming, however, is itself just as much the collapse within itself of its contradiction, the *immediately identical* and *existent* unity of both, namely, *Matter*"; Arnold V. Miller, trans.,

Hegel's Philosophy of Nature: Being Part Two of the Encyclopaedia of the Philosophical Sciences (1830) (Oxford: Clarendon,), 41.

18. "All that exists, all that lives on land and under water, exists and lives only by some kind of movement. Thus the movement of history produces social relations; industrial movement gives us industrial products, etc. Just as by dint of abstraction we have transformed everything into a logical category, so one has only to make an abstraction of every characteristic distinctive of different movements to attain movement in its abstract condition—purely formal movement, the purely logical formula of movement. If one finds in logical categories the substance of all things, one imagines one has found in the logical formula of movement the absolute method, which not only explains all things, but also implies the movement of things. Karl Marx, *The Poverty of Philosophy* (New York: International, 1963), 78.
19. Aristotle, *On the Soul*, in *Complete Works of Aristotle: Revised Oxford Translation*, ed. Jonathan Barnes (Princeton, NJ: Princeton University Press, 1984), 33–34, translation slightly modified.
20. See Kurt Gödel, *On Formally Undecidable Propositions of Principia Mathematica and Related Systems* (New York: Basic, 1962).
21. Geach used this phrase to describe Russell and McTaggart's theories of formal change. See P. T. Geach, *God and the Soul* (New York: Schocken, 1969), 71–72. See also Alfred North Whitehead's theory of change, in Alfred North Whitehead, *Concept of Nature* (Cambridge: Cambridge University Press, 1978), 73, 59. According to Whitehead, change is only "the difference between actual occasions comprised in some determined event" (73) and thus it is "impossible to attribute 'change' to any actual entity" (59). Change and motion thus relate to a succession of actual entities, and are constituted only by the differences among them. Every entity is simply "what it is" and it becomes with its whole set of relations to other entities inherent therein, thus *cannot change or move*. See also Leonard J. Eslick, "Substance, Change, and Causality in Whitehead," *Philosophy and Phenomenological Research* 18, no. 4 (June 1958): 503–13. Whitehead's transition "is not a real transition, not a flow or flux, and change so understood is merely a fact consequent upon the successive existence of a series of different unchangeable actual entities. *The very notion of change has been made incurably static*" (510).
22. See Nail, "Being and Motion," bk. I, part III.
23. Lucretius, *On the Nature of Things*, trans. Walter G. Englert (Newburyport, MA: Focus Classical Library, 2003), Book 1, lines 1–25.
24. Bergson, *Matter and Memory*, 193.
25. Paul Valéry, *The Collected Works of Paul Valery: Volume I* (New Jersey: Princeton University Press, 2015), 10–11. Translation modified slightly.

    De sa profonde mère, encor froide et fumante,
    Voici qu'au seuil battu de tempêtes, la chair
    Amèrement vomie au soleil par la mer,
    Se délivre des diamants de la tourmente.

    Son sourire se forme, et suit sur ses bras blancs
    Qu'éplore l'orient d'une épaule meurtrie,
    De l'humide Thétis la pure pierrerie,
    Et sa tresse se fraye un frisson sur ses flancs.

26. Bergson, *Matter and Memory*, 189.

27. Henri Bergson, *An Introduction to Metaphysics*, trans. T. E. Hulme (New York: Liberal Arts, 1949), 53.
28. By "whole," Bergson does not mean a "totality," because a totality cannot change or become other than it is. Bergson means something like an open and vibratory whole.
29. See Thomas Nail, *The Figure of the Migrant* (Stanford, CA: Stanford University Press, 2015), 11–20.
30. Even flows of light are composed of pedetic photons and bend around black holes.
31. The Clay Mathematics Institute (CMI), http://www.claymath.org/millennium-problems/navier–stokes-equation.
32. James Gleick, *Chaos: Making a New Science* (New York: Viking, 1987), 1–32; Steven Strogatz, *Sync: The Emerging Science of Spontaneous Order* (New York: Hyperion, 2014).
33. The quote is probably apocryphal, but the sentiment is striking.
34. Heisenberg ultimately adopts a theory of quantum interaction and abandons the idea of epistemological relativism often associated with his uncertainty principle.

    > Following a heated discussion wherein Bohr offers an important criticism of Heisenberg's analysis, Heisenberg acquiesces to Bohr's point of view. Though it is little discussed, Heisenberg includes an admission of these important shortcomings of his analysis in a postscript to his famous uncertainty paper. In an important sense, this postscript constitutes an undoing of the analysis that he presents in the body of the text, and yet this erroneous analysis has become the standard exposition on the reciprocity relations. The uncertainty principle continues to be taught to students and spoken of by physicists and non-physicists in accord with Heisenberg's account when by his own admission his account had been based on a fundamental error. Ironically, there is no mention of Bohr's account of the reciprocity relations, that is, the indeterminacy principle. Indeed, if Bohr's contributions to these discussions are mentioned at all, it is usually with a historically respectful nod to complementarity; but even this is seldom mentioned anymore. (Barad, *Meeting the Universe Halfway*, 301)

35. L. A. Rozema, A. Darabi, D. H. Mahler, A. Hayat, Y. Soudagar, and A. M. Steinberg, "Violation of Heisenberg's Measurement-Disturbance Relationship by Weak Measurements," *Physical Review Letters* 109, no. 10 (2012): 100404(1-5).
36. The "Boltzmann brain" is named after the nineteenth-century physicist Ludwig Boltzmann.
37. Gleick, *Chaos*, 1–32.
38. Philip Ball, "Van Gogh Painted Perfect Turbulence," *Nature*, July 7, 2006, http://www.nature.com/news/2006/060703/full/news060703-17.html.
39. The kinetic postulate of bifurcation is thus entirely consistent with contemporary astrophysics. From this continuum of motion, the universe rapidly bifurcated and split itself from itself through the big bang: moving from pure continuum to increasing heterogeneity, discontinuity, and entropy. Cosmologically, matter multiplies through bifurcation.
40. Paul Valéry, "The Graveyard by the Sea," trans. C. Day Lewis, http://unix.cc.wmich.edu/~cooneys/poems/fr/valery.daylewis.html.
41. See Nail, "Being and Motion," chap. 4.

42. Of course a neuronal flow can be imaged in a magnetoencephalography (MEG) machine, but only insofar the material flows of the machine's electrical paths are themselves—along with the material flows of the viewer of these patterns—in turn rendered insensible in an endless recession of material conditions for sensation.
43. David Biello, "How the First Plant Came to Be," *Scientific American*, February 16, 2012, http://www.scientificamerican.com/article/how-first-plant-evolved/.
44. See Peter Wohlleben, Tim F. Flannery, S. Simard, and Jane Billinghurst, *The Hidden Life of Trees: What They Feel, How They Communicate: Discoveries from a Secret World* (Vancouver: Greystone, 2016). Daniel Chamovitz, *What a Plant Knows: A Field Guide to the Senses* (New York: Scientific American/Farrar, Straus & Giroux, 2013).
45. Jakob Uexküll, *A Foray into the Worlds of Animals and Humans: With a Theory of Meaning* (Minneapolis: University of Minnesota Press, 2010), 44–51.
46. See Eric Schneider and Dorion Sagan, *Into the Cool: Energy Flow, Thermodynamics, and Life* (Chicago: University of Chicago Press, 2005).

CHAPTER 2

1. Michel Serres develops a similar theory of vortices: "The vortex conjoins the atoms, in the same way as the spiral links the points; the turning movement brings together atoms and points alike." Michel Serres, *The Birth of Physics* (Manchester: Clinamen, 2000), 16. Deleuze and Guattari then further develop this under the name of "minor science," in Gilles Deleuze and Félix Guattari, *A Thousand Plateaus: Capitalism and Schizophrenia*, trans. Brian Massumi (London: Continuum, 2008), 361–62.
2. The kinetic roots of the word *junction* come from the PIE root **yeug-*, "to join," "to yoke."
3. See Nail, "Marx: The Birth of Value," for a detailed theory of this abstraction process.
4. Jackson Pollock, Paul Falkenberg, Hans Namuth, and Morton Feldman, *Jackson Pollock 51*, NTSC color broadcast system, 2013, https://www.youtube.com/watch?v=CrVE-WQBcYQ
5. Simon Hantaï, "Letter of 9 July 1999," in Top of Form Geneviève Bonnefoi, *Hantai* (Montauban: Centre d'Art contemporain de l'Abbaye de Beaulieu, 1973), 23–24. http:/pmc.iath.virginia.edu/issue.503/translations.html#60.
6. The river rolls itself up like the periodicity of an electron shell.
7. Daniel Graham, *The Texts of Early Greek Philosophy: The Complete Fragments and Selected Testimonies of the Major Presocratics* (Cambridge: Cambridge University Press, 2010), 159–62
8. Virginia Woolf, *The Waves*, in *Selected Works of Virginia Woolf* (Hertfordshire: Wordsworth, 2005), 691.
9. Paul Valéry, *Cantate Du Narcisse* (Paris: Gallimard, 1944), scene II.
   Admire dans Narcisse un éternel retour
   Vers l'onde où son image offerte à son amour
   Propose à sa beauté toute sa connaissance:
   Tout mon sort n'est qu'obéissance
   A la force de mon amour.
10. See also Herbert Marcuse, *Eros and Civilization: A Philosophical Inquiry into Freud* (Boston: Beacon, 2015), 160–70, for an closer reading of the sensuous image of Narcissus's reflecting pool.

11. André Gide, *Le Traité Du Narcisse*, in Top of Form André Gide, *Romans: Récits Et Soties, Oeuvres Lyriques*, trans. Maurice Nadeau, Yvonne Davet, and J.-J Thierry (Paris: Gallimard, 1969). Bottom of Form.
12. Charles Baudelaire, *L'invitation Au Voyage: A Poem from the Flowers of Evil*, ed. and trans. Pamela Prince, Jane Handel, Richard Wilbur, and Carol Cosman (Portland, OR: Ma Nao, 2011).
    Là, tout n'est qu'ordre et beauté,
    Luxe, calme, et volupté
13. Virginia Woolf, *Orlando: A Biography* (New York: Harcourt, Brace, 1993), 323.
14. Woolf, *Orlando*, 323.
15. Uexküll, *Foray into the Worlds of Animals and Humans*, 49.
16. See Maurice Merleau-Ponty, *Phenomenology of Perception*, trans. Donald A. Landes (London: Routledge, [1962] 2012), 94. See also Maurice Merleau-Ponty, *The Primacy of Perception* (Evanston: Northwestern University Press, 1964), 163.
17. For example, the Koyukon of Alaska see "streaking like a flash of fire through the undergrowth," not a fox, and "perching in the lower branches of spruce trees," not an owl. The names of animals are not nouns but, rather, verbs. The sun is not understood as an object that moves across the sky but, rather, as the path of its movement through the sky, on its daily journey from the eastern to the western horizon. Richard Nelson, *Make Prayers to the Raven: A Koyukon View of the Northern Forest* (Chicago: University of Chicago Press, 1983), 108, 158. For similar examples, see Tim Ingold, *Being Alive: Essays on Movement, Knowledge and Description* (London: Routledge, 2011), 72.
18. Woolf, *The Waves*, 727.
19. Spinoza describes this as a kind of kinetic *affectus*. "Every body," he says, "is moved sometimes more slowly, sometimes more quickly." Affection is the capacity of flows to act and be acted on, thus increasing or diminishing their motion. Outside of this affective kinetic sensibility there is no transcendent essence of the thing. See Benedictus de Spinoza, *A Spinoza Reader: The Ethics and Other Works*, trans. Edwin M. Curley (Princeton, NJ, and Chichester: Princeton University Press, 1994), 154, Book II, Axiom II.
20. Victor Hugo, view from train window, August 1837. Quoted in Marc Baroli, *Le Train Dans La Littérature Française* (Paris: Éditions N.M, 1964), 58.
21. See Plato. *Philebus*, in *The Dialogues of Plato*, trans. Benjamin Jowlett (New York: Random House, 1937).
22. See Aristotle, *Categories*, in *The Basic Works of Aristotle*, trans. Richard McKeon (New York: Random House, 1941), line 254.
23. Gilles Deleuze, *Bergsonism*, trans. Hugh Tomlinson and Barbara Habberjam (New York: Zone, 1988), 87–88. "Qualities belong to matter as much as to ourselves: They belong to matter, they are in matter, by virtue of the vibrations and numbers that punctuate them internally."
24. Turin, *Secret of Scent*. Jason Palmer, "Quantum Biology: Do Weird Physics Effects Abound in Nature?," British Broadcasting Corporation, https://www.bbc.com/news/science-environment-21150047; Chandler Burr, *The Emperor of Scent: A True Story of Perfume and Obsession* (New York: Random House, 2003).
25. H. L. Klopping, "Olfactory Theories and the Odors of Small Molecules," *Journal of Agricultural and Food Chemistry* 19 (1971): 999–1004.
26. *Fourier Transform Processing with ImageMagick*, http://www.fmwconcepts.com/imagemagick/fourier_transforms/fourier.html.

27. This problem is also formalized in set theory. See Alain Badiou, *Being and Event*, trans. Oliver Feltham (London: Continuum, 2007), 267–68.
28. Friedrich Nietzsche, *The Will to Power*, trans. Walter A. Kaufmann (New York: Random House, 1967), 565.
29. See Karl Marx, *Capital* (New York: Penguin, 1976), 1:125–37.
30. Deleuze, *Bergsonism*, 74.
31. See Richard Liboff, *Kinetic Theory: Classical, Quantum, and Relativistic Descriptions* (New York: Springer, 2003).
32. For a full theory of the object, see Thomas Nail, "Theory of the Object," unpublished manuscript.
33. For a philosophical theory of diffraction, see Barad, *Meeting the Universe Halfway*.
34. See Anatoliĭ Burshteĭn, *Introduction to Thermodynamics and Kinetic Theory of Matter* (New York: Wiley, 1996).
35. As Hume had argued.
36. Plutarch, *Plutarch's Lives*, trans. John Dryden (New York: Modern Library, 2001), 14.

CHAPTER 3

1. See David Howes and Constance Classen, eds., *Ways of Sensing: Understanding the Senses in Society* (New York: Routledge, 2014) 152. James Porter, *The Origins of Aesthetic Thought in Ancient Greece: Matter, Sensation, and Experience* (Cambridge: Cambridge University Press, 2010), 64–66.
2. René Descartes, *The Philosophical Writings of Descartes*, trans. and ed. John Cottingham, Robert Stoothoff, and Dugald Murdoch (Cambridge: Cambridge University Press, 1985), 1:228, part II, sec. 13.
3. Isaac Newton, *A Treatise of the Method of Fluxions and Infinite Series: With Its Application to the Geometry of Curve Lines* (London: T. Woodman, 1737).
4. Descartes, *Philosophical Writings*, 236, part II, sec. 31.
5. Virginia Woolf, *The Waves* (New York: Harcourt, Brace, 1931), 126–27; excerpt from pp. 118–40.
6. I take this word *kinoumena* from Epicurus's "Letters," Book X of Diogenes Laërtius, *Lives of Eminent Philosophers* (London: W. Heinemann, 1925).
7. For a critique of this Kantian idea of phenomena and toward a phenomenology of motion, see Maxine Sheets-Johnstone, *The Primacy of Movement* (Amsterdam: John Benjamins, 1999).
8. For an expanded theory of social limit junctions, see Thomas Nail, *Theory of the Border* (New York: Oxford University Press, 2016).
9. These three motions are all developed at length in Nail, *Theory of the Border*.
10. The theory of knots is further developed in Nail, "Being and Motion," chap. 3.
11. Cited in Gilles Deleuze, *Francis Bacon: The Logic of Sensation*, trans. Daniel W. Smith (Minneapolis: University of Minnesota Press, 2005), 36.

SECTION A

1. Georg W. F Hegel, *Phenomenology of Spirit*, trans. Arnold V. Miller and J. N. Findlay (Oxford: Clarendon, 1977), 421.
2. Georg W. F. Hegel, *Aesthetics: Lectures on Fine Art*, trans. T. M. Knox (Oxford: Clarendon, 1975), 1:427.
3. Robert Goldwater, *Primitivism in Modern Art* (Cambridge, MA: Harvard University Press, 1987), 309.

CHAPTER 4

1. Marx, *Early Writings*, 353. translation modified.
2. André Leroi-Gourhan, *Gesture and Speech*, trans. Anna Bostock Berger (Cambridge, MA: MIT Press, 1993), 301.
3. Leroi-Gourhan, *Gesture and Speech*, 303.
4. For a full account with numerous other example, see Leroi-Gourhan, *Gesture and Speech*, 309. "One of the operating techniques of human beings from the earliest stages has been the application of rhythmic percussive movements repeated over prolonged periods. Indeed that is the only operation that marked the attainment of human status by the Australanthropians, whose only surviving traces are choppers made from splintered pebbles and polyhedral spherical objects produced by prolonged hammering. Manufacturing techniques developed from the beginning in a rhythmic setting—at once muscular, visual, and auditive—born of the repetition of impact-making gestures." See also Gary Tomlinson, *A Million Years of Music: The Emergence of Human Modernity* (New York: Zone, 2015), especially chap. 2 on the biface.
5. Merleau-Ponty, *Phenomenology of Perception*, 112.
6. Richard Menary, *The Extended Mind* (Cambridge, MA: MIT Press, 2012)
7. Alva Noë, *Strange Tools: Art and Human Nature* (New York: Hill and Wang, 2016); Humberto Maturana and Francisco J. Varela, *The Tree of Knowledge: The Biological Roots of Human Understanding* (Boston, MA: Shambhala, 2008).
8. Maturana and Varela, *Tree of Knowledge*, 242.
9. R. Swenson, "Galileo, Babel, and Autopoiesis (It's Turtles All The Way Down)," *International Journal of General Systems* 21, no. 1 (1992): 267–69. The autopoietic model is "miraculously decoupled from the physical world by its progenitors . . . (and thus) grounded on a solipsistic foundation that flies in the face of both common sense and scientific knowledge."
10. For theories of agential realism and networked agency, see Barad, *Meeting the Universe Halfway*; Bruno Latour, *Reassembling the Social: An Introduction to Actor-Network-Theory* (Oxford: Oxford University Press, 2008).
11. See "Deer Hunt," detail of a wall painting from Level III, Çatal Höyük, Turkey, ca. 5750 BCE, Museum of Anatolian Civilization, Ankara. Cited in Helen Gardner and Fred S. Kleiner, *Gardner's Art Through the Ages: The Western Perspective 13th edition* (Boston, MA: Wadsworth Cengage Learning, 2010), 12.
12. A similar argument has be developed by Michael Pollan with respect to plants: that plants use humans in order to reproduce themselves. Corn, in this respect, has been incredibly successful. Michael Pollan, *The Botany of Desire: A Plant's Eye View of the World* (New York: Random House, 2008). Donna Haraway has done something similar for companion species, specifically dogs. See Donna Haraway, *When Species Meet* (Minneapolis: University of Minnesota Press, 2010).

CHAPTER 5

1. This study is limited to human aesthetic history, but other animal histories of aesthetics are possible.
2. For these paragraphs, see Leroi-Gourhan, *Gesture and Speech*.
3. Leroi-Gourhan, *Gesture and Speech*, 113–14.
4. Leroi-Gourhan, *Gesture and Speech*, 363; italics mine.
5. Elias Canetti, *Crowds and Power* (New York: Viking, 1962), 31.
6. Shannon P. McPherron, Zeresenay Alemseged, Curtis W. Marean, Jonathan G. Wynn, Denne Reed, Denis Geraads, Rene Bobe, and Hamdallah A. Bearat,

"Evidence for Stone-Tool-Assisted Consumption of Animal Tissues Before 3.39 Million Years Ago at Dikika, Ethiopia," *Nature* 466 (2010): 857–60.

7. See Tomlinson, *Million Years of Music*, 23–50.
8. Steven James, "Hominid Use of Fire in the Lower and Middle Pleistocene: A Review of the Evidence," *Current Anthropology* 30, no. 1 (1989): 1–26.
9. Cited in Gaston Bachelard, *The Poetics of Space*, trans. Maria Jolas (Boston: Beacon, 2013), 34, "Tout ce qui brille voit."
10. Bachelard, *Poetics of Space*, 35.
11. Aeschylus, *Prometheus Bound*, in *Aeschylus: 1*, trans. Herbert W. Smith (Cambridge: Harvard University Press, 1973), 215, line 7.
12. Hesiod, *Theogony: And, Works and Days*, trans. Stephanie A. Nelson (Newburyport, MA: Focus Classical Library, 2009), lines 530–70. Aeschylus, *Prometheus Bound*, lines 1–50.
13. Heather Pringle, "Quest for Fire Began Earlier Than Thought," *ScienceNOW* ? (2012): https://www.sciencemag.org/news/2012/04/quest-fire-began-earlier-thought.
14. Elizabeth Pennisi, "Did Cooked Tubers Spur the Evolution of Big Brains?" *Science* 283, no. 5410 (1999): 2004–5.
15. Catherine Perles, "Feeding Strategies in Prehistoric Times," in *Food: A Culinary History from Antiquity to the Present*, ed. Jean-Louis Flandrin, Massimo Montanari, Albert Sonnenfeld, and Clarissa Botsford (New York: Columbia University Press, 2013).
16. H. Saul, M. Madella, A. Fischer, A. Glykou, S. Hartz and O. E. Craig, "Phytoliths in Pottery Reveal the Use of Spice in European Prehistoric Cuisine," *PLoS ONE* 8, no. 8 (2013): e70583, https://doi.org/10.1371/journal.pone.0070583. In the Neolithic, spices like mustard seed and garlic were first used to augment flavors around 6100 to 5750 BCE. And between 7000 and 4000 BCE, olive oil and sesame oil were used as medicinal and sensual ointments on the body.
17. Children were often present, and sometimes "fluted" their fingers over the images; see K. Sharpe and L. Van Gelder, "Evidence for Cave Marking by Paleolithic Children," *Antiquity* 80 (2006): 937–47.
18. John Noble Wilford, "In African Cave, Signs of an Ancient Paint Factory," *New York Times*, October 13, 2011, http://www.nytimes.com/2011/10/14/science/14paint.html
19. Thomas Rickert, "Rhetorical Prehistory and the Paleolithic," *Review of Communication* 16, no. 4 (2016): 352–373. This is what Rickert calls "bio-material-affect" (364).
20. David Lewis-Williams, *Mind in the Cave* (London: Thames & Hudson, 2011), 204.
21. See Dean R. Snow, "Sexual Dimorphism in European Upper Paleolithic Cave Art," *American Antiquity* 78, no. 4 (2013): 746–61.
22. Rickert, "Rhetorical Prehistory and the Paleolithic," 366.
23. Cited in Gardner and Kleiner, *Gardner's Art Through the Ages*, 8.
24. Gardner and Kleiner, *Gardner's Art Through the Ages*, 7.
25. Christopher Henshilwood et al., "Middle Stone Age Shell Beads from South Africa," *Science* 304 (2004): 404.
26. "The pregnant vegetation goddess . . . was one of the most-represented female figures depicted in Neolithic Old Europe"; Marija Gimbutas, *The Living Goddesses* (Berkeley: University of California Press, 1999), 15.

27. Thomas H. Maugh II, "Venus Figurine Sheds Light on Origins of Art by Early Humans," *Los Angeles Times*, May 14, 2009, http://articles.latimes.com/2009/may/14/science/sci-Venus14
28. Gimbutas, *Living Goddesses*, 8.
29. Marija Gimbutas, *The Language of the Goddess* (London: Thames & Hudson, 2001), 43.
30. Lewis Mumford, *The City in History: Its Origins, Its Transformations, and Its Prospects* (New York: Harcourt, Brace, & World, 1961), 13.
31. Genevieve Von Petzinger, *First Signs: Unlocking the Mysteries of the World's Oldest Symbols* (New York: Atria, 2017), 246–56.
32. See Gimbutas and Campbell, *Language of the Goddess*.
33. Gimbutas and Campbell, *Language of the Goddess*, 279.
34. Leroi-Gourhan, *Gesture and Speech*, 367–68.
35. Gimbutas and Campbell, *Language of the Goddess*, 277.
36. Woman Holding a Bison Horn, from Laussel, France, ca. 25,000–20,000 BCE. Painted limestone, 1' 6" high, Musée d'Aquitaine, Bordeaux.
37. Gimbutas and Campbell, *Language of the Goddess*, xix.
38. "Antithetic spirals whirl around this lidded vase, simulating the life power of the central egg, which is marked with a net design, 4200–4100 BCE." See plate 22, Gimbutas and Campbell, *Language of the Goddess*.
39. Crispin Sartwell, *Six Names of Beauty* (New York: Routledge, 2015), 50.
40. Tomlinson, *Million Years of Music*, 257: "individual pitch connect the perception of discrete pitch to capacities of late hominins for hierarchic cognition and its subset, discrete combinatoriality."
41. Steven Brown's "musilanguage" hypothesis, for example, aims to locate a common source for both music and language in a stage of "lexical tones," discrete pitches matched to meanings and combined into referential, "melodorhythmic" phrases. See Nils L. Wallin, Björn Merker, and Steven Brown, eds., *The Origins of Music* (Cambridge, MA: MIT Press, 2000). For a theory of "pitch continuum" and "vector grammar," see also M. Rohrmeier, W. Zuidema, G. A. Wiggins, and C. Scharff, "Principles of Structure Building in Music, Language and Animal Song," *Philosophical Transactions of the Royal Society of London, Series B: Biological Sciences* 370 (2015): 1664. Steven Mithen, *The Singing Neanderthals: The Origins of Music, Language, Mind and Body* (Cambridge, MA: Harvard University Press, 2006).
42. See Tomlinson, *Million Years of Music*, 258: "The winnowing of discrete pitches from the graded intonational contours of the calls of protodiscourse brought with it an abstraction, a distancing of the pitches themselves from meaning."
43. Dušan Borić, "First Households and 'House Societies' in European Prehistory," in *Prehistoric Europe: Theory and Practice*, ed. Andrew Jones (Chichester: Wiley-Blackwell, 2008), 109–42, 114.
44. Bachelard, *Poetics of Space*, 22.
45. Bachelard, *Poetics of Space*, 52–56.
46. The etymological origins of the word *fence* come from the PIE root **gwhen-*, "to strike."
47. Bernard Cache, *Earth Moves: The Furnishing of Territories*, trans. Anne Boyman and Michael Speaks (Cambridge, MA: MIT Press, 2010), 24.
48. "The wall delimits and the window selects"; Cache, *Earth Moves*, 28.
49. Mumford, *City in History* 7.
50. See Nail, *Theory of the Border*, 57–62, for a more developed account and typology of megaliths.

SECTION B

1. Porter, *Origins of Aesthetic Thought*, 21. Roger Fry describes art as "pure form . . . [art] sets free a pure and as it were disembodied functioning of the spirit," while in Clive Bell's terms, "the nature of the focus [of 'pure aesthetics'] is immaterial," which is to say, emphatically nonmaterial, because its object is what is "significant form" evokes: an emotional content (a kind of objective correlative) divorced from meaning and reference, but also from "the chatter and tumult of material existence" (71–72).
2. See Edmund Husserl, *Ideas*, trans. W. R. Gibson (New York: Humanities, 1976), part 1, sec. 74.
3. See Aristotle, *Politics: Translation, Introduction, and Glossary*, trans. Joe Sachs (Newburyport, MA: Focus Classical Library, 2012), bk. I, chap. 1, on the naturalization and teleological structure of the city.
4. See Marx, *Early Writings*, 419–23, "Theses on Feuerbach."
5. Lucretius, *De Rerum Natura*, in *On the Nature of Things*, trans. Walter G. Englert (Newburyport, MA: Focus Classical Library, 2003), 2:633–39.
6. "And he [Epicurus] says that the world began in the likeness of an egg, and the Wind [Khronos (Time) and Ananke (Inevitability) entwined?] encircling the egg serpent-fashion like a wreath or a belt then began to constrict nature. As it tried to squeeze all the matter with greater force, it divided the world into the two hemispheres [Ouranos and Gaia, heaven and earth]." Carl Holladay, *Orphica* (Chico, CA: Scholars, 1996), Epicurus fragment (from Epiphanius).
7. First was Chaos and Night, and black Erebus and vast Tartarus;

    And there was neither Earth, nor Air, nor Heaven: but in the boundless bosoms of Erebus.
    Night, with her black wings, first produced an aerial egg,
    From which, at the completed time, sprang forth the lovely Eros,
    Glittering with golden wings upon his back, like the swift whirlwinds.
    But embracing the dark-winged Chaos in the vast Tartarus.
    He begot our race (the birds), and first brought us to light.
    The race of the Immortals was not, till Eros mingled all things together;
    But when the elements were mixed one with another, Heaven was produced, and Ocean,
    And Earth, and the imperishable race of all the blessed Gods.
    (Aristophanes, *The Birds* [Newburyport, MA: Focus Classical Library, 1999], line 698)

8. First (I have sung) the vast necessity of ancient Chaos,

    And Cronus, who in the boundless tracts brought forth
    The Ether, and the splendid and glorious Eros of a two-fold nature,
    The illustrious father of night, existing from eternity.
    Whom men call Phanes, for he first appeared.
    I have sung the birth of powerful Brimo (Hecate), and the unhallowed deeds
    Of the earth-born (giants), who showered down from heaven
    Their blood, the lamentable seed of generation, from whene sprung
    The race of mortals, who inhabit the boundless earth for ever. (Holladay, *Orphica*, Arg. v. 12)

9. Today, we continue to use such kinetic so-called metaphors to describe speech-acts such as "to advance claims," "direct an argument," "gain/lose 'momentum' in speech," "gain/lose traction with an argument," "to stand for something," "direct an audience," "to be moved or touched by something someone said," "make a forceful statement," "to walk someone through something," "shift positions in an argument," "to drive home a point," "to stake a claim," "to settle a dispute."
10. Plato, *Timaeus*, trans. Peter Kalkavage (Newburyport, MA: Focus Classical Library, 2001), 74d.
11. Plato, *Timaeus*, 28a–b.

CHAPTER 7

1. Verse cannot be radically distinguished from prose in either speech or writing. Prose and verse each have their own rhythm, cadence, and relative grammatical structure, whether they are spoken or written. Historically, however, the birth of writing has an increasingly centrifugal kinetic effect on both—as we saw in the case of phonetic writing. Thus, the important kinesthetic distinction is not the difference between prose and verse but between the ways both are ordered by the material kinetics of writing. However, since the formal features of verse are clearer and more dramatic, we will focus on those here.
2. See Nail, "Being and Motion," chap. 15.
3. See Milman Parry and Adam Parry, *The Making of Homeric Verse: The Collected Papers of Milman Parry* (Oxford: Oxford University Press, 1987).
4. Atrahasis I, 70–73, in W. G. Lambert, A. R. Millard, and Miguel Civil, *Atra-ùhasâis: The Babylonian Story of the Flood* (Winona Lake, IN: Eisenbrauns, 1999), 47.
5. Walter J. Ong and John Hartley, *Orality and Literacy: The Technologizing of the Word* (New York: Routledge, 2013), 45–46.
6. Ong and Hartley, *Orality and Literacy,* 45–46.
7. John Maier, *Gilgamesh: A Reader* (Wauconda, IL: Bolchazy-Carducci, 1997), 183–84.
8. Parry and Parry, *Making of Homeric Verse,* 269; italics mine.
9. Eric Csapo and Margaret C. Miller, "General Introduction," in *The Origins of Theater in Ancient Greece and Beyond: From Ritual to Drama*, ed. Eric Csapo and Margaret C. Miller (New York: Cambridge University Press, 2007), 1–2. See Theodor Gaster, *Thespis: Ritual, Myth, and Drama in the Ancient Near East* (New York: Norton, 1977), 9–11.
10. See Francis Cornford, *The Origin of Attic Comedy* (Cambridge: Cambridge University Press, 2010), for a full application of the notion of the rebirth ritual to Hellenistic drama.
11. For an in-depth account of the ritual aspects of the various vegetation deities, see Jane Harrison, *Prolegomena to the Study of Greek Religion* (Cambridge: Cambridge University Press, 1903). See also Sir James Frazer, *The New Golden Bough: A New Abridgment of the Classic Work* (New York: S. G. Phillips, 1972).
12. Anne Baring and Jules Cashford, *The Myth of the Goddess: Evolution of an Image* (London: Arkana, 2000), 216.
13. Diane Wolkstein and Samuel N. Kramer, *Inanna, Queen of Heaven and Earth: Her Stories and Hymns from Sumer* (New York: Harper & Row, 1983), 71.
14. Baring and Cashford, *Myth of the Goddess,* 407.
15. Baring and Cashford, *Myth of the Goddess,* 410.

16. Baring and Cashford, *Myth of the Goddess*, 408.
17. Emily Teeter, *Religion and Ritual in Ancient Egypt* (Cambridge: Cambridge University Press, 2011), 58–66.
18. Tobin Nellhaus, Bruce A. McConachie, Carol F. Sorgenfrei, and Tamara L. Underiner, *Theatre Histories: An Introduction* (New York: Routledge, 2016), 57.
19. Teeter, *Religion and Ritual in Ancient Egypt*, 58–66.
20. The festival of Dionysus and Greek choral poetry from which Greek tragedy emerge are continuous with this long tradition of ritual and theatrical kinesthetics.
21. Quoted from Harrison, *Prolegomena to the Study of Greek Religion*, 341. "It regards Dionysus in this connection as an 'Eniautos-Daimon,' or vegetation god, like Adonis, Osiris, etc., who represents the cyclic death and rebirth of the Earth and the World, i.e., for practical purposes, of the tribe's own lands and the tribe itself. It seems clear, further, that Comedy and Tragedy represent different stages in the life of this Year Spirit; Comedy leads to his Marriage Feast . . . Tragedy to his death." See also Cornford, *Origins of Attic Comedy*.
22. Aristotle, *Poetics*, in *Complete Works of Aristotle: Revised Oxford Translation*, ed. Jonathan Barnes (Princeton, NJ: Princeton University Press, 1984), 1454a16-1454b14.
23. Aristotle, *Poetics*, 1449a10–30.
24. Friedrich Nietzsche, *The Birth of Tragedy and the Case of Wagner*, trans. Walter Kaufmann (New York: Vintage, 2010), 36.
25. Aristotle, *Poetics*, 1449b22–31.
26. Y. Garfinkel, F. Klimscha, S. Shalev, and D. Rosenberg, "The Beginning of Metallurgy in the Southern Levant: A Late 6th Millennium CalBC Copper Awl from Tel Tsaf, Israel," PLoS ONE 9, no. 3 (2014): e92591.
27. Miljana Radivojević et al., "On the Origins of Extractive Metallurgy: New Evidence from Europe," *Journal of Archaeological Science* 37, no. 11 (2010): 2776.
28. Radivojević et al., "On the Origins of Extractive Metallurgy," 2776–77.
29. Carol Mattusch, *Enduring Bronze: Ancient Art, Modern Views* (Los Angeles: J. Paul Getty Museum, 2014), 56–57.
30. P. R. S. Moorey, "The Chalcolithic Hoard from Nahal Mishmar, Israel, in Context," *World Archaeology* 20, no. 2 (1988): 173–82.
31. Robert Maddin, *The Beginning of the Use of Metals and Alloys* (Cambridge, MA: MIT Press, 1988), 28–33.
32. Jack Ogden, *Jewellery of the Ancient World* (New York: Rizzoli, 1982).
33. Amany El-Kheshen, "Glass as Radiation Sensor," InTechOpen, 2012, 580–81, at http://cdn.intechopen.com/pdfs/32110/InTech-Glass_as_radiation_sensor.pdf.
34. Mattusch, *Enduring Bronze*, 65–73.
35. Mattusch, *Enduring Bronze*, 59–65.
36. Mattusch, *Enduring Bronze*, 20–22.
37. Mattusch, *Enduring Bronze*, 87–88.
38. We have only a Roman marble copy of the Greek bronze original *Dying Gaul*, which must have been even more expressive in its original bronze.
39. Frank Carlson, "Vestiges of an Ancient Greek Art Form, Preserved by Catastrophe," PBS broadcast, January 25, 2016, http://www.pbs.org/newshour/bb/vestiges-of-an-ancient-greek-art-form-preserved-by-catastrophe/
40. Galen, *De placitis Hippocratis et Platonis*, 5, in *The Art of Ancient Greece: Sources and Documents*, by J. J. Pollitt (New York: Cambridge University Press, 1990), 76.
41. See Nail, "Being and Motion," 2, parts II and III.

42. Acts 7:48 (NIV). "However, the Most High does not live in houses made by human hands."

CHAPTER 8

1. This can be clearly seen in the creation of Roman *castra* (military encampments), which later would turn into towns and cities. The orientation of the entire grid was first and foremost a central point (*decumanus* and *cardo maximus*) from which the cardinal directions were measured and the *castra* divided. "Meanwhile the survey party was staking out the centre line of the camp (*decumanus maximus*) crossed by its axis (*Cardo maximus*). These two lines formed the basis of two path-ways bisecting the camp, the *via principalis* about 30 metros wide. All the various *strigae* (or rectangular spaces where the tents were to be erected) were marked off, as were the corners of the square of oblong camp." Martin Brice, *Stronghold: A History of Military Architecture* (London: Batsford, 1984), 48.
2. For an excellent literature review of theories of ancient urban planning, including their cosmological design, see Michael E. Smith, "Form and Meaning in the Earliest Cities: A New Approach to Ancient Urban Planning," *Journal of Planning History* 6, no. 1 (2007): 3–47.
3. Mumford, *City in History*, 16.
4. See Aurangzeb Khan and Carsten Lemmen, "Bricks and Urbanism in the Indus Valley Rise and Decline," *History and Philosophy of Physics*, http://arxiv.org/abs/1303.1426.
5. Lemmen, "Bricks and Urbanism," 6. The map clearly shows an explosion of brick usages after 3200 BCE.
6. See M. L. Smith, "The Archaeology of South Asian Cities," *Journal of Archaeological Research* 14, no. 2 (2006): 97–142.
7. Nail, *Theory of the Border*, 47–87.
8. Rosalie Baker and Charles Baker, *Ancient Egyptians: People of the Pyramids* (Oxford: Oxford University Press, 2001), 23.
9. It is the mobility of the columns themselves that gives them strength in earthquakes and high winds.
10. Charles Woolley, *The Sumerians* (New York: Norton, 2004), 191.
11. Paul Artus, *Art and Architecture of the Roman Empire* (New York: Bellona, 2006), 45–48.
12. F. W. Galpin, "The Sumerian Harp of Ur, C. 3500 B.C.," *Oxford Journal of Music and Letters* 10, no. 2 (1929): 108–23.
13. "History of the Harp | International Harp Museum," www.Internationalharpmuseum.org.
14. Mark Cartwright, "Lyre," in *Ancient History Encyclopedia*, http://www.ancient.eu/Lyre/.
15. See Nail, *Lucretius I*, chap. 8. See also Nietzsche, *Birth of Tragedy*.
16. The oldest known musical notation written in cuneiform was found at Ugarit, Syria. The song is in a 7-note diatonic scale and produces harmonies rather than a melody of single notes.
17. Plato, *Laws*, 812d, cited and discussed in Francesco Pelosi, *Plato on Music, Soul and Body* (Cambridge: Cambridge University Press, 2010), 338.
18. Quoted in Umberto Eco, ed., *History of Beauty*, trans. Alastair McEwen (New York: Rizzoli, 2004), 62.
19. Plato, *Timaeus*, 47d–e.

20. In Mesopotamia, early evidence of beer is in a Sumerian poem honoring Ninkasi, the patron goddess of brewing, which contains the oldest surviving beer recipe, describing the production of beer from barley via bread.
21. Rudolph Michel, Patrick E. McGovern, and Virginia R. Badler. "The First Wine & Beer: Chemical Detection of Ancient Fermented Beverages," *Analytical Chemistry* 65, no. 8 (1993): 408–13. There was also beer residue from a pottery vessel from Godin Tepe, in the Zagros Mountains of Iran, and Sumerian texts that contain recipes for beer production.
22. According to Lise Manniche and Werner Forman, *Sacred Luxuries: Fragrance, Aromatherapy, and Cosmetics in Ancient Egypt* (Ithaca, NY: Cornell University Press, 1999), 49–50:

    Directions: Places the items marked "*" in a mortar and grind them. Two-fifths of this will be in the form of liquid to be discarded. There remain three-fifths in the form of ground powder. Reduce the ingredients marked "*" to powder. Moisten all these dry ingredients with wine in a copper vessel. Half of this wine will be absorbed by the powder (the rest is to be discarded). Leave overnight. Moisten the raisins with oasis wine. Mix everything in a vessel and leave for five days. Boil to reduce by one-fifth. Place honey and frankincense in a cauldron and reduce volume by one-fifth. Add to the honey and frankincense the *kyphi* macerated in wine. Leave overnight. Grind the myrrh and add to the *kyphi*.

23. Roy Porter, *The Cambridge Illustrated History of Medicine* (Cambridge: Cambridge University Press, 2001) 54–55.
24. Hippocrates, *Collected Works*, ed. W. H. S. Jones (Cambridge, MA: Harvard University Press, 1995), I:Internal Affections, 51.
25. Martin Levey, "Some Chemical Apparatus of Ancient Mesopotamia," *Journal of Chemical Education* 32, no. 4 (1955): 180–81. "The most important find in recent years for the history of chemical apparatus has been in the excavations at Tepe Gawra, in northeast Mesopotamia, approximately fifteen miles from present-day Mosul, Iraq. Here, vessels have been unearthed that show evidence of having been used in heating processes and could very well have been utilized for the specialized purposes of extraction, sublimation, and distillation."
26. A. Lucas, *Ancient Egyptian Materials and Industries* (New York: Dover, 2012), 90. "At what date incense was first used in Egypt is uncertain, but the earliest references that can be traced are of the Fifth (2498–2345 BCE) and Sixth (2345–2181 BCE) Dynasties. Both incense and incense burners (censers) are mentioned the ancient records and the offering of incense is shown in the illustrations to the Book of the Dead and is one of the commonest subjects pictured in temples and tombs, and incense and incense burners have been found in graves."
27. Definition of φάρμακον in Henry G. Liddell, Robert Scott, Jones H. Stuart, and Roderick McKenzie. *Greek-english Lexicon* (Oxford: Clarendon Press, 2006). Bottom of Form
28. Levey, "Some Chemical Apparatus of Ancient Mesopotamia."
29. Exodus 30:25–26, "You should make of these a holy anointed oil, a perfume mixture, the work of a perfumer; it shall be a holy anointing oil."
30. Susan Harvey, *Scenting Salvation: Ancient Christianity and the Olfactory Imagination* (Berkeley: University of California Press, 2015), 15. "Incense offerings were prominent in the cultic system institutionalized in the First Temple." See Genesis 8:20–9:1, when Noah offered sacrifice "of every clean animal and every clean bird . . . the Lord smelled the pleasing odor."

31. Porter, *Cambridge Illustrated History of Medicine*, 58.
32. See Plato, *Timaeus*, 47d–e.
33. We have not dealt with ancient painting (on vases or frescos) or marble sculpture here because both arts, in their general form, were invented much earlier and are thus kinetically more similar to those of the centripetal functions of the prehistoric. The act of painting on walls, vases, and frescos goes back to the Neolithic, while rock sculpture and relief go back to the Paleolithic. The painting of sculpture and architectural forms also goes back to the Paleolithic—and it thus follows the kinetics described in the chapters on functional aesthetics.

SECTION C

1. In this sense, they are different from both the conjunctions and the knots defined in chapter 3, this volume.
2. Origen, *The Song of Songs: Commentary and Homilies*, trans. R. P. Lawson (Westminster, MD: Newman, 1957), 220.
3. Tertullian, *Adversus Marcionem*, bk. I, chap. 18.
4. Basil, *Hexameron*, bk. I, chap. 6 (FC 46, 11), in *Fathers of the Church* (Washington, DC: Catholic University of America Press, 1947), 46, 11.
5. Augustine, *Confessions*, bk. IV, chap. 15, sec. 25.
6. Augustine, *Confessions*, bk. XII, chap. 25, sec. 35.
7. Thomas Aquinas, *Somme Théologique; Vol. 3: Ia. 12–13, Knowing and Naming God: Latin Text and English Translation*, trans. Herbert McCabe (Cambridge: Cambridge University Press, 2010), 45.
8. Pseudo-Dionysius, *The Complete Works*, trans. Colin Luibheid (New York: Paulist, 1987).
9. Erwin Panofsky, *Renaissance and Renancences in Western Art* (Stockholm: Almquist and Wiksells; New York: Harper Torchbooks, 1960), 187; and Emile Male, *Religious Art, From the Twelfth to the Eighteenth Century* (New York: Pantheon, [1972] 1949), n. trans., 9.
10. Pseudo-Dionysius, *Complete Works*, 68ff.
11. Plotinus, I.6, in Plotinus, *The Enneads*, trans. Stephen MacKenna (Burdett, NY: Larson, 1992).
12. Robert Grosseteste, *On Light (De Luce)* (Milwaukee, WI: Marquette University Press, 2000), 10.
13. Grosseteste, *On Light*, 10.
14. Grosseteste, *On Light*, 11.
15. Grosseteste, *On Light*, 15.

CHAPTER 10

1. Cecil Gray, *The History of Music* (Abingdon: Routledge, 2009). The music historian Cecil Gray argues that the ancient arts were dominated by sculpture, while the medieval arts were dominated by music.
2. Liz James, *Light and Colour in Byzantine Art* (Oxford: Clarendon, 1996); Moshe Barasch, *Light and Color in the Italian Renaissance Theory of Art* (New York: New York University Press, 1978). Janetta Benton, *Materials, Methods, and Masterpieces of Medieval Art* (Santa Barbara, CA: Praeger, 2009).
3. "Iranian Visual Arts: History of Iranian Tile," *Iran Chamber Society*, http://www.iranchamber.com/art/articles/tile_history1.php
4. Katherine Dunbabin, *Mosaics of the Greek and Roman World* (New York: Cambridge University Press, 2012), 5.

5. Procopius, *De Aedificiis*, in William Richard Lethaby and Harold Swainson, *The Church of Sancta Sophia, Constantinople: A Study of Byzantine Building* (New York: Macmillan, 1894), 24–28.
6. James, *Light and Colour in Byzantine Art, 4.*
7. Richard Marks, Stained Glass in England During the Middle Ages (New York: Routledge, 2014), 108.
8. Hugh of Saint-Victor, *Speculum de mysteriis ecclesiae*, Sermon 2.
9. William Durandus, *Rationale Divinorum Officiorum*, 1.1.24, in *The Symbolism of Churches and Church Ornaments*, ed. John Mason Neale and Benjamin Webb (Leeds: T. W. Green, 1843), 28.
10. Barasch, *Light and Color*, 140. Suzanne Akbari, *Seeing Through the Veil: Optical Theory and Medieval Allegory* (Toronto: University of Toronto Press, 2015), 21; see "The Multiplication of Forms."
11. Perry Anderson, *Passages from Antiquity to Feudalism* (London: NLB, 1974), 134.
12. Clifford Backman, *The Worlds of Medieval Europe* (New York: Oxford University Press, 2003), 73.
13. "While the dominant strand of monasticism came from the Mediterranean, another strand came from the Celtic people of, what is today, northwestern France and the British Isles. For the Celts, monasticism offered an escape from rural misery and clan warfare. By 600 CE Ireland had well over a hundred thriving monasteries and abbeys—the most fully monasticized region in Europe"; Backman, *Worlds of Medieval Europe*, 76.
14. Since Saint Pachomius, the "father of cenobitic monasticism," also spent time in the Roman army, Marilyn Dunn suggests that this cellular structure may have even been partially inspired by Roman army barracks; Marilyn Dunn, "The Development of Communal Life," in her *The Emergence of Monasticism: From the Desert Fathers to the Early Middle Ages* (Malden: Blackwell, 2000), 29.
15. For floor plans and additional details of the cellular structure of Carthusian charterhouses, see Roger Palmer, *English Monasteries in the Middle Ages: An Outline of Monastic Architecture and Custom from the Conquest to the Suppression* (London: Constable, 1930).
16. In the Benedictine Rules, there are cells for the sick (chapter 36), cells for guests (chapter 1), cells for sleeping (chapter 22), a cell for the porter (chapter 66), and so on. According to the rule of Saint Pachomius, strict laws govern each cell: "It is prohibited to enter in the cell of the neighbor without having knocked first on the door," "When everybody separates to go to sleep, no one will be allowed to leave their cell, except in case of necessity," "Do not let anybody eat anything inside his cell," a monk "will not have a cell in which he can lock himself," and so forth. http://www.ben.edu/center-for-mission-and-identity/resources/rule-of-st-benedict.cfm.
17. Quoted in Erwin Panofsky, *Abbot Suger on the Abbey Church of Saint-Denis and Its Art Treasures*, 2nd ed. (Princeton, NJ: Princeton University Press, 1979), 101.
18. See Nail, *Theory of the Border*.
19. Robert James Forbes, *Short History of the Art of Distillation: From the Beginnings Up to the Death of Cellier Blumenthal* (Leiden: Brill, 1948), 57, 89; T. Fairley, *The Early History of Distillation* (London: Harrison and Sons, 1907).
20. Bruce Moran, *Distilling Knowledge: Alchemy, Chemistry, and the Scientific Revolution* (Cambridge, MA: Harvard University Press, 2005), 5.
21. See Moran, *Distilling Knowledge*.
22. Moran, *Distilling Knowledge*, 11.

23. Of course, the discourse on luminous ether begins much earlier in the ancient world. See Nail, "Being and Motion," chap. 24.
24. Jim Drobnick, "Sublime Essences," in *The Smell Culture Reader*, ed. Jim Drobnick and David Howes (Oxford: Berg, 2012), 387.
25. Quoted in Drobnick, "Sublime Essences," 388.
26. Quoted in Drobnick, "Sublime Essences," 389.
27. Quoted in Drobnick, "Sublime Essences," 389.
28. Forbes, *Short History of the Art of Distillation*, 57, 89.

CHAPTER 11

1. Erwin Panofsky, *Perspective as Symbolic Form*, trans. Christopher Wood (New York: Zone, 1991), 29–30.
2. Proclus, *Elements of Physics*, 142a, cited in Panofsky, *Perspective as Symbolic Form*.
3. Panofsky, *Perspective as Symbolic Form*, 67.
4. Other examples of square halos are found at Santa Prassede in Rome, in a mosaic of Pope Paschal I (ca. 820) and a mosaic that includes a woman specified as "Theodora, Bishop").
5. See Antoniazzo Romano, altarpiece of the confraternity of the *Annunciation*, ca. 1500, Santa Maria sopra Minerva, Rome.
6. The star halo sometimes appears in depictions of the Immaculate Conception. This type of halo refers to the description of the Virgin being crowned with twelve stars (Rev. 12:1). Several depictions of the Immaculate Conception appear in Counter-Reformation art, including Velásquez's *The Immaculate Conception*, ca. 1619, and Francesco Pacheco's *Immaculate Conception with Miguel Cid*, ca. 1621, Seville Cathedral.
7. J. F. L. Mérimée, *The Art of Painting in Oil and in Fresco, Being a History of the Various Processes and Materials Employed from Its Discovery to the Present Time* (London: Whittaker and Company , 1839), 14, 18. Mérimée mentioned that if a file were dressed over a painting by Jan van Eyck's hand, the surface of the paint took on the "shining appearance" of being painted with varnish. Mérimée also claimed that Ruben's teacher, Otho Venius (a.k.a. Otto Vaenius), painted with the original van Eyckian varnish and gave that device and the process for its use to his remarkably famous student (who, in his own turn, very likely handed that to his own remarkable student, Anthony van Dyke). "In the present state of knowledge it is not possible to say whether this special quality of Rubens' paint is due to the addition of some resin to the oil medium or to the mode of preparation of the oil itself"; Joyce Plesters, "Samson and Delilah: Rubens and the Art and Craft of Painting on Panel," *National Gallery Technical Bulletin* 7 (1983): 18. Vasari writes that

    [van Eyck] . . . began to devise means for preparing a kind of varnish which should dry in the shade, so as to avoid [the danger incurred by] placing his pictures in the sun. Having made experiments with many things, both pure and mixed together, he at last found that linseed oil and nut oil, among the many which he had tested, were more drying than all the rest. These [oils], therefore, boiled with other mixtures of his, made him the varnish which he, nay, which all the painters of the world, had long desired. Continuing his experiments with many other things, he saw that the immixture of the colours with these kinds of oils [Vasari must rather mean the invented varnish here, and not the oil] gave them a very firm consistence, which, when dry, was proof against wet; and, moreover, that the vehicle lit up the colours so powerfully, [and] that it gave a gloss of itself without [the need to have a final] varnish; and that which appeared to him still more admirable was that it allowed of blending [melding the paint-layers] infinitely better than [egg] tempera. (Giorgio Vasari, *Vasari on Technique: Being the Introduction to the Three Arts of Design, Architecture, Sculpture and Painting, Prefixed to the Lives of the Most Excellent Painters, Sculptors, and Architects*, trans. Louisa S. Maclehose [Mineola, NY: Dover, 2011], 294–95)

Rubens writes to Mayerne: "To use smalt so that it shall be beautiful and light, it is necessary to temper it with the hard varnish"; Théodore Mayerne, Donald C. Fels, Joseph H. Sulkowski, Richard Bedell, Rebecca A. McClung, and Ernst Berger. *Lost Secrets of Flemish Painting: Including the First Complete English Translation of the De Mayerne Manuscript, B.m. Sloane 2052* (Eijsden, the Netherlands: Alchemist, 2010), 10.

See also Charles Eastlake, *Methods and Materials of Painting of the Great Schools and Masters: Two Volumes Bound As One* (Mineola, NY: Dover, 2001).

8. Gardner and Kleiner, *Gardner's Art Through the Ages*, 400.
9. Philippi Villani, *De origine civitatis Florentie et de eiusdem famosis civibus*, ed. Giuliano Tanturli (Patavii: In Aedibus Antenoreis, 1997), 155.
10. Giovanni Boccacio, *Des femmes nobles et renommées* (Paris: Bibliothèque Nationale), ms. Fr. 12420, fol. 101v; and ms. Fr. 598, fol. 100v. Cf. Yiu, "Spiegel."
11. Filarete, *Filarete's Treatise on Architecture*, trans. John R. Spencer, 2 vols. (New Haven, CT: Yale University Press, 1965), 315.
12. Yvonne Yiu, "The Mirror and Painting in Early Renaissance Texts," *Early science and medicine*, 10(2) (2005): 187–210, 192.
13. Laura Snyder, *Eye of the Beholder: Johannes Vermeer, Antoni Van Leeuwenhoek, and the Reinvention of Seeing* (New York: W.W. Norton, 2016), 79–83.
14. Leon Battista Alberti, *On Painting and On Sculpture*, ed. Cecil Grayson (London: Phaidon 1972), 88–89, §46 (includes the Latin text).
15. Antonio Manetti and Howard Saalman, *The Life of Brunelleschi* (State College, PA: Pennsylvania State University Press, 1970), 42–43.
16. Leonardo da Vinci, *Leonardo's Notebooks: Writing and Art of the Great Master*, ed. H Anna Suh (New York: Black Dog & Leventhal, 2013), 18.
17. Leonardo da Vinci, *Treatise on Painting: Codex Urbinas Latinus 1270*, trans. A. P. McMahon (Princeton, NJ: Princeton University Press, 1979), §72.
18. Carlo Pedretti, *The Literary Works of Leonardo da Vinci*, 2 vols. (Oxford: Phaidon, 1977), 1:328.
19. Although textual evidence is lacking for the secretive Netherlandish painters, the appearance of convex mirrors in paintings like Jan van Eyck's *Giovanni Arnolfini and His Bride* (1434) and Petrus Christus's *A Gold smith in His Shop* (1449) suggests that they were used in some capacity.
20. Snyder, *Eye of the Beholder*, 129–30.
21. Gardner and Kleiner, *Gardner's Art Through the Ages*, 566–67. See also Tim Jenison Teller, Penn Jillette, and David Hockney, *Tim's Vermeer* (London: Sony Pictures Home Entertainment, 2014) for a demonstration of how Vermeer may have worked. David Hockney and physicist Charles Falco have argued that artists going back to 1430 were extensively using mirrors, lenses, and other optical instruments. They claim that these artists were simply tracing the images seen in them. However, given the quality of mirrors at the time, this level of accuracy would have been

impossible to merely trace. Rather, artists used mirrors and lenses to aid their painting in other ways than simple tracing; See Snyder, *Eye of the Beholder*, 91–93.

22. Translated from the original Latin of the *Docta sanctorum patrum* as given in *Corpus iuris canonici, ed. a. 1582*, 1256–57. Pope Clement VI, however, indulged in polyphonic music.
23. Giles Constable, *Letters and Letter-Collections* (Turnhout: Brepols, 1976), 31.
24. Alberico, *Alberici Casinensis Flores Rhetorici*, ed. Mauro Inguanez and Henry M. Willard (Montecassino: Arti grafiche Sansaini, 1938), 36–38.
25. Robert L. Benson, "Proto-humanism and Narrative Technique in Early Thirteenth-Century Italian 'Ars dictaminis," in *Boccaccio: Secoli di vita: Atti del Congresso Internazionale Boccaccio 1975, Università di California, Los Angeles 17–19 ottobre, 1975*, ed. Marga Cottino-Jones and Edward F. Tuttle (Ravenna: Longo, 1977), 32.
26. Alberico, *Alberici Casinensis Flores Rhetorici*, 36–38.
27. Ronald Witt, "Medieval 'ars Dictaminis' and the Beginnings of Humanism: a New Construction of the Problem." *Renaissance Quarterly* 35, no. 1 (1982): 1–35. See also note 20: "A speech (oratio) consisting of parts, harmoniously and clearly written, fully expressing the feeling of the speaker."
28. Cited in Benson, "Proto-humanism and Narrative Technique," 36n18.
29. Cited in Benson, "Proto-humanism and Narrative Technique," 98.
30. Cited in Benson, "Proto-humanism and Narrative Technique," 98.
31. Helen Waddell, *The Wandering Scholars* (London: General Books, 2009), 152–58.
32. Benson, "Proto-humanism and Narrative Technique," 44.
33. My argument here is not causal. I am not arguing that the love letter caused the chivalric romance but, rather, that they share the same kinetic condition and structure: a relation of distance traversed by amorous communication and quest.
34. Echevarría González, *Cervantes' Don Quixote* (New Haven, CT: Yale University Press, 2015), 39–40.

SECTION D

1. See Leroi-Gourhan, *Gesture and Speech*, 373–74.
2. On the modern importance of "the series," see Michel Foucault, *Security, Territory, Population: Lectures at the Collège de France* (Basingstoke: Palgrave Macmillan, 2009); and Deleuze, *Logic of Sense*.
3. The historical ontology of time that grants temporality ontological primacy is dealt with in Nail, "Being and Motion."

CHAPTER 13

1. H. Akanuma, "The Significance of the Composition of Excavated Iron Fragments Taken from Stratum III at the Site of Kaman-Kalehöyük, Turkey," *Anatolian Archaeological Studies* 14 (2005): 147–58.
2. Horace, *Odes*, I, 16.9, "*Noricus ensis*."
3. Quoted in Hans L. Jaffé, *De Stijl* (NewYork: Abrams, 1971), 185–88.
4. Mumford, *City in History*, 519.
5. Quoted in Vincent Scully Jr., *Frank Lloyd Wright* (New York: Braziller, 1960), 18.
6. Susan Sontag, *On Photography* (New York: Picador, 2010), 22–23.
7. Robert Hirsch, *Seizing the Light: A History of Photography* (New York: McGraw-Hill, 2000).
8. Anthony Feldman and Peter Ford, *Scientists & Inventors* (New York: Bloomsbury, 1989), 128.

9. Translated by Robin Fedden, in Elizabeth Gilmore Holt, ed., *From the Classicists to the Impressionists: Art and Architecture in the 19th Century* (New Haven, CT: Yale University Press, [1966] 1986), 406–409.
10. Margaret Anne Doody, *The True Story of the Novel* (New Brunswick, NJ: Rutgers University Press, 1997), 1.
11. See Ian Watt, *The Rise of the Novel: Studies in Defoe, Richardson and Fielding* (London: Bodley Head, [1957] 2015).
12. Henry Ward Beecher, "The Duty of Owning Books," *Friends' Intelligencer* 16 (1860): 747.
13. See Paul Saenger, *Space Between Words: The Origins of Silent Reading* (Stanford, CA: Stanford University Press, 2001).
14. Adalaide Morris and Thomas Swiss, *New Media Poetics: Contexts, Technotexts, and Theories* (Cambridge, MA: MIT Press, 2013), 2. "Instead of the beginnings, middles, and ends that structured nineteenth-century linear narratives, Stein constructed for her writing a continuous present as additive as a drive in the country, as iterative as the frames in a filmstrip, as collaged as the view from a plane."
15. Georg Lukacs, *The Theory of the Novel* (Cambridge, MA: MIT Press, 1974), 75.
16. Lukacs, *Theory of the Novel*, 125.
17. Lukacs, *Theory of the Novel*, 76.
18. Lukacs, *Theory of the Novel*, 81.
19. Lukacs, *Theory of the Novel*, 89.
20. Lukacs, *Theory of the Novel*, 84.
21. Lukacs, *Theory of the Novel*, 92.
22. See Alex Woloch, *The One vs. the Many: Minor Characters and the Space of the Protagonist in the Novel* (Princeton, NJ: Princeton University Press, 2014); Jeremy Rosen, *Minor Characters Have Their Day: Genre and the Contemporary Literary Marketplace* (New York: Columbia University Press, 2017); Julian Murphet, "The Mole and the Multiple: A Chiasmus of Character," *New Literary History* 42, no. 2 (2011): 255–76.
23. In 1714, Henry Mill, an engineer with the New River Water Co. in London, received his inconsequential British patent (no. 395) "for 'a machine or artificial method, to print letters continuously one after another while writing, in a fashion so clean and precise that they are indistinguishable from the printing of letters"; cited in Friedrich Kittler, *Gramophone, Film, Typewriter* (Stanford, CA: Stanford University Press, 1999), 187.
24. See full typewriter diagrams at http://maritime.org/doc/typewriter/part2.htm.
25. Kittler, *Gramophone, Film, Typewriter*, 183–265.
26. See Kittler, *Gramophone, Film, Typewriter*, 229.
27. Samuel Beckett, *Three Novels: Molloy: Malone Dies* (New York: Grove, 1965), 120.

CHAPTER 14

1. Simha Arom, *African Polyphony and Polyrhythm: Musical Structure and Methodology* (Cambridge: Cambridge University Press, 2004), 190, 196. "Souris testifies that in early mensural music, 'the beat is a neutral pulsation with no metric accentuation. . . ,' while even music as late as the sixteenth century, according to Emmanuel, 'conceive[d] of rhythm as based on beats, but not on beats marshalled into measures.'"
2. Victor Zuckerkandl, *Sound and Symbol: Music and the External World* (Princeton, NJ: Princeton University Press, 1973) 166–68.

3. David Landes, *Revolution in Time: Clocks and the Making of the Modern World* (Cambridge, MA, and London: Harvard University Press. 1983), 58.
4. Lewis Mumford, "The Monastery and the Clock," in *The City Cultures Reader*, ed. Malcolm Miles et al. (London: Routledge.), 121.
5. See Mark Abel, *Groove: An Aesthetic of Measured Time* (Chicago: Haymarket, 2015), 92–115.
6. Werner Bachmann, et al. "Bow," in *Grove Music Online* (Oxford: Oxford University Press, 2001). https://doi.org/10.1093/gmo/9781561592630.article.03753.
7. Anton Webern, *The Path to the New Music* (Bryn Mawr, PA: Theodore Presser, 1963), 47.
8. Cited in Ulrich Weisstein, *Expressionism as an International Literary Phenomenon: Twenty-one Essays and a Bibliography* (Amsterdam: J. Benjamins, 2011), 157. See page 389 in Arnold Schoenberg, *Theory of Harmony*, trans. Roy E. Carter and Walter Frisch (Berkeley, CA: University of California Press, 2010).
9. Pierre Boulez, *Relevés d'Apprenti* (Paris: Éditions du Seuil, 1966), 373–74.
10. Theodore Adorno, "On Popular Music," in *Adorno: Essays on Music*, ed. Richard Leppert (Berkeley, CA: University of California Press, [1941] 2002), 461. "Insofar as dance is synchronous movement, the tendency to march has been present in dance from the very beginning; thus jazz is connected in its origins with the march and its history lays bare this relationship." See also Abel's *Groove* for an excellent critique and alternative to Adorno's critique of jazz and metered music.
11. See the excellent work of Mark Abel, who has argued this point convincingly at length in his book *Groove*.
12. Robert Palmer, *Deep Blues* (New York: Penguin, 1981), 37.
13. Don Michael Randel, ed., *The New Harvard Dictionary of Music* (Cambridge, MA: Belknap Press, Harvard University Press, 1986), 216.
14. Se Abel, *Groove*, 61–91. On the question of whether groove is African or not, Abel offers a well-supported and cited argument for the nonmetricality of African and medieval musics.
15. Arom, *African Polyphony and Polyrhythm*, 180.
16. Abel, *Groove*, 51–52.
17. For a full definition of groove and a full defense of this definition of twentieth-century music, see Abel, *Groove*.
18. This is what Rosalind Krauss calls the "expanded field." See Rosalind Krauss, "Sculpture in the Expanding Field," *October* 8 (Spring 1979): 30. "Categories like sculpture and painting have been kneaded and stretched and twisted in an extraordinary demonstration of elasticity, a display of the way a cultural term can be extended to include just about anything."
19. This is well attested to in the scholarship. For a full literature review, see Matthew Gelbart, *The Invention of "Folk Music" and "Art Music": Emerging Categories from Ossian to Wagner* (Cambridge: Cambridge University Press, 2011). Matthew Gelbart argues that folk music and art music became meaningful concepts only in the late eighteenth and early nineteenth centuries, and only in relation to each other. He examines how cultural nationalists motivated the earliest classification of music by origins, and how the notions of folk music and art music followed—in conjunction with changing conceptions of nature, and changing ideas about human creativity.
20. From the Middle English *daunsen*, from Anglo-Norman *dancer* and *dauncer* ("to dance") (compare Old French *dancier*), from Frankish **dansōn* ("to draw," "to

pull," "to tretch out," "to gesture") (compare Old High German *dansōn* ["to draw," " to pull"]), from Proto-Germanic **þansōną*, from **þinsaną* ("to draw," "to pull").

21. Carol Lee, *Ballet in Western Culture: A History of its Origins and Evolution* (London: Routledge, 2002).
22. Paul Valéry, *Collected Works of Paul Valery, Volume 13: Aesthetics* (Princeton, NJ: Princeton University Press, 2015), 198.
23. Valéry, *Collected Works*, 114
24. Although the futurist implicitly rely on the continuum of motion, they also tend to valorize only the extensive variety of motion associated with technological motion and the speed of machines. Movement, for the futurists, is often defined by a series of states (as differences) and not by a continuum of waves. This fetishism of technological or differential motion is what allows them to affirm the beauty of modern warfare, machine automation, and state fascism. Perhaps if Boccioni were a closer reader of Bergson he would have realized this important conceptual distinction and focused less on trying to represent motion in painting by superposition and let the materiality of the paint move on its own as Pollock did.
25. Rose Lee Goldberg, *Performance Art: From Futurism to the Present* (London: Thames & Hudson, 1996), 11–30.
26. Goldberg, *Performance Art*, 28.
27. Luigi Russolo, *The Art of Noises* (New York: Pendragon, 1986).
28. Goldberg, *Performance Art*, 28.
29. Goldberg, *Performance Art*, 60–61. Hugo Ball and Emmy Hennings established the Cabaret Voltaire in Zürich in 1916, featuring sound poems, variety theatre, noise music, and cardboard tube costumes, and they were, like the futurists, met with violence from the audience.
30. Goldberg, *Performance Art*, 64–66.
31. Quoted in Sir Frederick Hopkins, "Nobel Lecture, 1929," http://www.nobelprize.org/nobel_prizes/medicine/laureates/1929/hopkins-lecture.html.
32. Thomas Carlyle, "The Dandiacal Body," in *Sartor Resartus* (Frederick A. Stokes Company, 1836), 239. "A Dandy is a clothes-wearing Man, a Man whose trade, office and existence consists in the wearing of Clothes. Every faculty of his soul, spirit, purse, and person is heroically consecrated to this one object, the wearing of Clothes wisely and well: so that the others dress to live, he lives to dress ... And now, for all this perennial Martyrdom, and Poesy, and even Prophecy, what is it that the Dandy asks in return? Solely, we may say, that you would recognize his existence; would admit him to be a living object; or even failing this, a visual object, or thing that will reflect rays of light."
33. Albert Camus, "II Metaphysical Rebellion," in *The Rebel: An Essay on Man in Revolt* (New York: Knopf/Doubleday, 2012), 51.

PART III

1. Octavio Paz, *The Other Mexico: Critique of the Pyramid* (New York: Grove, 1972), 36.

CHAPTER 15

1. See Eugene Thacker, *Biomedia* (Minneapolis: University of Minnesota Press, 2004).
2. Media studies often treats the images of aesthetics as a science of categories of objects instead of fields of kinetic processes.

3. Lev Manovich, *The Language of New Media* (Cambridge, MA: MIT Press, 2010). Manovich is one of the best sources on defining "new media," but often ends up with definitions so specific that they become circular. The whole problem stems from not starting with kinetic processes and flows and instead trying to define categories of things or objects. He privileges media instead of the more primary kinetic process of *mediation*. For a critique of this object-media fetishism, see Sarah Kember and Joanna Zylinska, *Life After New Media: Mediation As a Vital Process* (Cambridge, MA: MIT Press, 2015), xviii. "Chapter 1 makes a case for a shift from thinking about 'new media' as a set of discrete objects to understanding media, old and new, in terms of the interlocked and dynamic processes of mediation. It also outlines what is at stake in this shift from thinking about media solely as objects of use, to recognizing our entanglement with media not just on a sociocultural but also on a biological level."
4. Media studies, especially in the 1990s and 2000s, described the digital image as immaterial, virtual, or de-material. See Martin Lister, *New Media: A Critical Introduction* (London: Routledge, 2010); and the foreword to Hansen, *New Philosophy for New Media*. This trend is only starting to wane, and there have been a number recent books emphasizing the material dimensions of aesthetics and digital media. See Bernd Herzogenrath, ed., *Media Matter: The Materiality of Media, Matter As Medium* (New York: Bloomsbury Academic, 2017), introduction; Anna Munster, *Materializing New Media: Embodiment in Information Aesthetics* (Lebanon, NH: Dartmouth College Press, 2011); Hansen, *Bodies in Code*; Thacker, *Biomedia*; Zylinska, *Life After New Media*.
5. See Jay Bolter and Richard Grusin, *Remediation: Understanding New Media* (Cambridge, MA: MIT Press, 2003).
6. James C Maxwell, *A Dynamical Theory of the Electromagnetic Field* (Edinburgh: Scottish Academic Press, 1982).
7. Giuliano Pancaldi, *Volta: Science and Culture in the Age of Enlightenment* (Princeton, NJ: Princeton University Press, 2003).
8. See an image of Samuel Thomas von Sömmering's "Space Multiplexed" Electrochemical Telegraph (1808–1810), http://people.seas.harvard.edu/~jones/cscie129/images/history/von_Soem.html.
9. The mathematical formalization of the binographic process was put forward by George Boole in his books *The Mathematical Analysis of Logic* (1847) and *An Investigation of the Laws of Thought* (1854). Boolean algebra demonstrated for the first time that not only communication but all logical and graphic processes could be reduced to a single and fundamental quantitive difference between one and zero.
10. For detailed explanation of and contrast with classical mechanics of the transitory, see Richard P. Feynman, *The Feynman Lectures on Physics* (Reading, MA: Addison-Wesley, 1963), 3:chs. 13–14.
11. See Ray Kurzweil, *The Age of Spiritual Machines: When Computers Exceed Human Intelligence* (New York: Penguin, 1999), 280; Friedrich Kittler, *Gramophone, Film Typewriter*, trans. Geoffrey Winthrop-Young and Michael Wutz (Stanford, CA: Stanford University Press, 1999), 1–2.
12. See Henri Bergson, *Creative Evolution*, trans. Arthur Mitchell (New York: Modern Library, 1944), 296, for his critique of the cinematographic mechanism. See also Martin Heidegger, "The Question Concerning Technology," in his *Basic Writings: From Being and Time (1927) to the Task of Thinking (1964)* (London: Harper Perennial, 2008), 311–41. As Bernard Stiegler says, digital

decomposition "resides in the tendency of capitalism to hyper-synchronize the temporalities of consciousness, to eliminate their diachronies"; Bernard Stiegler, *The Decadence of Industrial Democracies* (New York: Blackwell, 2011), 49.

13. See Nail, "Theory of the Object."
14. Robert McMillan, "Your PC just Crashed? Don't Blame Microsoft," *Wired Magazine*, August 30, 2012, https://www.wired.com/2012/08/your-pc-just-crashed-dont-blame-microsoft/
15. Alan Seabaugh, "The Tunneling Transistor," *Institute of Electrical and Electronics Engineers Spectrum*, September 30, 2013, http://spectrum.ieee.org/semiconductors/devices/the-tunneling-transistor
16. See Manovich, *Language of New Media*; and Lister, *New Media*.
17. Manovich, *Language of New Media*.
18. Marshall McLuhan, "The Future of Man in the Electric Age," interview by Frank Kermode, British Broadcasting Corporation, 1965, http://www.marshallmcluhanspeaks.com/media/mcluhan_pdf_2_No2kVoY.pdf

CHAPTER 16

1. Carlo Rovelli, *Reality Is Not What It Seems: The Elementary Structure of Things*, trans. Erica Segre (New York: Riverhead, 2017), 136–37.
2. See Gleick, *Chaos*.
3. C. Janecke, *Kunst und Zufall: Analyse und Bedeutung* (Nurnberg: Verlag fur Moderne Kunst, 1995), 75–83.
4. Marcel Duchamp and Simonetta Rasponi, *Marcel Duchamp* (Milan: Bompiani, 1993), 12.
5. David Sylvester, *Brutality of Fact: Interviews with Francis Bacon* (London: Thames & Hudson, 2016), 30.
6. Sylvester, *Brutality of Fact*, 190.
7. Sylvester, *Brutality of Fact*, 83.
8. Hans Emons, *Komplizenschaften: zur Beziehung zwischen Musik und Kunst in der amerikanischen Moderne. Kunst-, Musik- und Theaterwissenschaft* 2. (Berlin: Frank & Timme, 2006, 87).
9. David Nicholls, "Brown, Earle (Appleton)," in *The New Grove Dictionary of Music and Musicians*, 2nd ed., ed. Stanley Sadie and John Tyrrell (London: Macmillan, 2001).
10. See Eva Jablonka, Marion J. Lamb, and Anna Zeligowski, *Evolution in Four Dimensions: Genetic, Epigenetic, Behavioral, and Symbolic Variation in the History of Life* (Cambridge, MA: Bradford, 2014).
11. Gleick, *Chaos, 117*.
12. Maurizio Bolognini, *Artists Website*. http://www.bolognini.org/intro.htm.
13. Maurizio Bolognini, "De l'interaction à la démocratie. Vers un art gènératif post-digital," in *Artmedia X Proceedings* (Paris: Bibliothèque Nationale de France, Institut National d'Histoire de l'Art, 2008).
14. Scott Draves, *Triangulation*, www.triangulationblog.com.
15. Noam Cohen, "He Wrote 200,000 Books (but Computers Did Some of the Work)," *New York Times*, April 14, 2008.
16. Katherine Hayles, ed., *Electronic Literature Collection* (Notre Dame, IN: University of Notre Dame, 2008), http://collection.eliterature.org/1/.
17. Curtis Roads, *Microsound* (Cambridge, MA: MIT Press, 2001), vii.
18. Barry Truax, *Riverrun* (1986), https://www.youtube.com/watch?v=u81IGEFt7dM.

19. Brian Eno, *Reflection* (2016), http://www.brian-eno.net.
20. Brian Eno, *Reflection* (2016), http://www.brian-eno.net.
21. Brian Eno, *Reflection* (2016), http://www.brian-eno.net.
22. Brian Eno, *Reflection* (2016), http://www.brian-eno.net.

INDEX

Page numbers followed by *f* indicate figures. Numbers followed by n indicate notes.